Advanced Information and Knowledge Processing

For other titles published in this series, go to
http://www.springer.com/4738

JingTao Yao

Web-based Support Systems

Dr. JingTao Yao
University of Regina
Dept. Computer Science
3737 Wascana Parkway
Regina SK S4S 0A2
Canada
jtyao@cs.uregina.ca

AI&KP ISSN: 1610-3947
ISBN: 978-1-4471-2546-4 e-ISBN: 978-1-84882-628-1
DOI: 10.1007/978-1-84882-628-1
Springer London Dordrecht Heidelberg New York

British Library Cataloguing in Publication Data
A catalogue record for this book is available from the British Library

Cover design: SPi Publisher Services

Printed on acid-free paper

Springer is part of Springer Science+Business Media (www.springer.com)

Preface

Web-based Support Systems (WSS) are an emerging multidisciplinary research area in which one studies the support of human activities with the Web as the common platform, medium and interface. The Internet affects every aspect of our modern life. Moving support systems to online is an increasing trend in many research domains. One of the goals of WSS research is to extend the human physical limitation of information processing in the information age.

Research on WSS is motivated by the challenges and opportunities arising from the Internet. The availability, accessibility and flexibility of information as well as the tools to access this information lead to a vast amount of opportunities. However, there are also many challenges we face. For instance, we have to deal with more complex tasks, as there are increasing demands for quality and productivity. WSS research is a natural evolution of the studies on various computerized support systems such as Decision Support Systems (DSS), Computer Aided Design (CAD), and Computer Aided Software Engineering (CASE). The recent advancement of computer and Web technologies make the implementation of more feasible WSS. Nowadays, it is rare to see a system without some type of Web interaction.

The research of WSS is classified into four groups.

- WSS for specific domains.
  - WSS for specific domains:
  - Web-based DSS
  - Enterprise-wide DSS
  - Web-based group DSS
  - Web-based executive support systems
  - Web-based business support systems
  - Web-based negotiation support systems
  - Web-based medical support systems
  - Web-based research support systems
  - Web-based information retrieval support systems
  - Web-based education support systems
  - Web-based learning support systems
  - Web-based teaching support systems

- Web-based applications and WSS Techniques
  - Web-based knowledge management systems
  - Web-based groupware systems
  - Web-based financial and economic systems
  - Web-based multimedia systems
  - Web information fusion
  - Internet banking systems
  - XML and data management on the Web
  - Web information management
  - Web information retrieval
  - Web data mining and farming
  - Web search engines
  - Information fusion
  - Web services
  - Grid computing
- Design and development of WSS
  - Design and development of WSS
  - Web-based systems development
  - CASE tools and software for developing Web-based applications
  - Systems analysis and design methods for Web-based applications
  - User-interface design issues for Web-based applications
  - Visualizations of Web-based systems
  - Security issues related to Web-based applications
  - Web engineering

This book can be viewed as an extended culmination to three international workshops on WSS. The First International Workshop On WSS was held on October 13, 2003 in Halifax, Canada. The Second International WSS was held on September 20, 2004 in Beijing, China. The proceedings were published by Saint Mary's University in Canada. There are 26 and 24 papers in each set of the proceedings respectively. The Third International Workshop On WSS was held on December 18, 2006 in Hong Kong, China. There are 14 papers presented in a volume of IEEE published proceedings.

In order to keep track of the research on WSS, a Web site devoted to the research of WSS has been set up at http://www.cs.uregina.ca/∼wss/. There are articles on the Bibliography page of the Web site. If you want your publications to be listed on the page or identify yourself as a researcher in the area of WSS, please send information to wss@cs.uregina.ca. Proceedings for the first two workshops are also online at http://www.cs.uregina.ca/∼wss/wss03/wss03.pdf and http://www.cs.uregina.ca/∼wss/wss04/wss04.pdf.

This book is intended to present research related to fundamental issues of WSS, frameworks for WSS, and current research on WSS. A key issue of WSS research is to identify both domain independent and dependent activities before selecting suitable computer and Web technologies to support them. We will also examine how applications and adaptations of existing methodologies on the Web platform benefit our decision-making and other various activities.

The selection of this book was started with call-for-chapter proposals. We received 33 chapter proposals. Authors of 26 proposals were chosen to submit full chapters. After receiving the full chapters, each chapter was reviewed by three reviewers. Including the authors, some domain experts were asked to review chapters. After a couple of rounds of revisions, we present 19 chapters in this book.

There are three parts of this book: WSS for specific domains, Web-based applications and WSS techniques, and Design and development of WSS.

The first part consists of seven chapters. Chapter 1 entitled "Context-Aware Adaptation in Web-based Groupware Systems" presents research on applying context-based filtering technology for a Web-based group decision support system and its mobile users. Chapter 2 entitled "Framework for Supporting Web-based Collaborative Applications" proposes a framework to support automated service management, in particular for a Web-based medical support system. Chapter 3 entitled "Helplets: A Common Sense Based Collaborative Help Collection and Retrieval Architecture for Web-Enabled Systems" presents an online intelligent query support system for replacing a traditional help desk. Chapter 4 entitled "Web-based Virtual Research Environments" presents a Web-based research support system for research collaborations. Chapter 5 entitled "Web-based Learning Support System" proposes an adaptive learning support environment that effectively accommodates a wide variety of students with different skills, background, and cognitive learning styles. Chapter 6 entitled "A Cybernetic Design Methodology for 'Intelligent' Online Learning Support" describes a Web-based learning support system that aims to evolve Web-based teaching environments into intelligent learning systems modelled on cybernetic, systems theory principles. Chapter 7 entitled "A Web-based Learning Support System for Inquiry-based Learning" employs a treasure hunt model for teaching and learning support.

Part two includes six chapters. Chapter 8 entitled "Combinatorial fusion analysis for meta search information retrieval" describes a combinatorial fusion methodology including a theoretical framework and illustrations of various applications of the framework using examples from the information retrieval domain. Chapter 9 entitled "Automating Information Discovery within the Invisible Web" discusses issues related to the deep Web information retrieval. Chapter 10 entitled "Supporting Web Search with Visualization" introduces information visualization as a means for supporting users for their search and retrieval tasks on the Web. Chapter 11 entitled "XML Based Markup Languages for Specific Domains" describes the need for domain-specific Markup Languages within the context of XML, provides a detailed outline of the steps involved in markup language development, and gives the desired properties of Markup Languages for WSS in incorporating human factors from a domain based angle. Chapter 12 entitled "Evaluation, Analysis and Adaptation of Web Prefetching Techniques in Current Web" presents a study on Web prefetching to reduce the user-perceived latency in three steps: evaluating Web prefetching techniques from the user's point of view, analyzing how prefetching algorithms can be improved, and exploring the performance limits of Web prefetching to know the potential benefits of this technique depending on the architecture in which it is implemented. Chapter 13 entitled "Knowledge Management System Based on Web 2.0 Technologies" demonstrates that Web 2.0 technologies could be used to design user interaction in a knowledge management system to improve online interaction with WSS in other application domains.

Part three consists six chapters. Chapter 14 entitled "A Web-based System for Managing Software Architectural Knowledge" presents and discusses the design, implementation, and deployment details of a Web-based architectural knowledge management system, called PAKME, to support the software architecture process. Chapter 15 entitled "CoP Sensing Framework on Web-based Environment" presents a Web-based social learning support system based on the concept of Community of Practice and theories of social constructivism. Chapter 16 entitled "Designing a Successful Bidding Strategy using Fuzzy Sets and Agent Attitudes" presents the implementation of an online biding system armed with intelligent agents using fuzzy strategy. Chapter 17 entitled "Design Scenarios for Web-Based Management of Online Information" discusses a scenario-based design process, and results thereof, used to examine how online communication management might be supported by a Web-based system. Chapter 18 entitled "Data Mining for Web-based Support Systems: A Case Study in e-Custom Systems" provides an example of a Web-based support system used to stream-line trade procedures, prevent potential security threats and reduce tax-related fraud in cross-border trade. Chapter 19 entitled "Service Oriented Architecture (SOA) as a technical framework for Web-based Support Systems (WSS)" discusses issues on applying service oriented technique to WSS.

Last but not least, I would like to thank all authors who contributed a chapter in this book as well as the reviewers who helped to improve the quality of chapters. I would like to thank Series Editors Drs. Lakhmi Jain and Xindong Wu and Springer editors Catherine Brett and Rebecca Mowat, for their assistance and help editing this book. Dong Won Kim, a graduate student of the University of Regina under my supervision, helped final compiling with LaTeX. He also spent a lot of his time on converting some chapters prepared in Microsoft Word. I thank him for his time and patience. Without everyone's effort, it is impossible to see the completion of this book.

Regina, Saskatchewan
Canada, June 2009

JingTao Yao

Contents

List of Contributors

Ali Babar, Muhammad
Software Development Group, IT University of Copenhagen, Rued Langaards, Vej 7, DK-2300, Copenhagen, S. Denmark
e-mail: malibaba@itu.dk

Allan, Robert J.
STFC Daresbury Laboratory, Warrington WA4 4AD, UK
e-mail: robert.allan@stfc.ac.uk

Barradas, Carlos
Center for Intelligent Computing and Robotics, Tecnológico de Monterrey, Av. E. Garza Sada 2501, 64849 Monterrey, NL, Mexico
e-mail: carlos.barradas@itesm.mx

Berbers, Yolande
Department of Computer Science, Katholieke Universiteit Leuven, B-3001 Heverlee, Belgium
e-mail: Yolande.Berbers@cs.kuleuven.be

Carrillo-Ramos, Angela
Department of Computer Science, Pontificia Universidad Javeriana, Carrera 7 # 40-62, Bogotá, Colombia
e-mail: angela.carrillo@javeriana.edu.co

Curran, Kevin
School of Computing and Intelligent Systems, University of Ulster, Londonderry, Northern Ireland, UK
e-mail: KJ.Curran@ulster.ac.uk

Dai, Wei
School of Management and Information Systems, Victoria University, PO Box 14428, Melbourne City MC, Victoria 8001, Australia
e-mail: wei.dai@vu.edu.au

Domènech, Josep
Universitat Politècnica de València, Camí de Vera, s/n 46022 València, Spain
e-mail: jdomenech@ai2.upv.es

Fahrenholz, Sally
Publishing and Content Management Professional, Northeast Ohio, OH 44073, USA
e-mail: spfahren@gmail.com

Fan, Lisa
Department of Computer Science, University of Regina, 3737 Wascana Parkway, Regina, Saskatchewan, Canada
e-mail: Lisa.fan@uregina.ca

Gensel, Jérôme
Grenoble Computer Science Laboratory, 38000 Grenoble, France
e-mail: Jerome.Gensel@imag.fr

Gil, José A.
Universitat Politècnica de València, Camí de Vera, s/n 46022 València, Spain
e-mail: jagilg@disca.upv.es

Goyal, Madhu Lata
Faculty of Information Technology, University of Technology, Sydney, PO Box 123, Broadway, NSW 2007, Australia
e-mail: madhu@it.uts.edu.au

Hepting, Daryl H.
Department of Computer Science, University of Regina, 3737 Wascana Parkway, Regina, Saskatchewan, Canada
e-mail: dhh@cs.uregina.ca

Hoeber, Orland
Memorial University of Newfoundland, St. John's, NL, Canada
e-mail: hoeber@mun.ca

Hsu, D. Frank
Department of Computer and Information Sciences, Fordham University, Bronx, NY 10458, USA
e-mail: hsu@cis.fordham.edu

Jimenez, Guillermo
Center for Intelligent Computing and Robotics Tecnológico de Monterrey, Av. E. Garza Sada 2501, 64849, Monterrey, NL, Mexico
e-mail: guillermo.jimenez@itesm.mx

Khan, Sanaullah
1-A, Sector E5, Phase VII, Hayatabad, Peshawar, Pakistan
e-mail: sanaullah@imsciences.edu.pk

Khan, Shahbaz
1-A, Sector E5, Phase VII, Hayatabad, Peshawar, Pakistan
e-mail: shahbaz@imsciences.edu.pk

Kim, Dong Won
Department of Computer Science, University of Regina, 3737 Wascana Parkway, Regina, Saskatchewan, Canada
e-mail: kim263@cs.uregina.ca

Kirchner, Kathrin
Department of Business Information Systems, Faculty of Business and Economics, Friedrich-Schiller-University of Jena, 07743, Jena, Germany
e-mail: kathrin.kirchner@uni-jena.de

Ma, Jun
Faculty of Information Technology, University of Technology, Sydney, PO Box 123, Broadway, NSW 2007, Australia
e-mail: junm@it.uts.edu.au

Maciag, Timothy
Department of Computer Science, University of Regina, 3737 Wascana Parkway, Regina, Saskatchewan, Canada
e-mail: maciagt@cs.uregina.ca

Mustapha, S.M.F.D. Syed
Faculty of Information Technology, Universiti Tun Abdul Razak, 16-5 Jalan SS6/12 Kelana Jaya, 47301 Petaling Jaya, Selangor, Malaysia
e-mail: syedmalek@unitar.edu.my

Nauman, Mohammad
1-A, Sector E5, Phase VII, Hayatabad, Peshawar, Pakistan
e-mail: nauman@imsciences.edu.pk

Pinheiro, Manuele Kirsch
Reseach Center on Computer Science, University of Paris 1 (Panthéon Sorbonne), 75013 Paris, France
e-mail: Manuele.Kirsch-Pinheiro@univ-paris1.fr

Pont-Sanjuán, Ana
Universitat Politècnica de València, Camí de Vera, s/n 46022 València, Spain
e-mail: apont@disca.upv.es

Quinton, Stephen R.
Technology Enhanced Learning and Teaching (TELT) Portfolio Leader, Learning & Teaching @ UNSW, Division of the Deputy Vice-Chancellor (Academic), Level 4, Mathews Building, The University of New South Wales, Sydney, NSW 2052, Australia
e-mail: s.quinton@unsw.edu.au

Razmerita, Liana
Center of Applied Information and Communication Technologies (CAICT), Copenhagen Business School, Denmark, Solbjerg Plads, 3 DK-2000 Frederiksberg
e-mail: liana.razmerita@cbs.dk

Rundensteiner, Elke
Department of Computer Science, Worcester Polytechnic Institute, Worcester, MA, USA
e-mail: rundenst@cs.wpi.edu

Sahuquillo, Julio
Universitat Politècnica de València, Camí de Vera, s/n 46022 València, Spain
e-mail: jsahuqui@disca.upv.es

Singh, Vishav Vir
Intersil Corporation, 1001 Murphy Ranch Road, Milpitas, CA 95035
e-mail: vsingh@intersil.com

Sweeney, Edwina
Computing Department, Letterkenny Institute of Technology, Letterkenny, Ireland
e-mail: Edwina.sweeney@lyit.ie

Taksa, Isak
Department of computer Information systems, Baruch College, New York, NY 10010, USA
e-mail: Isak.Taksa@baruch.cuny.edu

Varde, Aparna
Department of Computer Science, Montclair State University, Montclair, NJ, USA
e-mail: vardea@montclair.edu

Villanova-Oliver, Marlène
Grenoble Computer Science Laboratory, 38000 Grenoble, France
e-mail: Marlene.Villanova-Oliver@imag.fr

Xie, Ermai
School of Computing and Intelligent Systems, University of Ulster, Londonderry, Northern Ireland, UK
e-mail: Xie-e@email.ulster.ac.uk

Yang, Xue Dong
Department of Computer Science, University of Regina, 3737 Wascana Parkway, Regina, Saskatchewan, Canada
e-mail: yang@cs.uregina.ca

Yang, Xiaobo
IT Services, University of Birmingham, Edgbaston, Birmingham B15 2TT, UK
e-mail: x.yang.3@bham.ac.uk

Yao, JingTao
Department of Computer Science, University of Regina, 3737 Wascana Parkway, Regina, Saskatchewan, Canada
e-mail: jtyao@cs.uregina.ca

Part I
Web-Based Support Systems for Specific Domains

Chapter 1
Context-Aware Adaptation in Web-Based Groupware Systems

Manuele Kirsch Pinheiro, Angela Carrillo-Ramos, Marlène Villanova-Oliver, Jérôme Gensel, and Yolande Berbers

Abstract In this chapter, we propose a context-aware filtering process for adapting content delivered to mobile users by *Web-based Groupware Systems*. This process is based on context-aware profiles, expressing mobile users preferences for particular situations they encounter when using these systems. These profiles, which are shared between members of a given community, are exploited by the adaptation process in order to select and organize the delivered information into several levels of detail, based on a progressive access model. By defining these profiles, we propose a filtering process that considers both the user's current context and the user's preferences for this context. The context notion of context is represented by an object-oriented model we propose and which takes into account consideration both the user's physical and collaborative context, including elements related to collaborative activities performed inside the groupware system. The filtering process selects, in a first step, the context-aware profiles that match the user's current context, and then it filters the available content according to the selected profiles and uses the progressive access model to organize the selected information.

M.K. Pinheiro
Research Center on Computer Science, University of Paris 1 (Panthéon Sorbonne), 75013 Paris, France
e-mail: Manuele.Kirsch-Pinheiro@univ-paris1.fr

A. Carrillo-Ramos
Department of Computer Science, Pontificia Universidad Javeriana, Carrera 7 # 40-62, Bogotá, Colombia
e-mail: angela.carrillo@javeriana.edu.co

M. Villanova-Oliver and J. Gensel
Grenoble Computer Science Laboratory, 38000 Grenoble, France
e-mail: Marlene.Villanova-Oliver@imag.fr; Jerome.Gensel@imag.fr

Y. Berbers
Department of Computer Science, Katholieke Universiteit Leuven, B-3001 Heverlee, Belgium
e-mail: Yolande.Berbers@cs.kuleuven.be

J.T. Yao (ed.), *Web-Based Support Systems*, Advanced Information and Knowledge Processing, DOI 10.1007/978-1-84882-628-1_1,

1.1 Introduction

1.1.1 The Web and the Collaboration Issue in Mobile Environment

Nowadays, the Web is omnipresent in both our personal and professional lives. The recent development of Web 2.0 technologies makes it simpler for users to share information and collaborate with each other. The underlying notions of *social networks* and *communities* are a main concern for Web 2.0 [1] and systems based on the Web 2.0 principles, such as Wiki systems like XWiki[1] and MediaWiki,[2] can be considered as a new generation of Web-based groupware systems.

For this new generation of Web-based groupware systems, as well as for traditional ones, one expects from collaboration gains not only in terms of productivity (in the case of workers for instance), but also in terms of facility and efficiency in retrieving and accessing some information that matches the needs of each member of a collaborative group. Such gains depend on the capability of exploiting the social dimension not only of large communities of users, but also of smaller groups (for instance in some collaborative work). *Folksonomies*[3] built in social bookmarking approaches or the *Friend Of A Friend (FOAF)* initiative [5] are examples of this recent trend.

The social dimension of these communities can be exploited by making the users aware of the communities they belong to, of the goals of these communities, and of the activities performed inside them. The notion of *group awareness* [13, 30] refers to the knowledge a user has about her or his colleagues and their actions related to the group's work. By taking into consideration this knowledge, a Web-based groupware system can supply the community members with some content better related to their own activities inside this community.

At the same time, mobile technologies, such as Wi-Fi networks, PDA, and 3G cellular phones, make it now possible for users to access any Web-based system from various kinds of devices, *anytime* and *anywhere*. This is the underlying idea of the *Ubiquitous Computing* [32, 33], defined by the W3C [16] as the paradigm of "Personal Computing," which is characterized by the use of small wireless devices.

Combining mobile technologies together with some of the Web 2.0 principles has given a rise to a new generation of Web-based systems that allow *mobile users* to share data and contextualized information (e.g., location-aware annotations or photos exchanged by Flyckr[4] users) and to collaborate with other users (through wikis or blogs, for example). However, we are only at the beginning of this new mode of accessing and sharing information, and still many problems have to be overcome.

---

[1] http://www.xwiki.org/

[2] http://www.mediawiki.org/

[3] http://www.vanderwal.net/folksonomy.html

[4] http://flickr.com/

1.1.2 Adaptation to Web-Based Groupware Systems Mobile Users

In this chapter, we focus on the specificities of the adaptation that can be expected by mobile users who have to cope with the intrinsic limitations of mobile devices (MD) (such as battery lifetime, screen size, intermittent network connections, etc.) and with the characteristics of their nomadic situation (noisy or uncomfortable environment) when accessing Web-based collaborative systems. These users generally use MD in brief time intervals in order to perform urgent tasks or to consult a small, but relevant, set of information. They usually consult, through these devices, information that is needed regarding the current situation (e.g., when away from her or his office, a user may use her or his 3G cellphone to consult only high priority messages related to her or his group work). All these aspects have to be considered when searching for and displaying information. More than traditional users, mobile users need informational content that suits their current situation.

In order to provide any member of a community with some appropriate and relevant information when she or he uses an MD some adaptation mechanisms are required. Adaptation, as a general issue, has to be tackled from different perspectives: What has to be adapted (i.e., data, service, etc.)? To whom or to what is adaptation required? Which are the guidelines or strategies for the adaptation? How can adaptation be performed? Which are the subjacent technologies? In this chapter, we do not intend to address every aspect of adaptation. Rather, we limit our study of content adaptation to the case of Web-based groupware systems that are accessed through MD.

Adaptation mechanism can be guided by different criteria such as user's personal characteristics, background, culture, and preferences. Often, the user's interests and preferences differ according to the situation in which this user is interacting with the Web-based groupware system: While some information may be valuable when the user works at her or his office, it may well be completely useless when the same user travels even for professional reasons. The situation in which the user is accessing the system, including the device she or he is using, her or his location, the activities she or he is performing, and many other aspects, can be seen as the user's context.[5] The actions and expectations of the user directly depend on the context in which she or he interacts with the system [14]. Thus, in order to improve the adaptation process, both the user's current context and her or his preferences (in terms of services, data, presentation, etc.) for this, context should be considered by the adaptation process. For example, let us consider a student preparing with her friends a dissertation. She is organizing a meeting in order to manage their work (group *activity*). For this, she invokes the service "consult colleagues agenda." In this case, she is only interested in a content referring to her friends availability for the period during which they have to prepare the dissertation. In another situation, for example in a holiday period, this student may plan a picnic (a different group activity), the same service

[5] Context refers here to "any information that can be used to characterize the situation of an entity. An entity is a person, place, or object that is considered relevant to the interaction between a user and an application, including the user and applications themselves." [12]

can be used but only selecting friends who are in the same city during the holiday period. This example shows how a same service (here, "consult colleagues agenda") can be carried out differently for the same user according to the context in which it is invocated.

Moreover, since the user interacts with other users through the collaboration process she or he is involved in, this interaction must be taken into account when considering some contextual information [15] and integrated in decisions made during the adaptation process. Indeed, the goals and activities of a group usually highly influence the actions of each member. Consequently, information related to this collaboration process, such as the concepts of *group*, *role*, and *collaborative activity* must be considered as part of each user's context. We call this set of information the *user's collaborative context*. This notion of the user's collaborative context is related to the *group awareness* concept [13,30], since both refer to the same knowledge: the knowledge about the group and the collaborative process. In our previous example, the group awareness refers to the fact that the people whose agendas are consulted are not the same since in the two situations the groups are different from the point of view of the user's collaborative context.

1.1.3 A Context- and Preference-Based Adaptation for Web-Based Groupware Systems

An idea, more and more widely accepted, is that by considering the characteristics of the group a user is a member of (i.e., considering the group's common goal to be reached, the group's expertise and knowledge, etc.) helps in better selecting information she or he individually needs. Thus, we propose a context-based filtering process which combines an object-oriented context model [18], representing the user's physical and collaborative context, with shared profiles definition, describing the user's preferences for a given context. The underlying principle of this filtering process is to improve the group productivity by helping individual users who compose the group in their own tasks. The main idea is to allow Web-based groupware systems to provide an adapted content to each user in order to improve the group activities as whole. For example, when a geographically distributed group of specialist doctors is using a Web-based groupware system in order to make a diagnosis for a patient, they can access to it through different kind of access devices and use the system to exchange information about the patient. By supplying content adapted to any doctor's preferences and current context, the system helps this doctor to make her or his own diagnosis activities, which directly contributes to the diagnosis of the patient as a whole.

The proposed filtering mechanism, based on [21], is guided by context-aware profiles, which define filtering rules that apply in a specific context. We believe that the collaborative and social dimensions of the group can be exploited by offering users the opportunity to share their profiles and preferences so that they benefit from the experiences of each other. We achieve this by allowing the users to share, with their colleagues, the profiles they have defined for some specific situations they usually encounter when using a given Web-based groupware system. These shared profiles are associated to particular context situations to which the preferences in the profile refer to. Moreover, these profiles use the Progressive Access Model (PAM) [31] in order to organize the delivered information according to the user's preferences.

The proposed filtering mechanism is therefore performed in two steps: first, it analyzes the user's current context and selects, among the available predefined profiles, those that match the user's current context. Second, it filters and organizes available information according to the selected profiles. Besides, we implement this filtering mechanism in a framework, named BW-$\mathcal{M}$, which has been used to build up a Web service. By using this technology and a context model explicitly referring to collaborative context, we believe that the proposed filtering process better suits the new generation of Web applications that focus on communities user-generated content sharing.

1.1.4 Chapter Organization

This chapter is organized as follows: In Section 1.2, we discuss related work in order to identify open issues and drawbacks from the state of the art on context-aware adaptation mechanisms for Web-based systems. In Section 1.3, we introduce an object-oriented context model particularly designed for context representation in Web-based groupware systems. Section 1.4 presents the content model we propose in order to represent the informational content that is filtered following the adaptation mechanism. Section 1.5 proposes a context-aware profile model for representing the user's preferences in a Web-based groupware system. Section 1.6 introduces the filtering process we propose in order to adapt available informational content based on a set of context-aware profiles. Section 1.7 presents our implementation of the filtering process and discusses some experimental results. Finally, Section 1.8 gives our conclusions and future work.

1.2 Related Work

Content adaptation, which aims at providing user with a customized content, is not a new topic for Web-based support systems (WSS) [6, 9, 31]. In this chapter, we are particularly interested in how context-aware adaptation mechanisms are used for content adaptation in these systems. Context-awareness [8, 12, 24] can be defined as the capacity of the system to perceive and analyze the user's context and to adapt itself (i.e., its behavior, services, interface, etc.) accordingly. The concept of user's context refers to a very large notion, for which there is no single definition [4, 8, 12, 14, 24]. In one of the pioneer works, Schilit et al. [29] define context as "the location of use, the collection of nearby people and objects, as well as the changes to those objects over time." Another view is given by Moran and Dourish [24], "context refers to the physical and social situation in which computational devices are embedded." Dey [12] defines context as "any information that can be used to characterize the situation of an entity. An entity is a person, place, or object that is considered relevant to the interaction between a user and an application, including the user and the applications themselves." In this work, we adopt this largely accepted definition since it also applies particularly for designing context-aware systems.

In order to exploit this notion of user's context inside Web-based support systems (WSS), we have to represent somehow this notion. The literature proposes different models for representing context information [2, 18, 26, 29]. In their majority, these models cover concepts related to the user's location and to the characteristics of the mobile devices. Even if some propositions, such as [26], include the user as part of the contextual model, most of these models consider the user as an isolated individual and do not consider the activities performed by her or him. However, when considering the users of Web-based groupware systems, the social aspects related to the collaborative process that takes place in these systems should be considered, since the goals and the activities of a group influence the actions of each individual in the community. In this chapter, we consider users as members of a social network, of a given community of users, or simply people using a wiki system. As member of a community, a user generally shares, with her or his colleagues, activities, goals, interests, etc. Thus, Web-based groupware systems, which support a community of users, need a context model that takes into consideration the social aspects related to the collaborative activities. We propose an object-oriented context model that represents the user's collaborative context in addition to the user's physical context, traditionally used in context models.

In the same way that context model does, context-aware adaptation mechanisms usually focus on the user as individual, considering for instance her or his location and device for adaptation purposes [17,23,27,28,34]. When considering Web-based systems, these mechanisms often propose the transformation of an original content into another version that suits the capabilities of the access device (for instance, transforming a content from an unsupported format, dimension, or resolution to a supported one), or the selection of the most appropriate versions of a content (e.g., to choose a BMP version of an image instead of a JPEG version) [23, 29]. Besides the capabilities of the access device, other elements of the user's context can also be considered for adaptation purposes. For instance, Yang and Shao [34] use concepts such as network bandwidth, user accessibility (e.g., if the user is blind), and situation (if the user is driving, or in her or his office, or in a meeting, etc.) for proposing a rule-based content adaptation mechanism in which content presentation is adapted based on a dynamically selected set of rules.

As one may observe, works such as [23,29,34] are mainly concerned with adapting content presentation according to the user's current context. However, the context in which the users interact with a Web-based support system can affect the relevance of a given content. A content that is relevant in a particular situation can be completely irrelevant (even useless) in another situation. Some works in the literature, such as [17, 27, 28], propose to supply users with a content particularly adapted to their current location. However, the user's current activity or her or his possible interactions with the communities she or he belongs to are not considered by these mechanisms.

Finally, it is worth noting that traditional content adaptation mechanisms for WSS are guided by user's preferences represented in user *profiles*, which can also refer to user's interests, history, or information needs [11]. Several works propose user profiles regarding different aspects (e.g., the user's knowledge or goals) [6, 11, 22] for adaptation purposes. However, these works often do not consider contextual aspects important for mobile users, such as location and device constraints [6, 22]. Others works, such as [11], do not relate these profiles to particular situations. However, mobile users interests and preferences may vary according to the context in which these users access a Web-based system. By proposing a fixed set of preferences that applies in every situation, these profiles might not correspond to the user's expectations.

In order to overcome some of the drawbacks presented by the works above, we propose a content-adaptation mechanism based on a filtering process. This process is guided by a set of context-aware profiles that can be shared among a community of users, and which also allow users to represent their preferences for a given context.

1.3 Context Representation

Inspired by the context definition given by Dey [12], we adopt the object-oriented representation of context originally proposed in [20], which focuses on a mobile use of collaborative systems. Based mainly on a set of UML diagrams, this model represents both the user's physical context (including the concepts of location, device, and application) and the user's collaborative context (which includes the concepts of group, role, member, calendar, activity, shared object, and process). We claim that collaborative context should be taken into consideration since users are also involved in some collaborative activities. Web-based support systems that claim to belong to Web 2.0 often propose applications (or services) whose main goal is to allow collaboration among users. Systems that propose wiki applications are a good example of this tendency. Thus, since we consider users belonging to communities, some information related to the group, such as its composition, its activities, etc., can be considered as relevant for such users, and consequently have to be included into the user's context. Moreover, since these systems are now accessible through different kinds of mobile devices (*MD*), it is also important to take into consideration information related to these devices and to the physical environment that surrounds the user interaction with these systems.

In this model (see Figure 1.1), the concept of context is represented by a class *Context Description*, which is a composition of both physical (location, device, space and service) and collaborative elements (group, role, activity, shared object, etc.). These elements are represented by classes that are specializations of a common

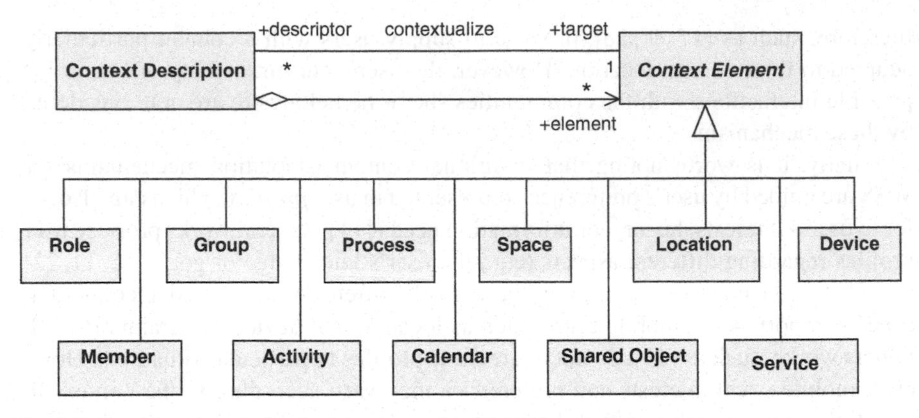

Fig. 1.1 A context description is seen as a composition of context elements

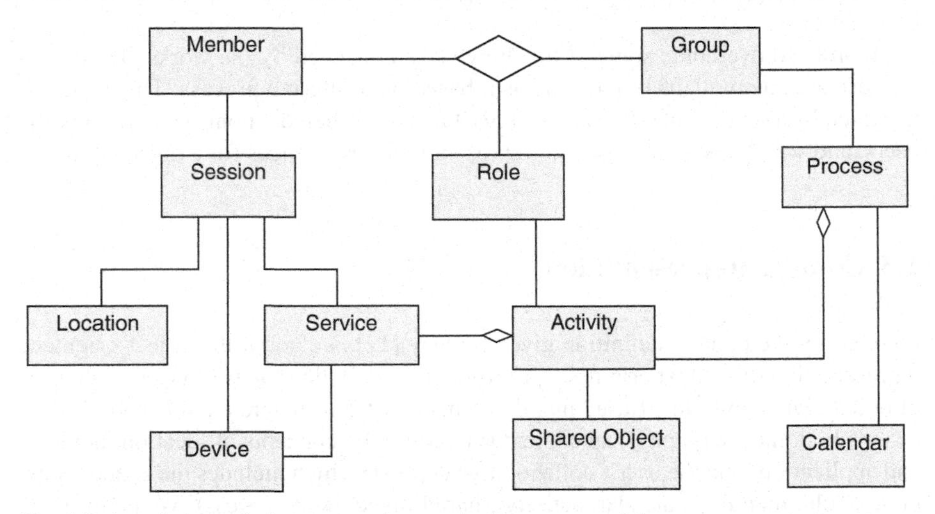

Fig. 1.2 Associations relating context elements in the context model

superclass, called *Context Element*. Furthermore, these context elements are related to each other, defining associations between the corresponding concepts. Each element of context is not isolated information, but does belong to a more complex representation of the user's situation. For instance, we consider that a user is the member of a group through the roles she or he plays in this group, and that a user handles a shared object through some application service that allows it, and so on. It is worth noting that this application service concept refers to system functionalities (e.g., weather previsions, agenda management, wiki edition, etc.), which often corresponds, in the case of new Web-based groupware systems, to Web services supplying particular functionalities. Moreover, from the system point of view, an activity can be considered as a set of services, which are executed in order to achieve this activity. Figure 1.2 presents a more complete description of these associations based on [20].

The context of a user (member of a group or community) is then represented in this model by an instance of the class *Context Description*, which is linked by composition to instances of the class *Context Element* and its subclasses (see Figure 1.1). Figure 1.3 illustrates an application of this context model. We consider a user ('Alice'), who is the coordinator ('coordinator' role) of a team ('administration' group) that uses a wiki service. Let us suppose that Alice is accessing this system through her PDA in order to consult the latest changes on the document that her team is writing. When Alice requests these changes, her current context can be represented by the context description object represented in Figure 1.3(b). This object is composed by the context elements representing Alice's location ("office D322" object of the location class) and device ("PocketPC" object of the device class), her team ("administration" object), her role in this team ("coordinator" object), and so on. These objects are related through a set of associations: Alice belongs to the administration group through the role coordinator; this role allows Alice to perform the

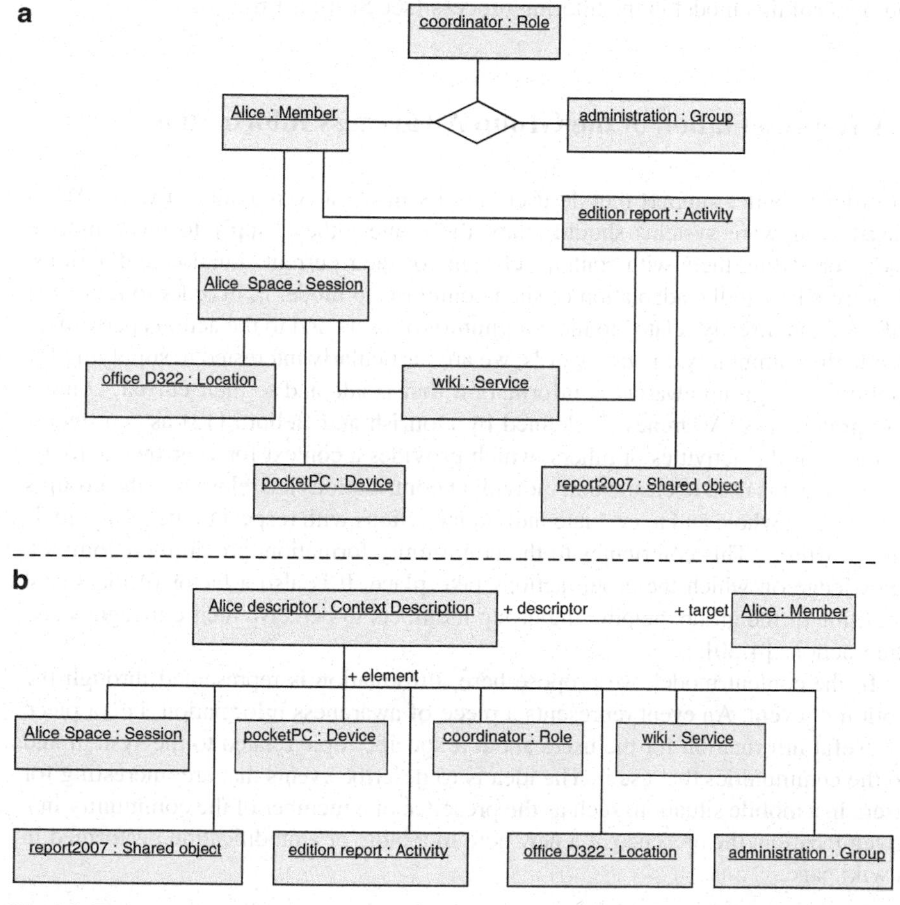

Fig. 1.3 Example of a context description for a given user (Alice). Global view on (**b**), association view on (**a**)

"report edition" activity, which handles the shared object "report 2007" through the "wiki" service, etc. All these associations, as well as the context elements objects connected by them, compose Alice's current context. In other words, according to this model, the context of a user (C_u) can be seen as a set of both context elements objects (O_u) and associations (A_u) between these objects, describing together the situation in which a user interacts with a Web-based groupware system:

$$C_u = O_u \cup A_u$$

Through this model, not only active users' current context can be represented, but also the descriptions of potential contexts established for the different users. In other words, we can describe a situation in which the user could potentially find herself or himself during her or his interaction with the system. Concerning the current context, instances of the model are created and dynamically updated by the system during each user's session, according to her or his behaviour. Regarding potential contexts, instances represent a knowledge that is permanently stored. We exploit the elements of this model in the filtering process (see Section 1.6).

1.4 Representation of the Group Awareness Information

In order to better support mobile user actions inside a community of users, Web-based groupware systems should adapt the content they supply to these mobile users, providing them with content relevant for their current situation and actions. A prerequisite to the adaptation of such content is to model it, in order to represent information directly related to the community of users and to the actions performed inside the community. In other words, we are particularly interested in supplying for mobile users group awareness information that is adapted to their current context and preferences. Awareness is defined by Dourish and Bellotti [13] as "an understanding of the activities of others, which provides a context for your own activity. This context is used to ensure that individual contributions are relevant to the group's activity as a whole and to evaluate individual actions with respect to the group goals and progress." This contributes to the community formation, providing a common knowledge on which the group actions take place. It is also a factor of the social coalition of the group, helping the group members to perceive their colleagues and their actions [4, 30].

In the content model, we propose here, information is represented through the notion of event. An event represents a piece of awareness information, i.e., a piece of useful information for the users about a specific topic related to the system and to the communities that use it. The idea is to describe events that are interesting for users in a mobile situation, such as the presence of a member of the community in a given location, the presence of a new note in a blog, or a modification performed in a wiki page.

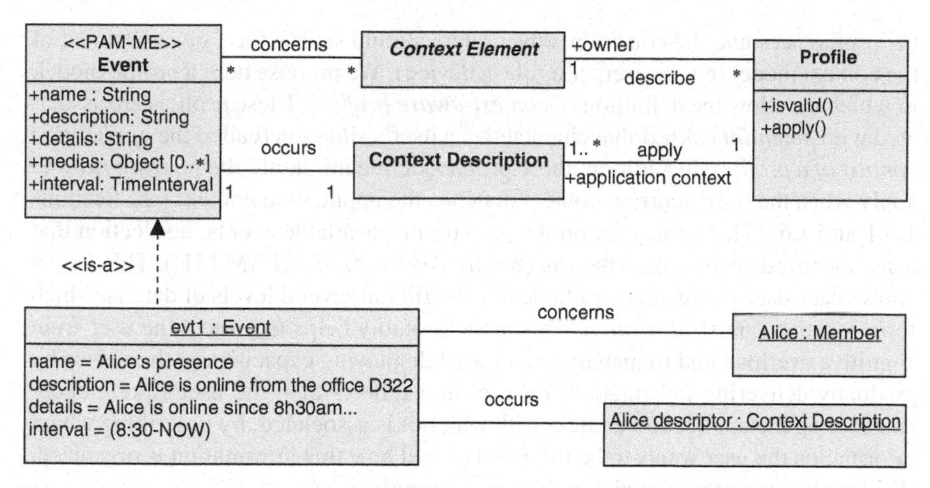

Fig. 1.4 UML class diagram describing the class `event` and an instance of this class

The `event` class has to be specialized by the system designer, when developing the system. She or he has to define the available event subclasses according to the users' interests toward the system. The `event` class contains some attributes that we consider as minimal in order to describe informational content (see Figure 1.4): an event `name` (for instance, "Alice's presence"), a `description` (e.g., "Alice is online from the office D322"), some `details` about it (i.e., a more detailed description, like "Alice is online since 8:30 a.m. from the office D322, third floor"), a `time interval` related to its content (for instance, "from 8:30 a.m. until now"), and some extra `media` that provide some complementary information (Alice's photo).

We also consider that an event can refer to one or more elements of the context model (see concerns association in Figure 1.4), since it can carry some information about a topic referring to these elements (for instance, the event referring to Alice's presence is linked to the object of the `member` class that refers to Alice). Additionally, each event instance is associated with a context description instance, representing the context in which the event is or has been produced (for instance, the `context description` object describing Alice's current context in Figure 1.3).

1.5 Representing User's Preferences

When considering mobile users accessing Web-based groupware systems, it is important to note that the user's interests and preferences may differ according to the context [14]: The same information may be interesting for a given user when she or he is at the office and completely useless when she or he is travelling. In this chapter, we represent user's preferences through the notion of *profiles*. A *profile* represents the preferences and the constraints the system should satisfy for a given element of the context model (e.g., a user, or a role, a device). We propose here a profile model, in which we allow the definition of *context-aware profiles*. These profiles are associated with *potential context* that characterize a user's situation (called the *application context of a profile*). In addition, these profiles define implicitly the *filtering rules* to apply when the user's current context matches the application context (see Sections 1.5.1 and 1.6). These rules act on the selection of available events, a selection that is personalized according to the *Progressive Access Model* (PAM [31]). This model allows each user to organize available information in several levels of details, which form together one *stratification*. This model notably helps to protect the user from cognitive overload and to manage the limited displaying capacities of the device by gradually delivering information. Each profile reflects then the user's preferences considering the application context with which it is associated, by describing which information the user wants to be informed of and how this information is organized. We describe this profile model in the next subsections.

1.5.1 Filtering Rules

This section presents a model for representing filtering rules based on the user preference model proposed in [7]. Filtering rules are expressed using a set of preferences, referring to user or group's preferences concerning content supplied by a system. In this chapter, we consider two types of preferences (see Figure 1.5): content and display preferences.

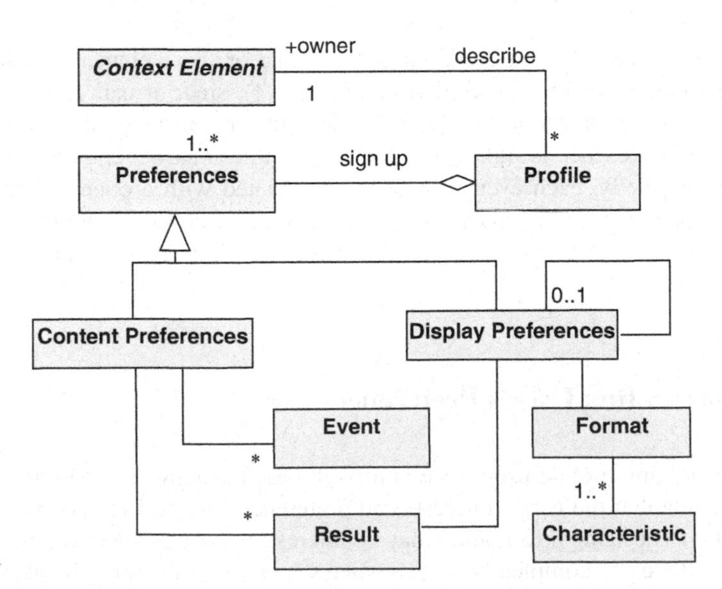

Fig. 1.5 UML class diagram of user preferences and their relations

The *content preferences* are built upon content selection queries (selection of events). They are defined as follows:

$$Content_{Pk} = Content\_Preference(\{content\}, \{condition\}, substitution)$$

where the *content* clause refers to a set of content classes (event subclasses) considered as relevant, *conditions* refers to query conditions for selection purposes (e.g., $event.TimeInterval > 24h$), and *substitution* correspond to a possible alternative content preference used when the first preference returns no content. This content preferences definition can be seems as a "*subscription operation*": a filtering rule subscribes to a particular content. Thus, we refer to these content preferences as the content *signed up* by a profile (see Section 1.5.2).

The *display preferences* describe the way the user wishes information to be displayed on her or his MD (e.g., image format). These preferences based on [7] are defined as follows:

$$Display_{Pk} = Display\_Preference(format, \{characteristics\}, substitution)$$

In these tuples, *format* can take as values: "*video*", "*text*", "*image*" or "*audio*" and *characteristics* specify the values taken by the attributes that characterize the format. The term *substitution* corresponds to another *display preference* that the system will try to use alternatively if the first display preference cannot be satisfied (substitution can take as value *nil*). The *display preference* P_1 given below corresponds to a preference for the display of a video, giving its dimensions and the file type (e.g., width, height, or type):

$$P_1 = Display\_Preference(video, \{200, 300, AVI\}, P_2)$$

where P_2 is the *substitution preference* of P_1 and contains the characteristics for the text (police, size, color, or file type):

$$P_2 = Display\_Preference(text, \{Arial, 10, bleu, .doc\}, nil)$$

In order to acquire user preferences, we can take into consideration information (according to Kassab and Lamirel [19]): (i) provided by the user by means of dedicated interfaces; (ii) defined like general user profiles; (iii) deduced from the history of the user, i.e., from her or his previous sessions (such as in [11]). In this chapter, we do not tackle directly this issue, considering that the preferences are provided by the users. Moreover, it is worth noting that other elements could compose a user profile (see, for instance [11, 22]), such as a user expertise level, sociocultural characteristics (e.g., gender, age, cultural background...), etc. These elements are not studied here since they are out of the scope of our proposal.

1.5.2 Context-Aware Profiles

The basic idea behind the profiles is to allow users, system designers, or administrators to define the profiles and the application contexts related to them. An application context is the description of a potential context, described using the context model (see Section 1.3). Profiles determine what information is considered as relevant given an application context, as well as how this information is organized in levels of detail. For instance, a system designer can create profiles whose application context describes the situation in which a given device is used. A system administrator can also define profiles for the situations in which users are playing some particular roles, like the coordinator or administrator roles.

Mobile users often access Web-based information systems in well-known situations, for which it is possible to identify elements that characterize them: from rail stations, from airports, from homes, using a particular PDA, a cellphone, a laptop, etc. By describing the application context corresponding to these potential situations and defining particular profiles for them, each user can express her or his needs and preferences for the most common situations, allowing the system to better adapt the content delivered in these situations. For instance, considering a system with wiki functionalities, its designer can predefine some event subclasses such as a "*document changed*" class, whose objects describe the changes performed in a document, or a "*new comment*" class, whose objects include the comments made by the group members about a document. In such cases, the user "Alice" can define a profile whose preferences sign up only the "comment" events, and associate this profile to the situations in which she is using her PDA (i.e., a profile that is valid when she is using this device).

Therefore, in the profile model we propose, each profile can be seen as a set composed by the following:

- An *owner*, for whom or what the profile is defined. The owner is represented by a context element object (association described in Figure 1.4), allowing the definition of a profile either for users or for any element of the context model.
- At least one *application context* to be considered, which represents the potential contexts in which this profile can be applied, i.e., the situations in which the profile is valid (association applied in Figure 1.4) and should be selected by the filtering mechanism (see Section 1.6).
- A set of content and display preferences (association *sign up* in Figure 1.6), representing the informational content considered as relevant for the owner.
- A set of *contextual conditions* that filters events instances of the signed-up classes.
- A set of *stratifications* defined for the owner (see Figure 1.8).

The stratifications together with the preferences and the contextual conditions define a complex set of filtering rules aiming at selecting the content related to the profile. The contextual conditions (see Section 1.5.3) and the content preferences (see Section 1.5.1) guide the filtering process, allowing the selection of information

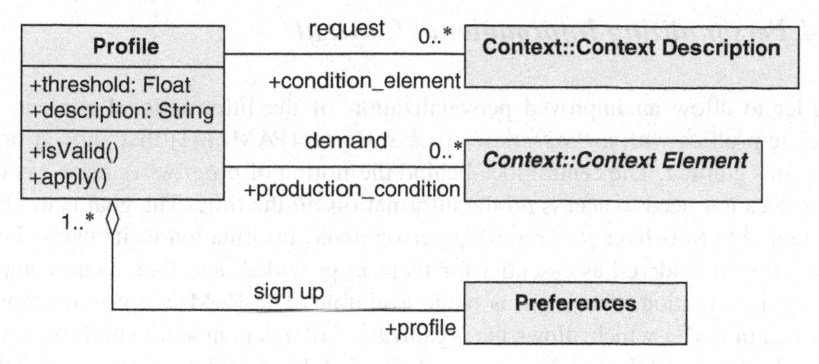

Fig. 1.6 The signed-up event classes and the contextual conditions that define the filtering rules of each profile

that matches only to the content (event classes) explicitly indicated as relevant by the preferences and that respects the contextual conditions imposed by the profile definition.

1.5.3 Contextual Conditions

In order to present the contextual conditions, let us consider first an illustration: Alice has defined a profile indicating that she is interested only in events linked to the "new comment class" that have been produced in a given location (her office) and that concern a given document. The first argument (events from "new comment" class) corresponds to the signed-up events (content preferences), the second one (event in a given location) is a *request condition*, whereas the third one (events concerning a document) correspond to a *demand condition*.

The contextual conditions defined in this profile model are represented by the request and the demand associations in Figure 1.6. These conditions analyze the context in which events have been produced, as well as the context elements concerned by the events. The definition of a "*request*" condition in a profile indicates that only events that are directly related to a given context should be selected, whereas the definition of a "*demand*" condition means that only events that concern some specific context elements should be selected. These contextual conditions exploit occurs and concerns associations proposed in the content model (see Figure 1.4), in order to select events based on this information. Thus, when applying a profile in which these conditions are expressed, these conditions are combined to the signed-up indication, recommending this way that only signed-up events whose occurs- and concerns-related objects match respectively the request and the demand conditions have to be selected. This matching is performed by the filtering process (see Section 1.6).

1.5.4 Personalizing Informational Content

In order to allow an improved personalization of the informational content, we associate profiles with a *Progressive Access Model* (PAM) [31] that aims at organizing this content. The central idea behind the notion of *progressive access* is that a user does not need to access *all* the information *all* the time. The goal is to make a system able to deliver *progressively* personalized information to its users: First, information considered as essential for them is provided, and then some complementary information, if needed, is made available. The PAM is a generic model, described in UML, which allows the organization of a data model in multiple levels of details according the user's interests. This model is based on some basic definitions that we present below (more details can be found in [31]).

The notion of progressive access is related to the one of *maskable entity*. A maskable entity (ME) is a set of at least two elements (i.e., $|ME| \leq 2$) upon which a progressive access can be set up. The progressive access to a ME relies on the definition of *representations of maskable entity* (RoME) for this ME. These RoME are subsets of the ME ordered by the set inclusion relation. Each RoME of a ME is associated with a level of detail. Thus, $RoME_i$ is defined as the *RoME* of a ME corresponding to the level of detail i, where $1 \leq i \leq$ max, and max is the greatest level of detail available for this ME.

Figure 1.7 illustrates a ME with three associated RoMEs. Some rules impose that a $RoME_{i+1}$ associated with the level of detail $i+1$ $(1 \leq i \leq max-1)$ contains at least one more element than $RoME_i$. A stratification for a ME is a sequence of RoME ordered by set inclusion as illustrated in the Figure 1.7. Please note that several and different stratifications can be defined for a given ME.

The progressive access relies then on several operations that allow to switch from a RoME to another within a given stratification: mask, unmask, previous, and next. The mask and unmask operations preserve the information of the previous levels when navigating forward or backward inside the stratification, whereas next and previous operations do not preserve this information, presenting the information

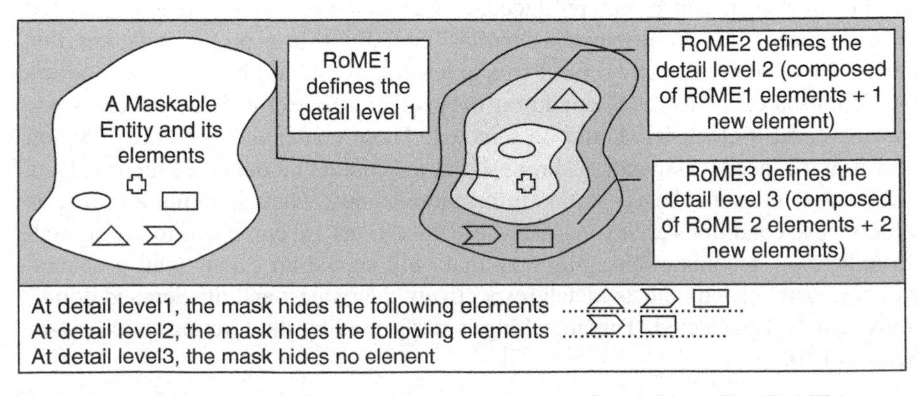

Fig. 1.7 A maskable entity (MEs) with three representatives of maskable entities (RoMEs)

of each level separately. These operations are defined as follow (a more detailed description is available in [21,31]):

- First of all, let us define a basic operation called `proper(i)`, which isolate the elements included at a level *i*:
  - `Proper`($RoME_i$) $= RoME_i$ if $i = 1$, otherwise `proper`($RoME_i$) $= RoME_i - RoME_{i-1}$, where $2 \leq i \leq \max$.
- From a $RoME_i$, at level of detail *i*, the *mask operation* gives access to the $RoME_{i-1}$, and to its possible predecessors as well, at level of detail $i-1$:
  - `Mask`$(RoME_i) = RoME_{i-1}$, $2 \leq i \leq \max$.
- From a $RoME_i$, at level of detail *i*, the *unmask operation* gives access to the $RoME_{i+1}$ at level of detail $i+1$:
  - `Unmask`$(RoME_i) = RoME_{i+1}$, where $1 \leq i \leq \max - 1$.
- From a $RoME_i$, at level of detail *i*, the *previous operation* gives access only to the elements of the $RoME_{i-1}$ at level of detail $i-1$:
  - `Previous`$(RoME_i) =$ `proper`$(RoME_{i-1})$, where $2 \leq i \leq \max$.
- From a $RoME_i$, at level of detail *i*, the *next operation* gives only access to the elements of the $RoME_{i+1}$ at level of detail $i+1$:
  - `Next`$(RoME_i) =$ `proper`$(RoME_{i+1})$, where $1 \leq i \leq \max - 1$.

In the proposed profile model, we consider the event class (and its subclasses) as a *maskable entity* when applying the *Progressive Access Model*. In other words, stratifications can be defined for the event class and its instances. Figure 1.8 shows the associations among `stratifications`, `profiles`, and `events` in the profile model. Through this model, we may then define different stratifications by associating, for instance, the attributes of the event class (and its subclasses) with the levels of detail in a stratification `s1 = {{name, interval}, {details}}`, or a stratification `s2 = {{name}, {interval, description}, {medias}}`. It is worth to note that this model allows the definition of many stratifications for a given event class,

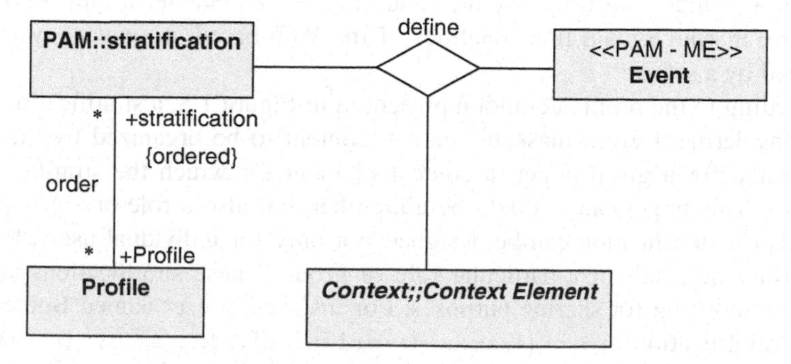

Fig. 1.8 Associations among stratifications, events, and profiles

by assigning it to different stratifications or to different profiles (and, consequently, to different application contexts). Besides, it is important to observe that stratifications can be defined for any context element (e.g., user, group, role, or device), not only users (see association definition in Figure 1.8). This is particularly interesting for considering the constraints of some mobile devices.

To sum up, the profile model we propose allows users, system administrators, and designers to define context-aware and personalized profiles. Each profile answers four important questions for the filtering mechanism: *what* information should be selected, for *who or what* is it selected, *when* should be it available, and *how* should be it organized.

1.5.5 Sharing Profiles

In a group or community, each member as an individual shares with other community members some goals or interests, and also contributes to the community by providing her or his particular knowledge and competences. Let us consider, for instance, researchers belonging to a research group or to a special interest community. If these researchers have similar interests, each of them has also her or his own speciality and her or his knowledge contributes to the group research activity. In this way, improving individual work inside a community benefits the community as a whole. Our approach aims at such enhancement: by adapting the content supplied to users, we are optimizing their chances of improving their own contribution to the community. As a step further, the definition of context-aware profiles, adapted to particular situations encountered by an individual user can reveal significance for the whole community (or at least for some of its members).

We claim that the possibility of sharing profiles inside a community represents an important knowledge-sharing mechanism for its members. Through this sharing, members of the community exchange with their colleagues stratifications they consider as interesting or particularly useful. This sharing can lead, over the time, to the improvement of the stratifications themselves (through the users' experience and feedback).

In order to allow user, to perform this knowledge sharing, all stratifications are stored in a central repository. We are assuming in this chapter a typical Web architecture, in which main functionalities of the Web-based groupware system are performed by a server.

According to the profile definition presented in Figure 1.8, a stratification is defined considering a given maskable entity (content to be organized by the stratification) and for a given target (a context element for which the stratification is designed). This target can not only be a member, but also a role or a group. This means that a stratification can be designed not only for individual users, but also considering the needs of a particular role or group. These stratifications are particularly interesting for sharing purposes. For instance, a user named Bob can define the stratification s_{137} = `{{name, description}, {details, interval}}`,

```
<pam>
    <stratification id="137">
        <me>
            <class name="BWM.Kernel.BWM_Event"></class>
        </me>
        <rome position="1">
            <element name="name"></element>
            <element name="description"></element>
        </rome>
        <rome position="2">
            <element name="details"></element>
            <element name="interval"></element>
        </rome>
        <usercategary>
            <role name="BWM_Kernel.BWM_Role" refid="coordinator"/>
        </usercategary>
    </stratification>
</pam>
```

Fig. 1.9 Example of shared stratification, expressed in XML

represented in XML in Figure 1.9, for the role "coordinator." This stratification is stored in the Web server common repository, allowing the user Alice (who also plays the role of coordinator) to use this stratification defined by Bob, by associating it with one of her profiles.

1.6 Filtering Process

The adaptation approach we propose here is based on a filtering process in *two steps*. The first step selects the profiles the system should apply in order to filter the available information, according the user's current context. The second step consists in applying the filtering rules defined by the selected profiles. These rules are based on the *signed up* event classes and their contextual conditions, and on the stratifications associated with a profile. We assume that for each user a set of predefined profiles are available (available profiles are those owned by the user). Thus, the first step selects, among the available profiles, those the system has to apply, and the second step applies the filtering rules defined in the selected profiles.

1.6.1 Selecting Profiles

The first step of the filtering process consists in selecting among the profiles those that are valid with regard to the user's current context. This selection is performed by comparing the *application context* related to the available user's profiles with the user's current context. Please note that these two kinds of contexts are both instances of the class context description of the context model. For each profile, we test if one of its *application contexts* has the same content or is a subset of the *user's current context description*. In other words, we verify if the situation described by the application profile occurs in the user's current context. If it is the case, then the profile is selected to be applied.

In order to identify this subset relationship, we consider that each `context description` instance and the `context element` instances associated with it define a graph, where the nodes represent the instances and the edges between them represent the tuples of associations involving these instances. Thus, a context C is a subcontext of a context C' whenever the graph corresponding to C is a subgraph of the graph corresponding to C'. The subgraph relationship is established using a pattern-matching algorithm, which is based on two operations (*equals* and *contains*), defined as follow:

- *Equals*: (*i*) a node N is considered as *equal* to a node N' if the object O represented by N belongs to the same class and defines the same values for the same variables that the object O' represented by N'. (*ii*) an edge E is *equal* to an edge E' if the associations they represent belong to the same type (tuples of the same association) and connect *equal* objects.
- *Contains*: a graph C contains a graph C if: (*i*) for each node N' that belongs to C'(called $N'_{C'}$), there is a node N belonging to C (N_C) for which $N'_{C'}\, equals N_C$; (*ii*) for each edge E' in C'(called $E'_{C'}$), there is an edge E in C (E_C) for which $E'_{C'}\, equals E_C$.

Thus, we consider that a context description C is a subset of a context description C if the graph defined by C *contains* the graph defined by C. For instance, let us consider the user *Alice* who is consulting the notes about a document, as shown in Figure 1.3. Let us suppose that Alice has defined two profiles: (*i*) a first one that includes in its *application context* an instance referring to her PDA (i.e., a profile for the situations in which she is using this device). (*ii*) And a second profile that is applicable only when she is involved in an `activity` which concerns a given document and when she is working on her desktop device. This means that the *application context* related to this profile includes the context elements corresponding to the `shared object` *report2007* and to the device *desktop*. When Alice finds herself in the situation described by the Figure 1.3, only the first profile is selected. The second one is rejected since the `context description` object representing the application context of this profile does not match the user's context description (the latter does not include a node referring to the desktop device that is present in the former, so it does not contain application context of this profile). Figure 1.10 represents the graphs defined by `context description` objects related to Alice's current context (Figure 1.10(a)) and her profiles (Figure 1.10(b) and (c) respectively). In this figure, we observe that the graph defined by the application context of the second profile is not a subgraph of the graph defined by Alice's context, while the graph defined by the application context of the first profile is a subgraph of the one defined by Alice's context.

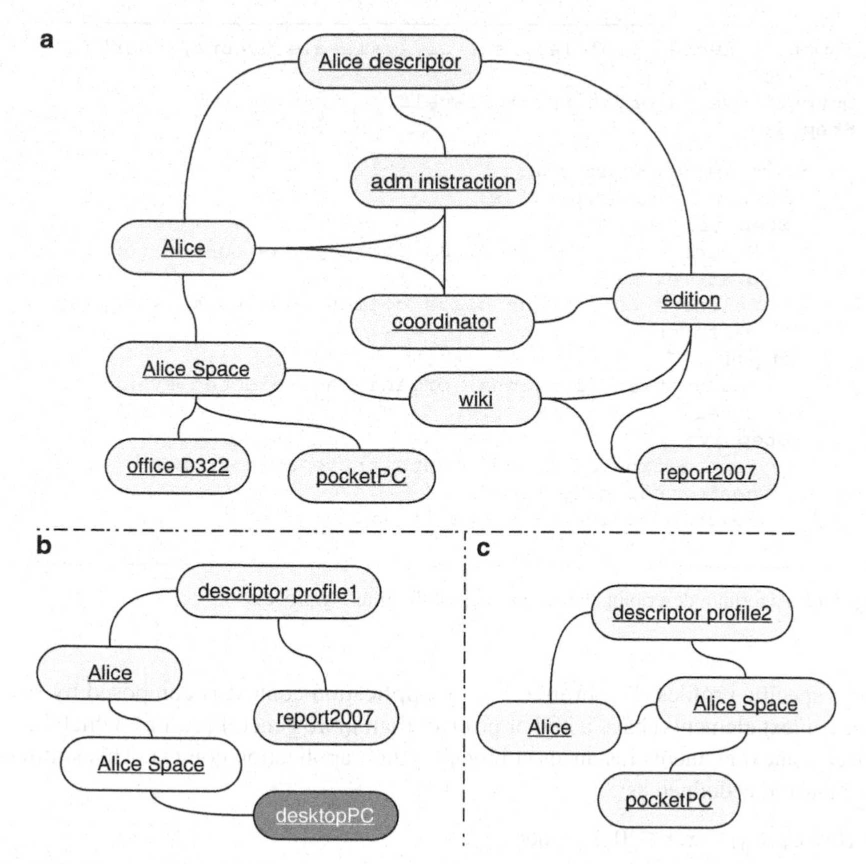

Fig. 1.10 The graphs defined by a user's current context (**a**), and the application context of two profiles (**b** and **c**)

1.6.2 Selecting and Organizing Content

Once the profiles have been selected, the second step of the filtering process applies the filtering rules defined in the selected profiles. This implies to filter the available set of events according to the contextual conditions (see Section 1.5.4) and to the content preferences indicated by the profile. Then the selected events have to be organized according to the stratifications defined in the profile. It is worth noting that, in order to respect the stratifications definition, we apply one profile at time by ordering the profiles. This is process achieved by the algorithm given in Figure 1.11.

The priority of a profile is defined by a similarity measure between the application context of the profile and the user's current context. This measure evaluates the matching degree of the application context with the user's context. In other words, it estimates the proportion of elements of the graph defined by the user's context that have equals elements in the graph defined by the application context. Therefore,

```
Input: selected profiles, set of available events, user's
context
Output: events organized in levels
Step i:
  Order selected profiles
  For each selected profile:
    Step ii:
      Select events whose class corresponds to the content
      preferences
      Select events whose class corresponds to the display
      preferences
    Step iii:
      Apply stratifications, organizing selected events in
      levels
    Step iv:
      Remove from selected events those that do not respect
      contextual conditions
      Remove duplicate events if any
```

Fig. 1.11 Algorithm describing the second step of the filtering process

more specific profiles (i.e., profiles whose application context is composed by several context elements) have a higher priority than more general profiles, which have fewer context elements instances composing their application context. This similarity measure is defined as:

- $Sim(C_u, C_p) = x, x \in [0,1]$, where:
  - $x = 1$ if each element of C_u has an equal element in C_p
  - otherwise $x = |X|/|C_u|$ with $X = \{x \mid x \text{ equals } y,\ x \in C_u,\ y \in C_p\}$

As an illustration, let us consider once again the example of a Web-based groupware system with wiki functionalities. The designer of such a system may have defined three event classes: *new comment*, *document changed* and *user presence* (indicating online users). Let us consider now that a user (Alice) has two selected profiles. The first profile, whose application context is illustrated by Figure 1.10, `signs up` the first `event` class, defining one stratification `S1 = {{name, interval}, {description}}`. The second profile `signs up` the third `event` class, defining the stratification `S2 = {{name}, {interval, description}, {medias}}`. Supposing that application context associated to the second profile refers to *Alice* using her *pocketPC* to access the *wiki* application, when comparing these two profiles, the second one has a higher priority than the first one, since $Sim(C_{Alice}, C_{P3}) = 0.42$ for the second one and $Sim(C_{Alice}, C_{P1}) = 0.38$ for the first one. This means that instances of the *user presence* `event` class will be available for Alice before instances of *new comment* `event` class. Moreover, instances of the *user presence* `event` class are organized according to S_2 stratification, whereas instances of the *new comment* `event` class are organized according to S_1 stratification.

Over this set of selected events, we apply the *contextual conditions* associated to each profile (request and demand conditions defined in Section 1.5.2). In order to evaluate *request condition*, we use *contains* operation: if the `context description` associated to the event does not contains the `context description` indicated in the *request condition*, then the event is removed from the selection. For evaluating *demand condition*, we use *equals* operation that allows us to compare the `context element` objects associated with each selected events and those pointed out in the *demand condition* associated with the profile. If one of the objects of the demand condition does not match any object related to the event, it is automatically removed from the selected events. Through these *contextual conditions*, we expect to reduce the set of selected events and to allow users to defined priority information related to a specific context. However, it worth noting that if a selected event is associated with no `context description` or `context element`, these contextual conditions do not apply, since we do not have enough knowledge about context to compare it with these conditions. In this case, event remains selected. At the end of the filtering process, an organized (in levels of details) set of events will be available for delivering to the mobile user, as illustrated by Figure 1.12. After that, the user can navigate into this set by using the operations defined by the *Progressive Access Model* (see Section 1.5.4). Besides, we believe that this set will probably better suit the current mobile user's context since it accords the user's preferences defined for her or his current context.

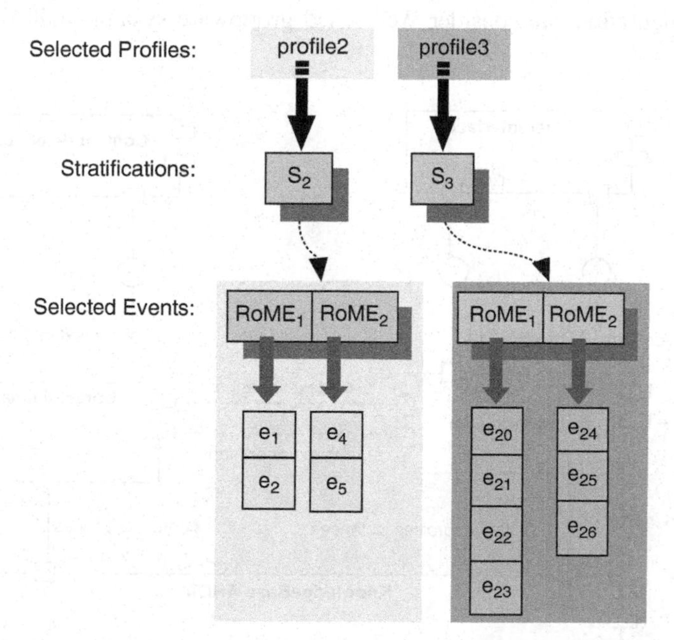

Fig. 1.12 Filtering results for Alice: selected events are organized according the stratifications defined by the selected profiles

1.7 Implementation

We have implemented the proposed filtering process in a framework called BW-$\mathcal{M}$, which uses an object-based knowledge representation system called AROM system [25]. We choose the AROM system since it proposes classes or objects and associations or tuples as main representation entities, allowing a quick translation of the proposed models into a knowledge base. Using AROM, we have created a knowledge base (KB) in which we keep the instances of the context model as well as the profiles and the events. We used the Java API supplied by the AROM system in order to manipulate these instances, as well as the corresponding models, from the BW-$\mathcal{M}$ framework. This allows the designer of a Web-based groupware system using this framework to easily adjust the context representation and the content model to the system and its purposes. Additionally, we have encapsulated this framework in a Web service, named `BWMFacade`, in order to facilitate the use of the BW-$\mathcal{M}$ framework into Web-based systems. Next subsections introduce the BW-$\mathcal{M}$ Framework and discuss some results.

1.7.1 BW-$\mathcal{M}$ Framework

The main structure of the BW-$\mathcal{M}$ framework is represented in Figure 1.13. For this implementation, we consider Web-based groupware systems built on the top

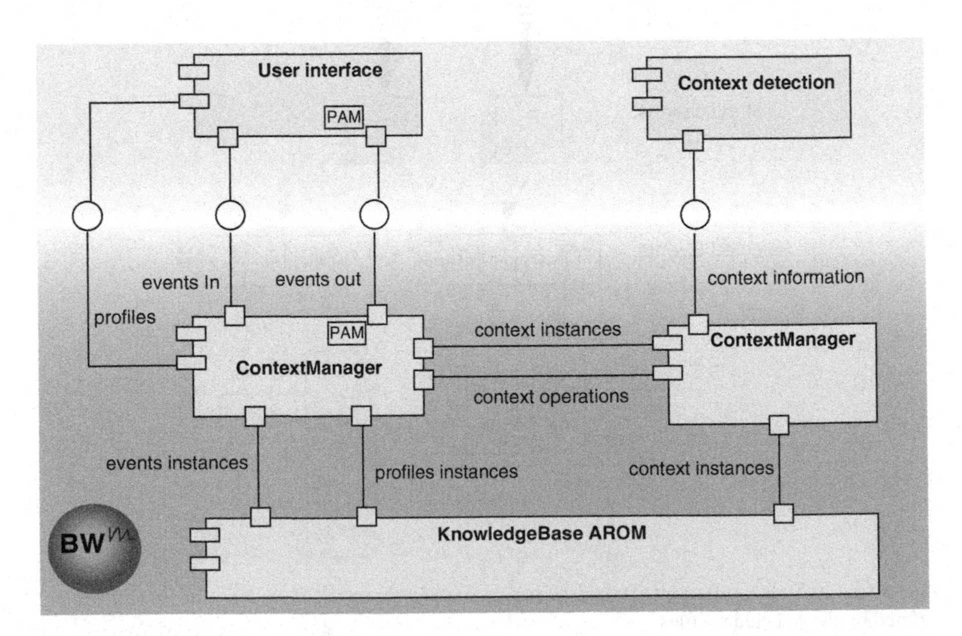

Fig. 1.13 BW-$\mathcal{M}$ framework architecture

of different independent components (or services) that interact while forming the system. The underlying assumption of this organization is that BW-$\mathcal{M}$ does not handle directly user's interface as well as context detection. We suppose that other components in the system, such as toolkits or middleware (e.g. [12, 17, 27, 28] are able to detect dynamically the elements of the user's context. Through these components, the system is able to identify the user and the activities she or he is performing, her or his physical location, her or his client device, etc. However, it is worth noting that the system cannot always determine all context elements. For instance, a system that uses the GPS technology may be unable to determine the *location* of a mobile user who is in an indoor environment (or whose GPS receiver battery level is too low). The BW-$\mathcal{M}$ framework (and the AROM knowledge base) allows the system to handle these situations by omitting unknown elements and attributes (that remain unvalued in this case) from the user's context description. This omission represents the fact that the system does not have enough knowledge to represent an element (or parts of it).

The BW-$\mathcal{M}$ framework is composed by three main components (see Figure 1.13): the *content manager*, the *context manager*, and the *AROM knowledge base*, which keeps the instances of the profile, content, and context models described in this chapter. The *content manager* component handles and filters the events that represent the awareness information. In our experiments, we have used a set of predefined events stored in the *knowledge base*. The content manager also receives the filtering requests from the Web-based groupware system. These requests trigger the filtering mechanism described in Section 1.6. Finally, the *context manager* handles the context information it receives from the context detection components. It transforms this information in the corresponding instances of the context model and stores them in the knowledge base. Moreover, the context manager also implements the equals, $\mathit{contains}$ and Sim operations, which are used by the *content manager* in order to compare the instances of the context model (see Section 1.6).

1.7.2 Implementation Issues

Using the BW-$\mathcal{M}$ framework, we have performed a set of tests, simulating situations against a set of predefined profiles and events. These tests have raised some interesting points, mainly considering the equals and $\mathit{contains}$ operations, which are key aspects of the filtering process. The tests have demonstrated that the performance of the filtering process strongly depends on these operations. Tests have also showed that these operations, as defined in this chapter, are too restrictive. For instance, equals operation compares all attributes that have a knowledge value in both instances. Consequently, all attributes are equally important when comparing two objects, and if one attribute has different values in the objects, these will be considered as distinct. When applied inside the $\mathit{contains}$ operation, this may lead to the nonselection of a profile or an event (when analyzing the contextual conditions) based only in the value of one attribute.

Table 1.1 Results of a BW-$\mathcal{M}$ execution test case

Operation	Contains		
Number of runs	88	Average memory	59.7613636 KB
Number of compared objects	28	Average time	43.4545455 ms
Number of total operations $n \times (n-1) \times runs$	66,528		

Based on our tests results, we have implemented new versions of the *equals* and *contains* operations that are more flexible than the original ones. These new versions include a *weight equals*, in which we associate each attribute a_i with a weight w_i ($\Sigma w_i = 1$). Thus, we consider that some attributes are more relevant than others (e.g., the `precision` attribute, which is used to indicate the precision of the information represented by the object, may get a reduced weight since it does not represent a content properly speaking). We have also implemented a new version of *contains* operation in which we allow indicating a list of objects, classes, or associations that should be ignored when comparing two graphs. Through this *ignore list*, we reduce the size of the graphs generated by the comparing objects and we prevent comparing instances that are associated with the context instances, but do not belong to the context description (e.g., profile and event objects related to `context description` objects).

With these new versions, our tests pointed out that the execution of the filtering process does not represent a significant bottleneck for Web-based groupware systems. Table 1.1 summarizes one of the test cases we have performed. In this test case, we have compared n objects in the knowledge base using a given operation and evaluated the memory and time consumption for executing $n \times (n-1)$ times the operation. We observe that the pattern-matching algorithm implemented by contains operation is not time consuming, even if it establishes subgraph relationships. However, it is memory consuming, since AROM system keeps the knowledge base active in the memory. Thus, we believe that this implementation fits the server implementation we choose by using the Web service `BWMFacade` encapsulating the BW-$\mathcal{M}$ framework. This architecture corresponds to a client-server solution (typical for Web-based system) in which the filtering process is performed by the server. However, the use of this implementation in the client side may represent some performance risk for constrained MD, such as PDAs. Nevertheless, with the recent evolution of MDs and their processor and memory capacities, memory consuming should not represent a problem in the near future.

1.8 Conclusion

In this chapter, we have presented a context-based filtering process that proposes to adapt the information delivered to mobile user by filtering it according to the current user's context and to the user's preferences for this context. In this approach we propose to associate the user's preferences with the user's context, allowing the expression of the user's preferences for specific situations. We believe that by allowing a direct participation of the users into the adaptation process, this process might provide them with results that suit more their expectations. Moreover, we allow users to share these preferences with other members of the community, allowing a progressive refinement of the profiles description by the reuse of profiles (and stratifications) that are considered as appropriate by the community members.

Besides, by using the *Progressive Access Model*, we not only filter the informational content delivered to the user (such as in [6, 34] but we also organize it according to its relevance for the user. This represents a clear improvement, since mobile users usually do not dispose of enough time or resources to analyze all information. Organizing the delivered information by relevance becomes then necessary, as pointed out by Billsus et al. [3] and by Coppola et al. [10].

We expect to extend the proposed filtering process by refining the pattern-matching algorithm used to compare instances of the context description class. We are interested in calculating an acceptable semantic distance between the objects, in order to check if they are sufficiently similar (and not necessarily equal) to establish a subgraph relationship. We are also interested in studying how to simplify the profile definition in order to reduce problems that may be caused by an incorrect definition.

References

1. Ankolekar, A., Krotzsch, M., Tran, T., Vrandecic, D., The two cultures: mashing up web 2.0 and the semantic web. In: Proceedings of the 16th international conference on World Wide Web (WWW '07), pp. 825-834. ACM Press, New York, NY, USA (2007). DOI http://doi.acm.org/10.1145/1242572.1242684
2. Bardram, J., The java context awareness framework (jcaf)- a service infrastructure and programming framework for context-aware applications. In: H. Gellersen, R. Want, A. Schmidt (eds.) Third International Conference in Pervasive Computing (Pervasive'2005), *Lecture Notes in Computer Science*, vol. 3468, pp. 98–115. Springer (2005)
3. Billsus, D., Brunk, C., Evans, C., Gladish, B., Pazzani, M., Adaptative interfaces for ubiquitous web access. Communications of the ACM **45**(5), 34–38 (2002)
4. Borges, M.R., Brézillon, P., Pino, J.A., Pomerol, J.C., Groupware system design and the context concept. In: W. Shen, Z. Lin, J.P.A. Barthes, T. Li (eds.) 8th International Conference Computer Supported Cooperative Work in Design – Revised Selected Papers (CSCWD 2004), *Lecture Notes in Computer Science*, vol. 3168, pp. 45–54. Springer (2004)
5. Brickley, D., Miller, L., Foaf vocabulary specification 0.91. http://xmlns.com/foaf/spec/ (2007). URL http://xmlns.com/foaf/spec/
6. Brusilovsky, P., Adaptive hypermedia: From intelligent tutoring systems to web based education. In: G. Gauthier, C. Frasson, K. VanLehn (eds.) 5th Int. Conference on Intelligent Tutoring Systems (ITS 2000), *Lecture Notes in Computer Science*, vol. 1839, pp. 1–7. Springer (2000)
7. Carrillo-Ramos, A., Villanova-Oliver, M., Gensel, J., Martin, H., Contextual user profile for adapting information in nomadic environments. In: Personalized Access to Web Information (PAWI 2007) – 8th International Web Information Systems Engineering (WISE 2007), *Lecture Notes in Computer Science*, vol. 4832, pp. 337–349. Springer (2007)

8. Chaari, T., Dejene, E., Laforest, F., Scuturici, V.M., Modeling and using context in adapting applications to pervasive environments. In: IEEE International Conference on Pervasive Services 2006 (ICPS'06). IEEE (2006)
9. Colajanni, M., Lancellotti, R., System architectures for web content adaptation services. IEEE Distributed Systems On-Line, Invited paper on Web Systems (2004). URL http://dsonline.computer.org/portal/pages/dsonline/topics/was/adaptation.xml
10. Coppola, P., Mea, V., Gaspero, L.D., Mizzaro, S., The concept of relevance in mobile and ubiquitous information access. In: F. Frestani, M. Dunlop, S. Mizzaro (eds.) International Workshop on Mobile and Ubiquitous Information Access (Mobile HCI 2003), *Lecture Notes in Computer Science*, vol. 2954, pp. 1–10. Springer (2004)
11. Daoud, M., Tamine, L., Boughanem, M., Chabaro, B., Learning implicit user interests using ontology and search history for personalization. In: Personalized Access to Web Information (PAWI 2007) – 8th International Web Information Systems Engineering (WISE 2007), *Lecture Notes in Computer Science*, vol. 4832, pp. 325–336. Springer (2007)
12. Dey, A., Understanding and using context. Personal and Ubiquitous Computing **5**(1), 4–7 (2001)
13. Dourish, P., Bellotti, V., Awareness and coordination in shared workspaces. In: Proceedings of ACM Conference on Computer-Supported Cooperative Work (CSCW 1992), pp. 107–114. ACM Press (1992)
14. Greenberg, S., Context as a dynamic construct. Human-Computing Interaction **16**(2-4), 257–268 (2001)
15. Grudin, J., Desituating action: digital representation of context. Human-Computing Interaction **16**(2-4), 269–286 (2001)
16. Heflin, J., Owl web ontologie language use cases and requirements. Tech. rep., W3C (2004). URL http://www.w3.org/TR/webont-req/. W3C Recommandation
17. Hightower, J., LaMarca, A., Smith, I.E., Pratical lessons from place lab. IEEE Pervasive Computing **5**(3), 32–39 (2006)
18. Indulska, J., Robinson, R., Rakotonirainy, A., Henricksen, K., Experiences in using cc/pp in context-aware systems. In: M.S. Chen, P. Chrysance, M. Sloman, A. Zas-lavsky (eds.) 4th Int. Conference on Mobile Data Management (MDM 2003), *Lecture Notes in Computer Science*, vol. 2574, pp. 247–261. Springer (2003)
19. Kassab, R., Lamirel, J., An innovative approach to intelligent information filtering. In: H. Haddad (ed.) ACM Symposium on Applied Computing 2006 (SAC 2006), pp. 1089–1093. ACM Press (2006)
20. Kirsch-Pinheiro, M., Gensel, J., Martin, H., Representing context for an adaptative awareness mechanism. In: 10th Int. Workshop on Groupware (CRIWG 2004), *Lecture Notes in Computer Science*, vol. 3198, pp. 339–348. Springer (2004)
21. Kirsch-Pinheiro, M., Villanova-Oliver, M., Gensel, J., Martin, H., A personalized and context-aware adaptation process for web-based groupware systems. In: 4th Int. Workshop on Ubiquitous Mobile Information and Collaboration Systems (UMICS'06), CAiSE'06 Workshop (2006), pp. 884–898 (2006)
22. Kurumatani, K., Mass user support by social coordination among citizen in a real environment. In: Proceedings of the International Workshop on Multi-Agent for Mass User Support (MAMUS 2003), *Lecture Notes in Artificial Intelligence*, vol. 3012, pp. 1–16. Springer (2003)
23. Lemlouma, T., Layaida, N., Context-aware adaptation for mobile devices. In: IEEE International Conference on Mobile Data Management, pp. 106–111. IEEE Computer Society (2004)
24. Moran, T., Dourish, P., Introduction to this special issue on context-aware computing. Human-Computer Interaction **16**(2-3), 87–95 (2001)
25. Page, M., Gensel, J., Capponi, C., Bruley, C., Genoud, P., Ziébelin, D., Bardou, D., Dupierris, V., A new approach in object-based knowledge representation: The arom system. In: L. Monostori, J. Váncza, M. Ali (eds.) 14th International Conference on Industrial and Engineering Applications of Artificial Intelligence and Expert Systems (IEA/AIE 2001), *Lecture Notes in Artificial Intelligence*, vol. 2070, pp. 113–118. Springer (2001)

26. Preuveneers, D., Vandewoude, Y., Rigole, P., Ayed, D., Berbers, Y., Context-aware adaptation for component-based pervasive computing systems. In: Advances in Pervasive Computing 2006, Adjunct Proceedings of the 4th International Conference on Pervasive Computing, pp. 125–128 (2006)
27. da Rocha, R., Endler, M., Context management in heterogeneous, evolving ubiquitous environments. IEEE Distributed Systems Online **7**(4) (2006). URL http://dsonline.computer.org/
28. Rubinsztejn, H., Endler, M., Sacramento, V., K. Gon c., Nascimento, F., Support for context-aware collaboration. In: A. Karmouch, L. Korba, M.E. L. (eds.) 1st International Workshop on Mobility Aware Technologies and Applications – MATA 2004, *Lecture Notes in Computer Science*, vol. 3284, pp. 37–47. Springer (2004)
29. Schilit, B., Adams, N., Want, R., Context-aware computing applications. In: Proceedings of the IEEE Workshop on Mobile Computing Systems and Applications, pp. 85–90 (1994)
30. Schmidt, K., The problem with 'Awareness': introductory remarks on 'Awareness in CSCW'. Computer Supported Cooperative Work **11**(3-4), 285–298 (2002)
31. Villanova, M., Gensel, J., Martin, H., A progressive access approach for web based information systems. Journal of Web Engineering **2**(1 & 2), 27–57 (2003)
32. Weiser, M., Some computer science issues in ubiquitous computing. Communication of the ACM **36**(7), 75–84 (1993)
33. Weiser, M., The futur of ubiquitous computing on campus. Communication of the ACM **41**(1), 41–42 (1998)
34. Yang, S., Shao, N., Enhancing pervasive web accessibility with rule-based adaptation strategy. Expert Systems with Applications **32**(4), 1154–1167 (2007)

Chapter 2
Framework for Supporting Web-Based Collaborative Applications

Wei Dai

Abstract The article proposes an intelligent framework for supporting Web-based applications. The framework focuses on innovative use of existing resources and technologies in the form of services and takes the leverage of theoretical foundation of services science and the research from services computing. The main focus of the framework is to deliver benefits to users with various roles such as service requesters, service providers, and business owners to maximize their productivity when engaging with each other via the Web. The article opens up with research motivations and questions, analyses the existing state of research in the field, and describes the approach in implementing the proposed framework. Finally, an e-health application is discussed to evaluate the effectiveness of the framework where participants such as general practitioners (GPs), patients, and health-care workers collaborate via the Web.

2.1 Introduction

Services account for large part of the world economy in the developed nations and increasingly large part of developing economies. Services science [1, 13] focuses on theories and methods from many different disciplines at problems that are unique to the service sector.

The dynamic nature of services increases the complexity of service management that attracted various related efforts offering solutions from certain aspects. For example, in an e-health environment, there can be new services arriving from a new primary care provider, while due to the workload some existing services become temporally unavailable. Another example is when a patient goes to a GP with a

W. Dai
School of Management and Information Systems, Victoria University, PO Box 14428, Melbourne City MC, Victoria 8001, Australia
e-mail: wei.dai@vu.edu.au

J.T. Yao (ed.), *Web-Based Support Systems*, Advanced Information and Knowledge Processing, DOI 10.1007/978-1-84882-628-1_2,

fever, responses (i.e., services) from the GP can range from a simple straightforward solution to a complicated one depending on the information collected on the patient.

Many firms are under increasing pressure from globalization, automation, and self-service, with which service chains can be assembled to form quickly strong competitive threats [14]. Providing services cost-effectively and rapidly especially in today's challenging e-business environment forces companies to interchange documents and information with many different business partners. The new competitors have created network organizational models that increase their innovative capacity and sensitivity to users in defining new products and applications.

2.1.1 Barriers and Obstacles

In a rapidly changing and increasingly complex world, service innovation requires new skills and associated underpinning knowledge to marshal diverse global resources to create value. Frequently, these resources are accessed using advanced information technology. Most academics and government policy makers are still operating in a manufacturing production paradigm rather than in a services paradigm. Change is slow, and this has a negative impact on service innovation rates.

There are many reasons why the shift to a new logic based on services has been slow to happen [26]. Nevertheless, pioneers in service research have reached initial satisfactory results and are calling for a wider range in service research [23].

The research presented here aims to explore the possibility of bringing coherence into the emerging strands of knowledge through a framework supporting Web-based collaborative applications. It builds on the theoretical foundations of service science where services are essential operations within the framework.

2.1.2 Research Motivations

There are several motivations for this research. First, it aims to incorporate knowledge management techniques [2, 22] into the proposed framework to support the automated service management. Second, it proposes mechanisms supporting intelligent use of resources based on context-aware computing [7] within the framework. It will thus enhance the framework's intelligence to process users' request by generating solution plans with the assistance of context-aware computing techniques, and dynamically managing services arrival, departure, and emergency through service management.

2.1.3 Benefits

The framework aims to provide benefits to the participants in the following way. For service providers, knowledge of services can be conveniently and effectively acquired which is going benefit a wider community of users through shared knowledge repositories. For service requestors, a dynamic service management mechanism is to be provided to meet any new requirements through cost-efficient solutions as only the required services are used during the solution process.

2.1.4 Research Questions and Aims

The main objectives therefore will prompt the following questions to be answered:

"How can service systems be understood in terms of a small number of building blocks that get combined via the Web to reflect the application requirements?" How can service systems be optimized to dynamically support services availability and autonomously reacting to changes in the application environments? The research aims to benefit users across different industry sectors and include business owners, providers, and customers who can engage with each other regardless of boundaries and locations. It also aims to make academic contributions toward real-time semantic services research through integration of knowledge management and services computing. The trend of semantic services research in the context of services oriented architecture (SOA) has been comprehensively studied in Martin et al. [17,18].

2.2 Research Background

2.2.1 Service System

The research attempts to bring together knowledge from different disciplines through a services management framework. Without a clear understanding of service system and its service management infrastructure linking different domains, knowledge will continue to be fragmented. A service system can be defined as a dynamic value co-creating sets of resources including people, technology, organization, and shared information, all connected internally and externally by value propositions, with the aim of meeting customer's needs better than competing alternatives. For instance, Fuchs [9] may have been the first to define services effectively as coproduction. Service management and service operations were proposed and extensively discussed by Fitzsimmons and Fitzsimmons [8] on service management and by Sampson [24]. Tien and Berg [25] recently demonstrated the need for service systems engineering.

Within a service system, interactions among resources take place at the service interface between the provider and the customer. Service science aims to provide a clear and common understanding of service system complexity. Hofmann and Beaumont [11] proposed an architecture to support connecting content services across network. The work focused on a fast personalized content delivery between two Internet endpoints with the architectural support to authorize, invoke, and trace application-level services. The architecture promotes open pluggable services that have compatible concepts in dynamic service management with the proposed research framework. However, it is focused on efficient user-specific content delivery between two Internet endpoints rather to explore the efficient use of existing resources (that are in the form of services) as proposed in this research. The success of modern service system development depends on integrating and delivering high-quality services within budget (i.e., the cost and quality factors). In this environment, there will be a great need to deliver robust and reliable application in less development time. To meet this challenge, software developers will seek using existing resources and methods to automate software design and deployment [16].

2.2.2 Dynamic Re-configurable System

The proposed research adopts the concepts of dynamically re-configurable systems by Hofmeister [12]. These systems use explicit architecture models and map changes into the existing models to the application implementation [20], which showed the limitations on flexibility, e.g., the existing models might not cover all the application requirements. The research on self-adaptation systems made further advancement and improvement on the dynamic aspects of the reconfiguration. Self-adaptation through dynamic reconfiguration is studied in Kon and Campell [15] and Georgiadis et al. [10]. Georgiadis et al. [10] explored the possibility of constructing self-assembling and self-adaptive applications, however, their approach required the adaptation logic written by programmers. To improve the usability of the adaptive technology and reduce development cost, research in software glue code [6] provided insights on the requirements for a complete integration solution through Integration Completion Functions (ICF), which is used as a metric in managing the effectiveness of our adaptive services toward business-oriented problems solving. However, the dynamic aspects of the solution processes in response to external requirements were not addressed. To help close this gap, Dai and Liu [4] proposed a self-adaptive system based on dynamic pluggable services supported by the OSGi framework (url: http://www.osgi.org). The OSGi specification defines a service platform that includes a minimal component model, a small framework for managing the components, and a service registry, which opens up the further services computing research as proposed.

2.3 Solution Approach

The outcome of studying service science from information technology (IT) and information systems perspectives [19] leads to a basic configuration of the proposed research framework including its fundamental components, i.e., service requestors, service providers, service registry, and repository.

2.3.1 Dynamic Services Management

The dynamic nature of services (e.g., arriving, departing, and emergency events) increases the complexity of service management, which attracted services from various aspects. Service providers, business owners, and service consumers interact through a service management infrastructure provided by the framework across the Web.

The infrastructure consists of Service Management, Service Bus, Service Registry and Repository, and Business Logic Modules interacting with each other as described in Figure 2.1. In this figure, service providers publish online services information as well as services meta data (such as those leading to services policy) into the Service Registry and Repository through the Service Management module. The Service Requestor receives the composed services via Service Bus after sending request to Service Management. Service Management responses to service request by invoking the relevant business logic modules provided by Business Owners to assist services discovery, composition, and execution before delivering results to Service Requestor.

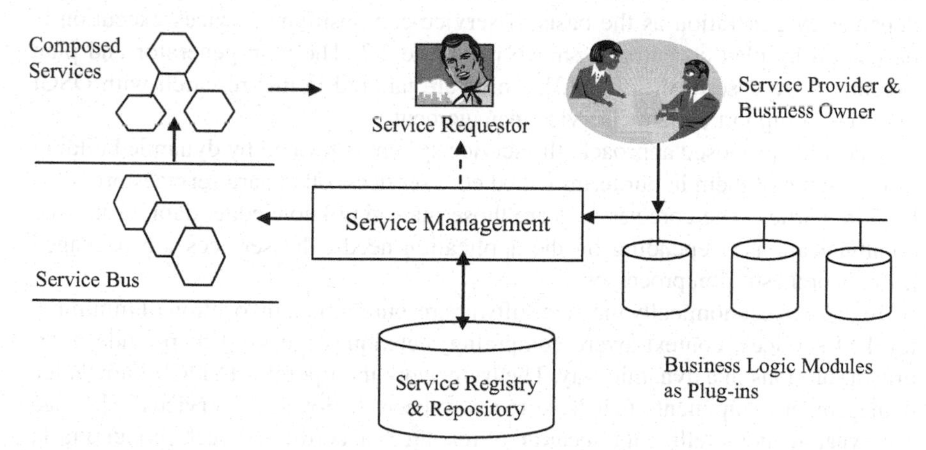

Fig. 2.1 Service framework supporting Web-based collaborations

2.3.2 Service Availability

In order to provide context-aware services management, services discovery, composition, and execution eventually need to be explored. Existing knowledge is limited to serve specific users who are knowledgeable about the services being provided. The new demands from customers may require service providers to frequently change (or modify) their services to fully satisfy the need of the customers. However, existing practice can only offer information solutions by using currently available services with predefined process or procedures [21].

One challenge here is to offer dynamic service availability where services may appear or disappear at anytime according to the sensitivity of task context. Service failures may occur, for example, when a server crashes or when a user simply walks out of wireless network range. These types of occurrences require that applications using the failed services deal with their dynamic departure. Likewise, applications may have to deal with dynamic service arrival when servers or network connections are restored or when completely new services are discovered. An important aspect of service dynamic availability is that it is not under application program control, which requires that applications be ready to respond at any time to services arrival and/or departure.

2.3.3 Services Invocation and Execution

The essential tasks, therefore, come down to service discovery to allow service management to pick up potential services, followed up by services composition and execution. The framework uses a plan generator for services-discovery and service-dependency generation as the basis of service composition. Services execution is performed by plan executor as shown in Figure 2.2. The plan generator and plan executor were described in Dai [3], which are tailored in this research with OSGi facilities to support dynamic service management.

Under the proposed approach, the service system is formed by dynamic building blocks, some of them in our terms called core services, others are services provided by third parties. The core services are those managing knowledge, data, tasks, and communications. Depending on the application needs, the services are packaged through orchestration processes.

In order to economically manage software resources, i.e., invocation of minimum level of services, context-aware computing techniques are used to provide goal-driven solutions in a dynamic way. The framework incorporates INDEX knowledge management components that have recently been deployed as services [5] to assist dynamic and intelligent allocation of resources according to each inbound task. The task description module and business logic modules of INDEX are reused and tailored to meet the needs of context-aware processing.

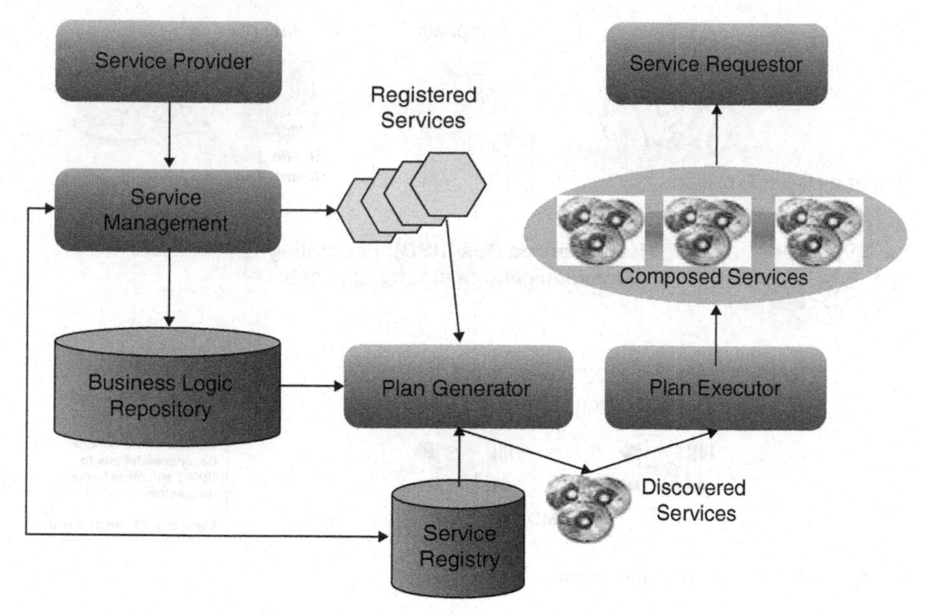

Fig. 2.2 Key process of service management

2.4 Application – Web-based Solution for e-Health

The dynamic collaborative framework as proposed is generic to different sectors. It depends on the sources of the domain knowledge as plug-ins, which will make the model domain dependent. Again, the plug-ins of domain knowledge sources depend on the tasks received. In this section, we use the e-health domain as an application example for the framework.

The main goal of the e-health application is to provide diseases management by assisting GPs generate care plans for patients and monitor the progress of those care plan, in anticipation with a number of health-care providers such as hospitals and third-party care providers. The solution is delivered through the Web where all the operations take the form of services. Services work with each other through a Health Service Bus (HSB) based on a vendor's product. Patients' progress of the disease management is reported and monitored remotely through devices.

In the application environment, as described in Figure 2.3, there are key players and technical entities, specifically, service providers including chronicle diseases management operator (CDMO), hospitals and third-party providers; service requestor including GPs and patients.

Requestors and providers work together through a service registry along the HSB. Service request (e.g., originated by a GP) can be issued at any time. CDMO responses as one of the providers by generating a suitable care plan which is to be progressed along with other service providers (such as hospitals) through care plan monitoring.

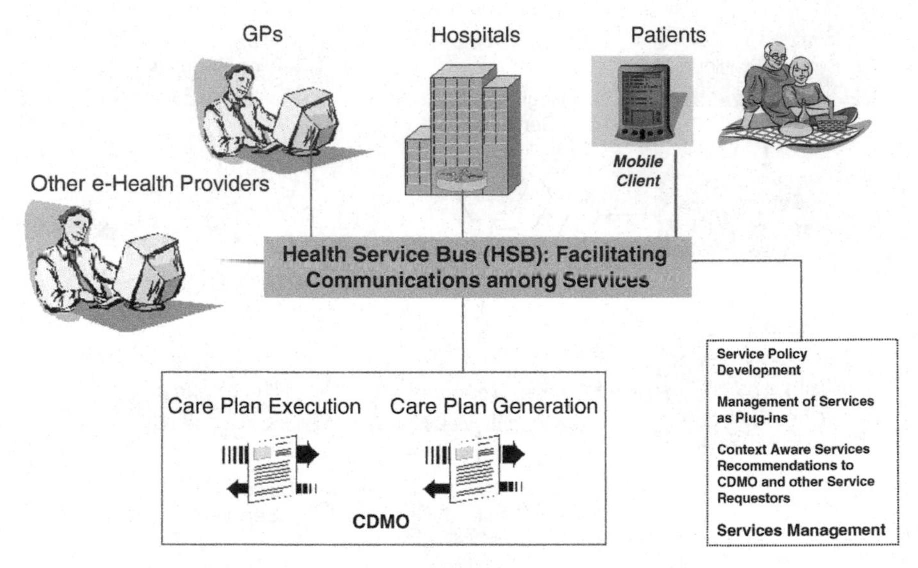

Fig. 2.3 Web-based e-health application

Service management is based on service policy in conjunction with service registry. Services can arrive and leave and can become unavailable. The dynamic services management is supported by plug-and-run paradigm. The application supports the HSB as a service provider specializing in plug-ins management for anticipating services. Therefore it is beneficial to the users to gain transparency in terms of the relevance and availability of other services.

A unique feature of the application is its context-aware capability to dynamically identify the relevant services according to the current focus of attention within the e-health framework. The service management functionality of the framework provides feedback to CDMO via the HSB to facilitate various providers (or members including new members) utilizing the services provided by the health network via the Web.

There are several contributions of this application to the e-health framework, which are: (1) providing Web-based context-aware solutions to CDMO to make effective use of available resources; (2) providing technical support in service policy development and management of dynamic services as plug-ins; (3) knowledge sharing from services providers to consumers.

Figure 2.4 gives a snapshot of a Web-based client receiving medical recommendations through relevant services working behind the scene.

The application is implemented on the OSGi platform (url: www.osgi.org) where the dynamically required services are invoked as plug-ins. Currently, the implementation state of the framework is a research prototype. To demonstrate the generic aspects of the proposed research, the prototype of the framework was also applied in the supply-chain sector [5].

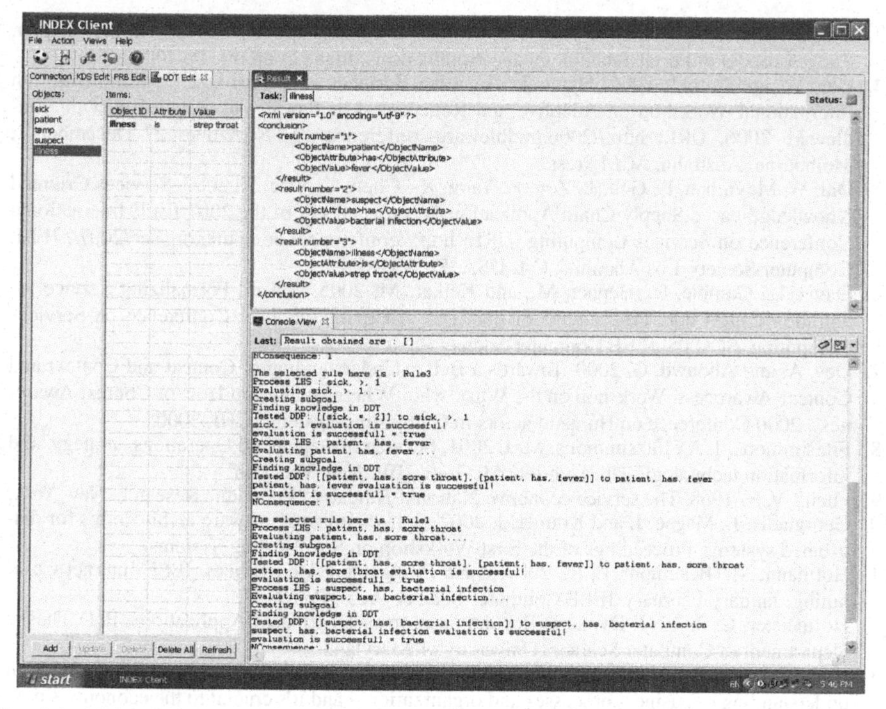

Fig. 2.4 A Web-based client program receiving medical advice

2.5 Conclusion

The article discussed the current state and emerging trend of service science and proposed an innovative approach of integrating and utilizing information resources in the form of service through the Web. It opens up an opportunity for service consumers and service providers in general to collaborate across the Web. Users of various needs including owners of business logic can make economical and flexible use of the services provided by the proposed solutions, therefore significantly increased the use of Internet to support each individual's need.

Acknowledgments The work is part of the PHOENIX research program at Victoria University, Australia. The leadership of Jonathan Liu in OSGi application development is hereby acknowledged.

References

1. Chesbrough, H. 2005. Toward a science of services. Harvard Business Review, 83, 16-17.
2. Cheung, L., Chung, P., Stone, R., Dai, W. 2006. A Personal Knowledge Management Tool that Supports Organizational Knowledge Management. Proceedings of 2006 Asia Pacific International Conference on Knowledge Management. Hong Kong, December 11–13.

3. Dai, W. 2007. Collaborative Real-Time Information Services via Portals. Book Chapter. Encyclopaedia of Portal Technology and Applications. Idea Group Inc. Pp 140 -145.
4. Dai, W. and Liu, J. 2006. Dynamic Adaptive Services in e-Business Environments. 5th International Workshop on Adaptive and Reflective Middleware, in association with Middleware 2006, URL: http://2006.middleware-conference.org/ November 27–December 1, Melbourne, Australia. ACM Press.
5. Dai, W, Moynihan, P., Gou, J., Zou, P., Yang, X., Chen, T., Wan, X. 2007. Services Oriented Knowledge-based Supply Chain Application. In proceedings of the 2007 IEEE International Conference on Services Computing, URL: http://conferences.computer.org/scc/2007/. IEEE Computer Society. Los Alamitos, CA. USA. Pp 660–667.
6. Davis, L., Gamble, R., Hepner, M., and Kelkar, M. 2005. Toward Formalizing Service Integration Glue Code, Proceedings of the 2005 IEEE International Conference on Services Computing, SCC'05. IEEE Computer Society Press.
7. Dey, A. and Abouwd, G. 2000. Towards a Better Understanding of Context and Context and Context-Awareness. Workshop on the What, who, Where, When and How of Context Awareness. 2000 Conference on Human Factors in Computing Systems, CHI. 2000.
8. Fitzsimmons, J. A., Fitzsimmons, M. J. 2001. Service management: Operations, strategy, and information technology. Third edition. McGraw-Hill: New York, NY.
9. Fuchs, V. R. 1968. The service economy. National Bureau of Economic Research. New York.
10. Georgiadis, I., Magee, J. and Kramer, J. 2002. Self-organising software architectures for distributed systems. Proceedings of the First Workshop on Self-Healing Systems.
11. Hofmann, M., Beaumont, L. R., 2007. Open Pluggable Edge Services. IEEE Internet Computing. January-February. IEEE Computer Society Press. Pp 67–73.
12. Hofmeister, C. R. 1993. Dynamic Reconfiguration of Distributed Applications. PhD Thesis. Department of Computer Science. University of Maryland.
13. Horn, P. 2005. The new discipline of Services Science: It's a melding of technology with an understanding of business processes and organization – and it's crucial to the economy's next wave. Business Week. January 21. URL: http://www.businessweek.com/technology/content/jan2005/tc20050121-8020.htm.
14. Karmarkar, U. 2004. Will you survive the services revolution? Harvard Business Review (June): 100-107.
15. Kon, F. and Campell, R. H. 1999. Supporting automatic configuration of component-based distributed systems. Proceedings of the 5th USENIX Conference on Object-Oriented Technologies and Systems.
16. Lowry, M. R. 1992.Software engineering in the twenty-first century. AI Magazine Fall: 71-87.
17. Martin, D., Domingue, J., Brodie, M. L., and Leymann, F. 2007. Semantic Web Services, Part1. IEEE Intelligent Systems, September/October. Pp 12-17.
18. Martin, D., Domingue, J., Sheth, A., Battle, S., Sycara, K., Fensel, D. 2007. Semantic Web Services, Part1. IEEE Intelligent Systems, November/December. Pp 8-15.
19. McAfee, A. 2005. Will web services really transform collaboration? MIT Sloan Management Review, 6 (2).
20. Oreizy, P. and Taylor, R. N. 1998. On the Role of Software Architectures in Running System Reconfiguration. IEEE Software. Vol. 145, No. 5.
21. Rebeca P. Díaz Redondo, Ana Fernández Vilas, Manuel Ramos Cabrer, José Juan Pazos Arias, Jorge García Duque, and Alberto Gil Solla. 2008. Enhancing Residential Gateways: A Semantic OSGi Platform. IEEE Intelligent Systems. January/February. Pp 32–40. IEEE Computer Society.
22. Rubin, S. H. and Dai, W. 2003. An integrated knowledge management environment: a study of counter terrorism technologies. Journal of Information and Knowledge Management, Vol. 3, No. 1, pp 1-15. World Scientific Publishing.
23. Rust, R. 2004. A call for a wider range of service research. Journal of Service Research. 6 (3).
24. Sampson, S. E. 2001. Understanding service businesses: Applying principles of unified systems theory. Second edition. John Wiley and Sons: New York, NY.
25. Tien, J. M., D. Berg. 2003. A case for service systems engineering. The Journal of Systems Science and Systems Engineering. 12 (1), 113-28.
26. Vargo, S. L. and Lusch, R. F. 2004. Evolving to a new dominant logic for marketing. Journal of Marketing. 68, 1-17.

Chapter 3
Helplets: A Common Sense-Based Collaborative Help Collection and Retrieval Architecture for Web-Enabled Systems*

Mohammad Nauman, Shahbaz Khan, and Sanaullah Khan

Abstract All computer software systems, whether online or offline, require a help system. Help texts are traditionally written by software development companies and answer targeted questions in the form of how-tos and troubleshooting procedures. However, when the popularity of an application grows, users of the application themselves start adding to the corpus of help for the system in the form of online tutorials. There is, however, one problem with such tutorials. They have no direct link with the software for which they are written. Users have to search the Internet for different tutorials that are usually hosted on dispersed locations, and there is no ideal way of finding the relevant information without ending up with lots of noise in the search results. In this chapter, we describe a model for a help system which enhances this concept using collaborative tagging for categorization of "helplets." For the knowledge retrieval part of the system, we utilize a previously developed technique based on common sense and user personalization. We use a freely available common sense reasoning toolkit for knowledge retrieval. Our architecture can be implemented in Web-based systems as well as in stand-alone desktop applications.

M. Nauman
e-mail: nauman@imsciences.edu.pk

S. Khan
e-mail: shahbaz@imsciences.edu.pk

S. Khan
e-mail: sanaullah@imsciences.edu.pk

Security Engineering Research Group, Institute of Management Sciences,
1-A, Sector E5, Phase VII, Hayatabad, Peshawar, Pakistan
e-mail: http://serg.imsciences.edu.pk.

* This is an extended version of work previously published in WI'07 under the title, "*Using Personalization for Enhancing Common Sense and Folksonomy based Intelligent Search Systems*".

J.T. Yao (ed.), *Web-Based Support Systems*, Advanced Information and Knowledge Processing, DOI 10.1007/978-1-84882-628-1_3,

3.1 Introduction

Help systems have traditionally followed the same pattern. They are arranged in topics, subtopics, and sections with unstructured text. Typical help systems include frequently asked questions (FAQs), user manuals, guides, and tutorials. In all these approaches, help texts remain unstructured static entities which are either manually categorized into sections or crawled, indexed, and retrieved without any understanding of what their content is about.

We note that help texts are usually written to describe *steps*, which need to be performed in order to solve a particular problem. A larger problem may be broken down into smaller ones and each of these addressed in turn. This is reminiscent of the artificial intelligence technique known as *problem decomposition* [24]. In traditional help systems, there is no automatic way of decomposing a problem into subproblems.

Another issue with help texts is that they are usually disconnected from the applications or software that they address. While context-sensitive help texts [13, 18, 30] solve this problem to some extent, they have never really scaled up to address larger, more organic problems. Also, tutorials written by third parties cannot take advantage of context-sensitive help and other such techniques currently in use. These third-party and community-written help texts reside on Web sites, blogs, social networks, etc. There is one way to link these help texts to their target applications.

In this chapter, we present an architecture for a help system which aims to address both of these problems. In this help collection and retrieval architecture, help texts are organized in *helplets* – small fragments of texts (or even photos, images, videos, and audio files) addressing either specific or general issues in a particular application or system. These helplets are stored in a central repository and organized using collaborative tagging [11] – a flexible, easy-to-use, and popular categorization technique used on the Web. Freely chosen keywords or *tags* are used to describe each helplet. The helplets are written from within the target application and can be linked to specific parts of the application itself.

For help retrieval, we use a mechanism previously developed [19], which uses machine common sense for search. Whenever a tutorial is written, it targets a specific problem. However, every solution is composed of smaller solutions addressing subproblems of the original target problem. The solution to the larger problem might be helpful for those who are looking for help regarding any of the subproblems. Also, any user looking for a solution to the larger problem might find a help text addressing the subproblem helpful. We use machine common sense to relate problems to subproblems. The result is a search system that "understands"[1] the target problem and can intelligently return any helplet which might be useful in solving the problem. The helplet submission and retrieval mechanisms both reside within the application and provide an ubiquitous Web-based help support system [33].

[1] Here, *understanding* refers to the structure of the problem and has nothing to do with the issue of whether a machine can actually reflect on the problems as humans can.

The rest of the chapter is organized as follows: In Sections 3.2 and 3.3, we describe the problems being addressed and background information on some of the concepts in use. Section 3.4 details our helplet architecture, content collection, organization, and retrieval in detail. Section 3.5 discusses some of the related help architectures in brief and the contribution is concluded in Section 3.6.

3.2 Issues in Contemporary Help Systems

Documentation, manuals, and help texts form an essential component of any software. A quick search on any research paper indexing site [7, 12] shows that help systems have received little attention from the research community throughout the history of software development. The result of this negligence has been that architectures of help texts have changed little over time since the inception of the F1 key and the association of help to it. Almost all softwares come bundled with a help system, which does little more than arranging the documentation in sections and subsections, providing an index, and embedding a search capability. These help systems, while useful for beginners, do little more than references for the more experienced users.

It takes expert users of any software to harness the true power of its capabilities. When a software grows in popularity, expert users start to feel the inadequacy of the help systems and start writing their own texts and tutorials for several tasks they performed using the software. A classic example of this phenomenon is Adobe's [6] popular photo- and graphics-editing software Photoshop. The Internet is full of tutorials written for all types of tasks, which can be performed in Photoshop. These tutorials are an example of user-generated content – a pillar of Web 2.0. With an increase in Web 2.0's blogs and other social networks, this corpus of information is ever increasing.

Another issue is the way this organic corpus of information is categorized, arranged, and accordingly retrieved effectively. Over the past few years, *collaborative tagging* [11] or *folksonomy* has emerged as the most popular [1–4, 10, 31] means of content categorization. In folksonomy, content is tagged with freely chosen keywords. These keywords describe the content and help arrange the content in a non-hierarchical, nonexclusive ontology [32]. The popularity of folksonomy is due to the reason that it is flexible, easy-to-use, and requires little expert knowledge on the part of the users.

The ease-of-use and flexibility of folksonomy, however, is not without some shortcomings. For one, there is the issue of lexical ambiguities in folksonomy based search [20]. These are related to how different people use words to describe concepts. A single word may be used to describe different concepts and in another situation different words may mean different concepts altogether. These are the issues of synonymy and polysemy (or homonymy) respectively. So, to summarize, the issues being addressed in this chapter follow.

3.2.1 Tutorial Embedding

There is a need to develop an architecture, which allows application developers to embed, in their applications, an interface for collecting help and allowing users to collaboratively write help articles and contribute to the help corpus. Such an architecture will allow the community to contribute to a central (or even possibly distributed) repository of help through an interface they are already familiar with. Through such an architecture, the users will not only be able to submit their content easily, but will also be able to associate their help texts with individual components of the application, thus allowing for a more dynamic and interactive help system.

3.2.2 Tutorial Decomposition

The second aspect of the problem of help systems being addressed in this chapter is tutorial or help text decomposition. This issue is analogous to the artificial intelligence concept of problem decomposition [24]. As described in Section 3.1, help texts usually address a complex problem involving many sub-steps. Help systems should be able to intelligently find out what these sub-steps are, what solutions form parts of a larger solution, and how the larger problems are linked to the smaller problems.

Such an architecture should be able to retrieve content related to subproblems if a user is searching for a solution to a larger problem. Also, a mechanism needs to be devised in which this problem–subproblem relation can be represented easily and using natural language. A person may write a tutorial for a complex procedure which covers many simpler steps. Each of these smaller problems can be a "component" of the bigger picture. A user, if searching for the smaller component, should be able to retrieve this larger picture. Otherwise, she might miss some help content that is relevant.

In order to overcome these problems, we devise an architecture based on machine common sense. An architecture for a system which is better representative of how the human brain works – a system with common sense. For this purpose, we utilize the machine common sense corpus called the Open Mind Common Sense [26].

3.3 Machine Common Sense

Open Mind Common Sense is a project started by Singh [26] at MIT Media Lab. The aim of the project and the underlying assumption is that common sense is information shared by all human beings and, therefore, expert involvement should be kept to a minimum during the collection of a corpus of this information. The project started collecting commonsense information from the general public through the

World Wide Web. The type of sentences input by the general public as *commonsense information* is depicted by the following examples [21].[2]

- Hockey is a sport.
- A chair is for sitting on.
- A goldfish is a carp.
- A boat can floats on water.

The basic architecture of OMCS was enhanced by Singh et al. to create OMCS-2 [29]. This Web-based front end could not only gather information from the users but also perform inference on that input and generate more commonsense information. This inferred information was then presented to the user who input the original information and was asked to verify the correctness of the inference. The form of input was also enhanced from a simple text entry to a fill-in-the-blanks-type entry, where the user was expected to fill in the blanks in a predefined sentence structure or template. Housekeeping functions such as spell-checking and disambiguation were also performed by the inference engine.

Our inspiration for using machine commonsense comes from the findings of Singh and Barry [27]. They note

> The most important lessons [they] learned from building the original OMCS site [are that] ...we do not need to wait until we have complete and perfect commonsense knowledge bases and inference techniques to begin exploring how to incorporate commonsense into applications. [27]

Different commonsense toolkits have been extracted from the corpus of commonsense information gathered through the OMCS project. The first one of these was ConceptNet. ConceptNet is based on rule-based reasoning. Another toolkit extracted from the corpus is LifeNet which is based on probabilistic reasoning. Using statistical techniques, LifeNet, allows for uncertainty in knowledge and in rules [28]. ConceptNet forms a central pillar of the retrieval part of our helplet architecture. A brief discussion of ConceptNet is thus necessary here.

ConceptNet [17] is a freely available commonsense-reasoning toolkit comprising of over 250,000 elements of commonsense knowledge built using *semistructured* natural language fragments. It builds on the simplicity of WordNet's[3] ontology of *semantic relations* (described using natural language) [9] and on Cyc's [4] [14] diversity of relations and concepts [16]. In essence, it combines the best of both worlds by creating a powerful and expressive ontology while still keeping the reasoning and *inference logic* simple.

The current version of ConceptNet is 2, consisting of 1.6 million *assertions*. These consist of a semantic network of concepts associated together using links

[2] Note that the grammatical and spelling mistakes are of the users entering the information. These mistakes are removed as part of a cleaning process during the extraction of information from this corpus.

[3] An ontology which contains the words of English language organized in synonym sets, which are linked together using relations.

[4] Another commonsense formalization, which is built around axioms, rules, and facts. It is a highly specialized ontology which requires expert input for its creation and use.

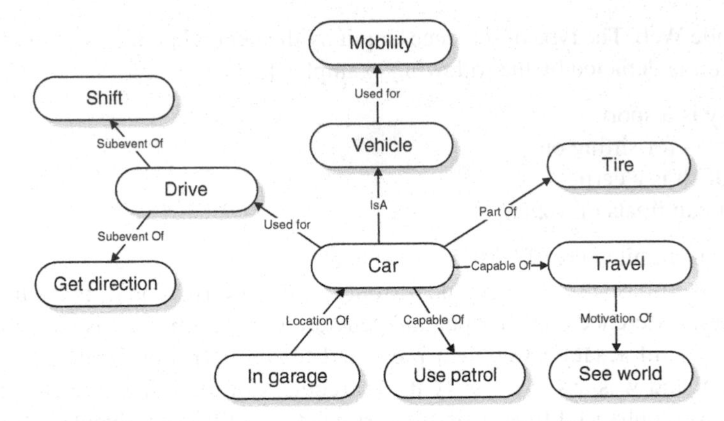

Fig. 3.1 A fragment of ConceptNet showing links from *Car*

such as *IsA*, *SubeventOf*, *MotivationOf*, *PartOf*, etc. [28]. There are 20 *relation-types* in ConceptNet. These are categorized in *thematics* called *k-lines*, *things*, *agents*, *events*, *spatial*, *causal*, *functional*, and *affective*.

Each instance of a relation is assigned a *score*. This score describes how *obvious* this relation is based on common sense. Scores are assigned to relations based on two criteria:

1. Number of utterances in OMCS corpus. This is the direct result of the user input to the Open Mind site. The number of utterances of a specific relation between two concepts is assigned a value called *f value*.
2. Inference from other facts in ConceptNet. This is the number of times the inference engine deduces the relation from the facts input by the users. This value is called *i value* in ConceptNet [17].

For the purposes of our helplet architecture, the most important relation types of ConceptNet are:

1. Subevent-Of
2. FirstSubEvent-Of
3. Prerequisite-Of

In Section 3.4.3, we describe how these relation types can help retrieve related help contents in an intelligent way using these relations of ConcepetNet.

Figure 3.1 shows an example of concepts and links as represented in ConceptNet.

3.4 Helplet Architecture

In order to address the issues of incorporating help texts within the target applications, organizing them in a flexible manner and retrieving them in a user-friendly, easy-to-use way, we propose an architecture built around the notion of helplets.

Helplets are pieces of help in the form of plain or rich text, audio, video, blog post, podcast, or any other form of media with the aim of solving a problem in a target application. It is an important strength of our architecture that is not limited to text-based help alone but can also accommodate the collection, organization, and retrieval of help based on rich media.

We note that it is not sufficient to develop any one aspect of the problem described in Section 3.2 and leave the rest. For providing a useful and flexible help system, it is necessary that the helplets be easily created, organically organized and searchable for the average user of the application without the intricate knowledge of the actual representation of the helplets within the application. For this purpose, we breakdown our help architecture in three parts: knowledge collection, knowledge organization, and knowledge retrieval.

1. *Knowledge Collection:* Knowledge collection (or submission by the users) is an important part of our helplet architecture. It has been noted through experiences in Web 2.0 applications that "network effects from the users are the most important aspect of Web 2.0" [22]. We therefore base our helplet architecture on a collaborative technique embedded within the target application.
2. *Knowledge Organization:* We propose the use of folksonomy (or collaborative tagging) as the underlying ontology for the organization of helplets. Recent growth in popularity and use of folksonomy as the means of content categorization on the Web has shown that it is a flexible and powerful mechanism for content organization – especially in knowledge bases built through collaboration among different users.
3. *Knowledge Retrieval:* The most important aspect of our proposed architecture is the use of machine common sense for the purpose of helplet retrieval.

Figure 3.2 shows the proposed architecture in brief. In the following sections, we discuss these three aspects of our architecture in detail.

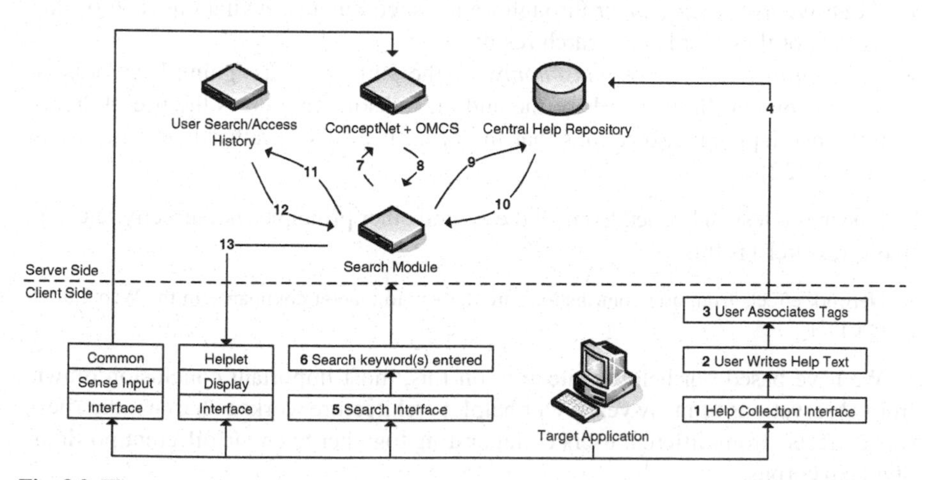

Fig. 3.2 The proposed helplets architecture

3.4.1 Knowlege Collection: Helplets

The major contribution of this chapter is the development of an architecture in which help texts are organized in "helplets". Helplets are very small to very large and very specific to very general. Our helplet architecture depends on the embedding of the knowledge collection into the target application for the purpose of providing an ubiquitous environment for writing helplets. Note that for this part of our architecture, the target application might be a stand-alone desktop-based application, a Web-enabled desktop application or a completely Web-based system.

The actual implementation details of embedding helplet collection system is left unspecified here as it is dependent heavily on the target application, scope of the application, and the existing interface.

3.4.2 Knowledge Organization: Folksonomy

One of the most challenging problems with search is huge amount of information that the search engines have to cope with. It was initially believed that the Web was overhyped and would not grow much after its initial burst of acceptance but in a conference brainstorm session between O'Reilly and MediaLive, the concept of Web 2.0 emerged [22]. This new concept has led to a faster growth rate of the Web than ever before.

O'Reilly [22] lists some principles common in all successful services of Web 2.0 nature:

- Extensive *hyperlinking* between content generated by different users.
- Inclusion of a *portal* to showcase the best works of the users.
- *Organic growth* of the content in response to user activity.
- Extensive *user engagement* through reviews, comments, invitations, and/or utilization of these for better search results.
- *Collaborative tagging* or *folksonomy* for the purpose of assigning keywords or *tags* to content. In the words of the author, folksonomy allows the use of "multiple, overlapping associations that the brain itself uses, rather than rigid categories" [22].

One major lesson learned from all these (and other principles not directly relevant to our research) is this:

> Network effects from user contributions are the key to market dominance in the Web 2.0 era. [22]

We have based our help architecture on this, most important conclusion drawn from observations of many years. Our helplet architecture works to harness the "network effects" from different users collaborating together to create different portions of a help corpus.

One of the most important aspects of this listing is *folksonomy*. *Folksonomy* (from *folk* and *taxonomy*) is a *structural representation* created by the general public [25]. The amount of information present in the Web 2.0 corpuses is extremely difficult to organize and search through. In Web 2.0, another problem has risen: most of the data generated by the general public in Web 2.0 is nonindexable. These include content such as photos, images, audio files, video podcasts, and the like [25].

Folksonomy is the most widely accepted solution to content categorization in Web 2.0 services. It refers to the assigning of descriptive keywords, called *tags* or *labels*, to the content for the purpose of categorization. Riddle [23] identifies several benefits of using tags for categorization:

- *Tagging is useful for personal recall:* Users creating tags are more likely to recall these tags in the future.
- *Tagging supports social effects:* Folksonomy-based services usually give users the ability to informally convey or convince others to tag similar content using the same tags.
- *Tagging promotes serendipity:* Since tags are assigned to content by a wide audience and not by some domain experts, users are likely to find useful content by browsing tags of interest.

In the same paper [23], Riddle has also identified some shortcomings of using tags for categorization. These are:

- *Precision and recall: Precision* is measure of those retrieved documents which are relevant to the search. *Recall* is the measure of relevant documents which are retrieved. Tagging is not good for either of these because users may use different words to describe a piece of content and because they may use the same words to describe different concepts. (These are the problems of *synonymy* and *polysemy* respectively.)
- *Tagging is not good for* ontologies*:* Tags are an *uncontrolled vocabulary*, meaning they are not created by experts but by the organic and dynamic keyword assignment by the general public. They therefore tend to be unstructured and lack compliance with other (formal) ontologies.

The major problem with tagging information is the way different individuals link real-world concepts with textual representation of these concepts. This is the problem of polysemy (using the same words to describe different concepts), synonymy (using different words to describe the same concept), and variation in basic level of categorization. Not only is this a problem with different users, it is also a problem of the same user recalling the tag after a long time. Users may not be able to recall correctly how they tagged the information a month earlier [11].

Xu et al. [32] define folksonomy as a *nonhierarchical* and *nonexclusive* ontology. This means that there is no parent–child relation between different sets of entities and that if an object belongs to one category, it does not necessarily have to be excluded from all others. Due to these reasons, the allocation of tags by different users creates a fuzzy and ambiguous boundary between sets of different concepts, thus leading to the issues of lexical ambiguity discussed earlier.

This is a major problem in our case since we want to be able to link helplet systems in a way that there can be a relation between problems and subproblems. It seems that this link cannot directly be established in a purely folksonomy-based system. We shall address the issue of linking problems with subproblems in Section 3.4.3. Here, we describe our motivation for using folksonomy, or collaborative tagging, for the purpose of content categorization of helplets:

- The most important motivation for using folksonomy is that the amount of information relating to any subject is huge. This size and the organic nature of collaborative systems, stemming from the network effects due to collaboration among users makes it extremely implausible for a single authority to organize the content. The best way to handle this organic corpus is to let the contributors submit not only data but also the meta-data associated with their submission.
- Folksonomy is easy to use and extremely flexible. There is hardly any learning curve involved in the use of folksonomy as it is based on "freely chosen" keywords attached to the content being submitted. This does lead to certain issues when retrieving content but we have developed a technique, which addresses these issues.
- The corpus of knowledge in a folksonomy-based system is generated by the users who create data. Content categorization comes from the same place as the content and is thus more likely to be both accurate and relevant.
- Folksonomy-based systems do not restrict users to a specific type of folder hierarchy. Instead it is more like the way humans normally associate things. Different, unrelated words are usually linked to concepts for the purpose of description.
- A vast majority of Web 2.0 content is already organized through folksonomy. By using folksonomy in our helplet architecture, we may be able to harness this shared information already present.

3.4.3 Knowledge Retrieval: Common Sense

The simplicity of the OMCS project and the freely available toolkit for common sense reasoning in the form of ConceptNet provides ample opportunities to anyone aiming to use textual processing in their applications. On the other hand, community-based content is also a vast source of information. The *tags* of folksonomy are already being utilized for the purpose of search and organization of content in Web 2.0. We believe that in combining these two sources of arrangement and inference, a powerful mix can be developed for an efficient search system for Web 2.0.

Also, the source of OMCS project's commonsense information and that of tags in folksonomy is the same. Users of the World Wide Web have contributed to both knowledge bases. Here, we describe how to apply the commonsense information present in OMCS through the ConceptNet toolkit for addressing the problem of searching through folksonomy.

3.4.3.1 Machine Common Sense and Helplets

We note that in each tutorial, there are some subprocesses. A solution to a larger problem can have many steps. Consider the ConceptNet's relation-type "SubEvent-of." Now imagine a user searching for helplets that address the issue of "connecting to the Internet". In machine common sense, there might be relation to the effect that "*installing a network card* is a subevent-of *connecting to the internet*". We note that helplets tagged with "installing a network card" might also be of use to the user since it might also guide her in finding a solution to her problem. Moreover, the relation-type "FirstSubEvent-Of" is also especially useful in our helplet architecture since it might provide the first step in solving a problem. Moreover, common sense has been shown [19] to be very useful in addressing the problems of lexical ambiguities in folksonomy.

3.4.3.2 Central Problems of Folksonomy

We argue that there are two central problems in folksonomy:

1. Lack of contextual information and
2. Limited inference capabilities due to its loose ontology.

We propose a framework [19, 20] in which both these issues are addressed through the application of a commonsense ontology. Figure 3.3 summarizes this mapping.

3.4.3.3 Flow of Control

An outline for using common sense on folksonomy is given below.

1. The user enters a search keyword and is presented with search options. These would decide the generality or specification of the concept association and the relation types used for expansion of concepts.
2. ConceptNet toolkit is used for concept expansion based on user's preferences.

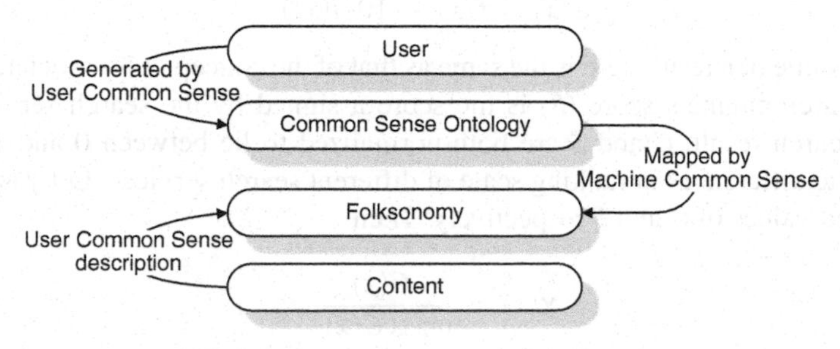

Fig. 3.3 Architecture of proposed framework

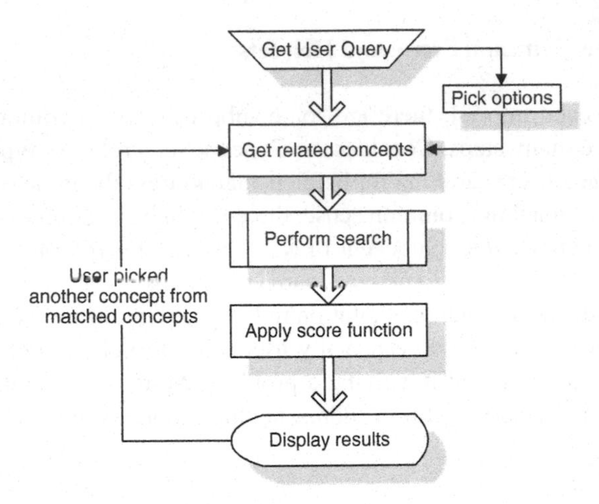

Fig. 3.4 Flow of control in proposed framework

3. A search is performed to get a set of *result items* for each resulting concept.
4. A *score function* is applied to the union of these sets to get a final listing of results.
5. The results are displayed to the user labeled with the associated concept and the user has the option of picking one of the expanded concepts for further search.

Figure 3.4 describes the flow of control in this scenario.

3.4.3.4 User Preferences

Generality or specification

Let the *Conceptual Similarity* (*C*) of a concept be the closeness of the concept to the original keyword. This closeness of concept is decided by the *f* and *i* values returned by ConceptNet. The *i* values are multiplied by 10 because they are usually an order of magnitude smaller than the *f* values but are still very important.

$$C(x) = f(x) + (10 \cdot i(x)) \tag{3.1}$$

The *C* value of a result item is the same as that of the concept it is associated with.

The search engine's score (*S*) is the score assigned by the search service to a specific search result. *C* and *S* are both normalized to lie between 0 and 1 to accommodate differences in ranking scale of different search services. Let γ and σ be normalized values of *C* and *S* respectively. Then

$$\gamma(x) = \frac{C(x)}{\max(C(x))} \tag{3.2}$$

$$\sigma(x) = \frac{S(x)}{\max(S(x))} \tag{3.3}$$

The user enters the required level of generality (G) of the combined results. A more specific preference would give more weight to the items conceptually more similar to the original keyword, i.e., those with higher γ values. A general preference would rank those items higher which have high ranking in the search service, i.e., those with higher σ values. This preference is implemented through the score function.

Relation Types

The user can also specify the relation types of ConceptNet for advanced control over concept expansion. The user may decide to expand concepts along a limited set of relations for better targeting of the results. For example, the user may prefer to search along the *IsA* relation type along with *PartOf* and not along *MotivationOf* relation type.

3.4.3.5 Score Vector

At the heart of the search results presentation is the score function. It is a function which assigns relative score to each individual result item based on conceptual similarity and service rank of the item and on user preferences of generality.

$$Score(x) = \sum_{i=1}^{n} inst\_score(x_i) \tag{3.4}$$

where x is an individual item returned by the search service, x_i are instances of the item x returned as responses to different concepts and n is the total number of these instances. The summation ensures that if an item appears as a result of more than one concepts, it gets a higher score in the final listing.

The score of a single instance of a result item is given by *inst_score*. *inst_score* is a function of γ, σ, and G where γ and σ are normalized values of conceptual similarity and service rank respectively and G is the level of generality specified by the user.

$$Inst\_score(x_i) = s(\gamma(x), \sigma(x_i), G) \tag{3.5}$$

In our framework, we leave "*s*" undefined so that it may be tailored according to specific situations. An example of this function we used in our prototypes is:

$$Inst\_score(x_i) = (G \cdot \sigma(x_i)) + (1 - G) \cdot \gamma(x) \tag{3.6}$$

3.4.3.6 Issues in the Basic Approach

The basic commonsense- and folksonomy-based approach described above has shown positive results in finding content tagged with related keywords. It successfully addresses the issues of synonymy and variances in basic levels of categorization. When it comes to the problems related to polysemy, however, the basic technique falls short. The following two aspects of this technique are identified as the root of the problem:

- The basic CS&F-based search technique uses only conceptual similarity for concept expansion, which is the problem of polysemy. Different concepts may be represented using the same term and will therefore be regarded by the system as being conceptually similar when in fact they are not. For example, a user searching for "architecture" may be interested in "building architecture" and not in "computer architecture." Both of these concepts will be considered similar to the original keyword due to the shared lexical representation of the base concepts but for the user they are not similar.
- Similarly, the score function ranks results according to conceptual similarity and level of generality. This leads to the same problem as described above when the user associates a keyword with only one of many possible categories to which the keyword may belong.

One solution to this problem is proposed here in which personalized Web search can help alleviate these problems by anticipating the user's categories of interest [20]. The expanded concepts and ranked results can be tailored for an individual user based on his or her search and access history while requiring little effort on the user's part. The history is maintained in a matrix that maps *terms* to *categories*. The semantics are that a user normally associates a single term with a specific category or domain. The access and usage history of a user can be used to create a matrix (called Matrix M), which defines a user's preference in ambiguous terms. Table 3.1 shows a sample of this matrix. We refer the reader to [15] for details regarding the construction of this matrix. For utilizing this matrix in our basic approach, some modifications will have to be made to the underlying personalization technique, which we describe in Section 3.4.3.7.

3.4.3.7 Modifications to Personalized Web Search

For the purpose of using personalized Web search technique with commonsense and folksonomy-based search systems, the following alterations are proposed:

Table 3.1 User preference stored in a matrix

Cat/term	Apple	Recipe	Pudding	Football	Soccer	Fifa
Cooking	1	0.37	0.37	0	0	0
Soccer	0	0	0	1	0.37	0.37

- Each single unit of content in a folksonomy-based Web 2.0 system is considered to be a document; this can be a post in a blog, a photo in a photo-sharing service, or a page in a wiki. In our helplet architecture, this single unit of content is a helplet.
- Tags of a service are considered index terms. This is because most content in Web 2.0 services is categorized by tags. For some types of content, such as images and maps, it is not possible to index by the content itself.
- Since index terms have been changed to tags, all occurrences of "B appears in A" in the original technique should be read as "A is tagged with B" in the modified approach.
- User-query inference part of the original technique is not utilized. The values of matrix M will be used only as weights to enhance the query expansion and score functions of commonsense- and folksonomy-based search systems.
- Since, in Web 2.0, the search service is also usually the content-provider service, *access history* can be used alongside *search history*. The procedure for using access history in the modified approach is exactly the same as that for search history.
- The original personalized Web search describes four algorithms for calculating the user profile: LLSF, pLLSF, kNN, and aRocchio. For our purposes, aRocchio is sufficient because it is efficient in both space and computation and is adaptive [15].

3.4.3.8 Enhancing the Basic Technique

Using the personalized Web search approach described in Section 3.4.3.7, we suggest the following changes in the basic CS&F-based search technique. The modifications target both of the problems identified in Section 3.4.3.6. There are two major proposed changes to the base technique: concept expansion enhancement and score function improvement. Note that in the following discussion "concept," "tag" and "keyword" are used interchangeably.

Concept Expansion Enhancement

The original algorithm for concept expansion is based on conceptual similarity decided only by the f and i values. We propose the use of category term matrix M for concept expansion. This would result in personalized search results for each user based on search and access history. Two alternatives for using the search history for concept expansion are described here. One only expands concepts that are in the same category as the original keyword and the other assigns weights to all expanded concepts based on category similarity. Both methods require that the category the user associates with the search keyword be known. The category (Φ_x) associated with a keyword x is that for which the column (T_x) representing the keyword has the highest value.

More precisely, let

Φ_o = Category of the original keyword
T_o = Column representing the original keyword
M_u = Matrix M for user u

then

Φ_o is that row for which

$$M_u(\Phi_o, T_o) = \max(M_u(i, T_o)) \tag{3.7}$$

where i ranges over all rows of matrix M.

After the category of the original keyword is known, either of the two methods described below can be used for enhancing concept expansion.

Method I

This method removes all concepts that, according to the user profile, are not in the same category as the original search keyword. First the category of the original keyword (in context of the user profile) is decided. Then, ConceptNet is used to generate related concepts. For each expanded concept $(e_1, e_2, \ldots, e_n)$, the same procedure is used to calculate the category as for the original keyword. Φ_{e_k} is calculated for each expanded concept e_k where k is the index of the expanded concept. Then, only those expanded concepts are kept in the list for which $\Phi_o = \Phi_{e_k}$.

Method II

This method does not remove any concepts. It assigns a *category similarity* (Θ) to all expanded concepts. The value of Θ for each expanded concept e_k is given by:

$$\Theta_{e_k} = M_u(\Phi_o, T_{e_k}) \tag{3.8}$$

where

Φ_o is the category of the original keyword and
T_{e_k} is the column representing the concept e_k

Θ_{e_k} is used as a weight for calculation of conceptual similarity of the expanded concepts. For the basic conceptual similarity defined in Equation (3.1), the *personalized conceptual similarity* (C') is given by:

$$C'(e_k) = C(e_k) \cdot \Theta_{e_k} \tag{3.9}$$

The value of personalized category similarly is normalized to lie between 0 and 1. The normalized value of personalized conceptual similarity is given by γ' using the function:

$$\gamma'(e_k) = \frac{C'(e_k)}{\max(C'(e_k))} \tag{3.10}$$

The use of Θ_{e_k} as a weight ensures that if a user associates a keyword with a category different than that of the original keyword, it is considered to be of less conceptual similarity. This means that search results will be tailored according to the preferences of an individual user.

Personalized Score Function

For improving the score function based on user profile and returning each user with personalized results, we utilize matrix M again. Most of the content has more than one associated tags. These *related tags* can be used to find similarity of a result item to the user's intended category. For example, if a result item is tagged with "apple" and is also tagged with "pepper," it is probably in the intended category if the user associates "apple" with the category "cooking". Whereas, if the item is tagged with "software," it is probably not of interest to the user. We utilize this extra information for improving the search result personalization.

For each search result item x, first the base score is calculated as in Equation (3.4). Then, for each tag $(r_1, r_2, \ldots, r_3)$ associated with the result item, category similarity to the original keyword is computed:

$$\Theta_{r_k}(x) = M_u(\Phi_o, T_{r_k}) \tag{3.11}$$

where

Φ_o is the category associated with concept x and
T_{r_k} is the column representing the concept r_k.

The *personalized score* ($score'$) is a function of the basic *score* and Θ_{r_k} given as:

$$Score'(x) = \frac{score(x) + \sum_{k=1}^{n} \Theta_{r_k}(x)}{n+2} \tag{3.12}$$

The use of Θ_{r_k} serves dual purpose. It gives preference to those documents that are tagged with keywords belonging to the same category as the original search keyword. It also ensures that if a document is tagged with irrelevant keywords, the score is penalized.

Figure 3.5 shows the working of personalized common sense and folksonomy-based search technique. Concept Z of the figure is not used during search because it belongs to a category different than that of the original search keyword. Search result r shown in the figure is assumed to be tagged with some keywords belonging to different categories than the original search keyword and is therefore penalized due to the use of Θ_{r_k} as a weight in final calculation of the score. An algorithmic representation of this personalized commonsense-based search is given in Figure 3.6.

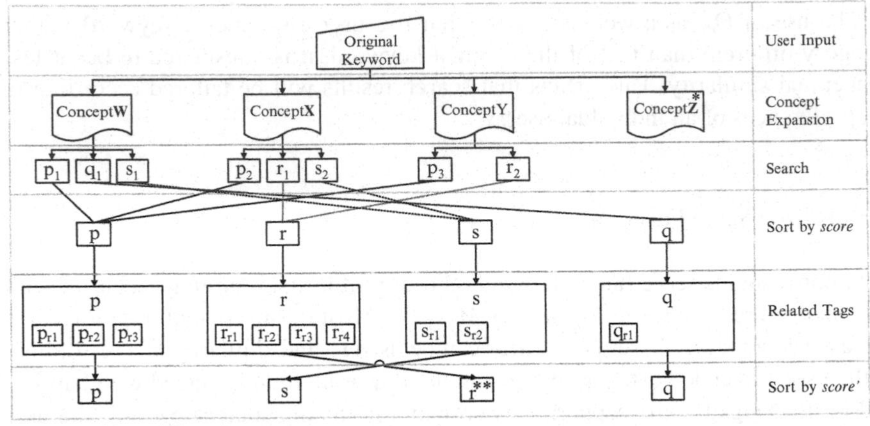

Fig. 3.5 Working of personalized CS&F-based search

Get search *keyword* from user
$\Phi_o :=$ getCategory(*keyword*)
$e :=$ expandConcepts(*keyword*)
$exConcepts := \{\}$
for each e_k in e
    $\Phi_{e_k} :=$ getCategory(e_k)
    if $\Phi_{e_k} = \Phi_o$ then $exConcepts$.add(e_k)
for each e_k in $exConcepts$
    $\gamma'(e_k) :=$ getConceptualSimilarity(e_k)
    $results :=$ performSearch(e_k)
    for each r_i in $results$
        $\text{inst\_score}(r_i) := G \cdot \sigma(r_i) + (1 - G) \cdot \gamma'(e_k)$
        addtoInstScores($inst\_score(r_i)$)
$scores[x] := \sum_{i=1}^{n} inst\_score(x_i)$
for each x in $scores$
    $relTags :=$ getRelatedTags(x)
    for each r_k in $relTags$
        $\Theta_{r_k} :=$ getCategorySimilarity(Φ_o, r_k)
    $scores'[x] := \frac{score[x] + \sum_{k=1}^{n} \Theta_{r_k}}{n+2}$
Sort by $scores'$ descending

Fig. 3.6 Algorithm using Method I

3.5 Related Work

The two help mechanisms most closely matching the objectives of this contribution are the embedded help architectures provided by Adobe's Photoshop [5] application and the Eclipse [8] platform. Adobe provides a framework that allows users to enter

Ps
Saving and exporting images / Placing Photoshop images in other applications
Create transparency using image clipping paths
You can use image clipping paths to define transparent areas in images you place in page-layout applications. In addition, Mac OS users can embed Photoshop images in many word-processor files.
You may want to use only part of a Photoshop image when printing it or placing it in another application. For example, you may want to use a foreground object and exclude the background. An image clipping path lets you isolate the foreground object and make everything else transparent when the image is printed or placed in another application.
Note: Paths are vector-based; therefore, they have hard edges. You cannot preserve the softness of a feathered edge, such as in a shadow, when creating an image clipping path.

Fig. 3.7 How-tos in Adobe Photoshop

how-tos related to the target application. Figure 3.7 shows an example of how-tos in Photoshop. There is however, a serious limitation in this mechanism. There is no way of organizing the how-tos except through a title. Users cannot collaborate with each other on a single how-to and the means of retrieving these how-tos are rudimentary.

The *cheatsheets* mechanism provided with the popular Eclipse [8] platform is another system similar to ours. In this help system, the users can create a series of steps that are aimed at solving a single problem. Figure 3.8 shows an example cheatsheet. Again, the problem remains that there is no way of linking cheatsheets together. There is no way of collaborating with other users for the creation of a cheatsheet.

Our helplet architecture offers many advantages over both these techniques. It uses a collaborative mechanism that can be built around all types of content besides simple (or hyper) text. It allows users to implicitly collaborate with each other by contributing to the larger corpus of helplets which can be searched in an efficient and effective manner. Since the search and retrieval mechanism is based on machine common sense, different helplets implicitly related to each other can be retrieved, thus decreasing the amount of noise and increasing the recall of help information.

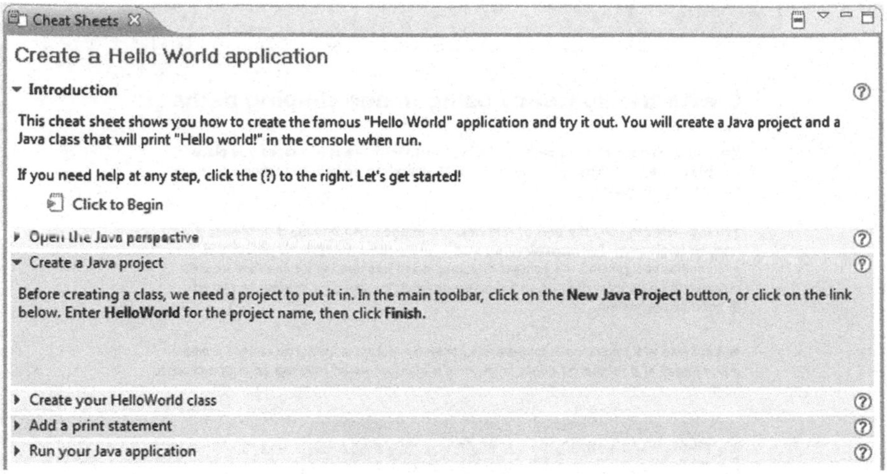

Fig. 3.8 Cheatsheets in eclipse

3.6 Conclusion

Help and documentation form an important part of any software. Traditional help systems have mostly relied on text-based corpuses, which can, at most, be indexed and searched. Complex information-retrieval mechanism designed to retrieve information from the social web have not been incorporated in help systems yet. The advent of Web 2.0 and the creation of collaborative systems that allow users to chapter to the corpus of information have opened new vistas in help and support systems. In this contribution, we have discussed how a collaborative help system can be developed around any Web-enabled system. Our helplet architecture allows different users to interactively submit help and tutorial texts, videos, screencasts, and images to a central repository. It embeds help collection and retrieval mechanism within the target application allowing for seamless integration with the application. Helplets are organized using collaborative tagging which allows for a flexible and user-friendly work flow. Content retrieval is automated through the use of machine common sense. The end result is a system which allows different users to contribute to a larger solution of a problem by submitting small bits of help. The small helplets are linked to each other using a corpus of machine common sense. The framework also allows users to contribute to the machine commonsense corpus, thus allowing for an adaptable system which can be tailored to fit many different domains.

Currently, our architecture is focusing on a general approach which does not favor any specific domain. However, it might be beneficial to consider specialized situations of this architecture for specific domains. For example, a geological survey software may develop a training architecture based on videos and images based on our architecture. This would require a deeper study of how the existing "common sense" corpus can be specialized to suit this specific domain. Moreover, the mechanism for incorporating this architecture in a distributed repository is being considered in our ongoing work. Such a distributed repository might be useful in cases where different applications share a corpus of knowledge. For example, a photo-editing softwares might make use of tutorials written for another image-manipulation software. We believe that the possibilities for the utilization of this helplet architecture are endless and there are many directions which need to be explored at length to fully harness the power of this architecture.

Acknowledgments This work has been supported by the Institute of Management Sciences. The authors would like to acknowledge the other members of Security Engineering Research Group for their input in this work.

References

1. What is del.icio.us? http://del.icio.us/about/ (Retrieved on February 24, 2007)
2. Basecamp. Accessed at: http://www. basecamphq. com/ (Retrieved on November 13, 2007)
3. Gmail. Accessed at: http://www. gmail. com/ (Retrieved on November 13, 2007)
4. TypePad. Accessed at: http://www. typepad. com/ (Retrieved on November 13, 2007)
5. Adobe Photoshop: Accessed at: http:// www. adobe. com/ products/ photoshop/ (2008)
6. Adobe Systems: Accessed at: http:// www. adobe. com/ (2008)
7. Association of Computing Machinery: ACM Digital Library. http://portal.acm.org (2007)
8. Eclipse: Accessed at: http:// www. eclipse. org/ (2008)
9. Fellbaum, C.: Wordnet: An Electronic Lexical Database. MIT Press (1998)
10. Flickr: About flickr. http://www. flickr. com/ about/ (Retrieved on February 24, 2007)
11. Golder, S., Huberman, B.: The Structure of Collaborative Tagging Systems. Arxiv preprint cs.DL/0508082 (2005)
12. Google Inc.: Google Scholar. http://scholar.google.com (2007)
13. Hoeber, A., Mandler, A., Cox, N., Shea, T., Levine, R., et al.: Method and apparatus for displaying context sensitive help information on a display (1992). US Patent 5,157,768
14. Lenat, D.: CYC: A Large-Scale Investment in Knowledge Infrastructure. COMMUNICATIONS OF THE ACM **38**, 33–38 (1995)
15. Liu, F., Yu, C., Meng, W.: Personalized Web Search by Mapping User Queries to Categories. Proceedings of the eleventh international conference on Information and knowledge management pp. 558–565 (2002)
16. Liu, H., Singh, P.: Commonsense Reasoning In and Over Natural Language. Proceedings of the 8th International Conference on Knowledge-Based Intelligent Information & Engineering Systems (KES-2004) (2004)
17. Liu, H., Singh, P.: ConceptNet: A Practical Commonsense Reasoning Tool-Kit. BT Technology Journal **22**(4) (2004)
18. Miller, J., Ganapathi, J.: Method of and system for displaying context sensitive and application independent help information (1996). US Patent 5,535,323
19. Nauman, M., Hussain, F.: Common Sense and Folksonomy: Engineering an Intelligent Search System. In: Proceedings of ICIET'07: International Conference on Information and Emerging Technologies, Karachi, Pakistan. IEEE (2007)
20. Nauman, M., Khan, S.: Using Personalized Web Search for Enhancing Common Sense and Folksonomy Based Intelligent Search Systems. In: Proceedings of WI'07: IEEE/WIC/ACM International Conference on Web Intelligence, Silicon Valley, USA (November 2007)
21. OMCS: The Open Mind Common Sence Project. Accessed at: http://openmind.media.mit.edu/
22. O'Reilly, T.: O'Reilly – What is Web 2.0. Accessed at: http://www. oreilly. com/ pub/ a/ oreilly/ tim/ news/ 2005/ 09/ 30/ what-is-web-20.html (Retrieved on October 22, 2007)

23. Riddle, P.: Tags: What are They Good For? http://prentissriddle. com(2005)
24. Russell, S., Norvig, P.: Artificial intelligence: a modern approach. Prentice-Hall, Inc. Upper Saddle River, NJ, USA (1995)
25. Schmitz, C., Hotho, A., Jaschke, R., Stumme, G.: Mining Association Rules in Folksonomies. Proceedings of the IFCS 2006 Conference
26. Singh, P.: The public acquisition of commonsense knowledge. Proceedings of AAAI Spring Symposium: Acquiring (and Using) Linguistic (and World) Knowledge for Information Access (2002)
27. Singh, P., Barry, B.: Collecting commonsense experiences. Proceedings of KCAP03 (2003)
28. Singh, P., Barry, B., Liu, H.: Teaching Machines about Everyday Life. BT Technology Journal **22**(4), 227–240 (2004)
29. Singh, P., Lin, T., Mueller, E., Lim, G., Perkins, T., Zhu, W.: Open Mind Common Sense: Knowledge acquisition from the general public. Proceedings of the First International Conference on Ontologies, Databases, and Applications of Semantics for Large Scale Information Systems (2002)
30. Sukaviriya, P., Foley, J.: Coupling a UI framework with automatic generation of context-sensitive animated help. ACM Press New York, NY, USA (1990)
31. WordPress: Wordpress.com. Accessed at: http://www. wordpress. com/ (Retrieved on November 13, 2007)
32. Xu, Z., Fu, Y., Mao, J., Su, D.: Towards the Semantic Web: Collaborative Tag Suggestions. Collaborative Web Tagging Workshop at WWW2006, Edinburgh, Scotland, May (2006)
33. Yao, J., Yao, Y.: Web-based Support Systems. In: Proceedings of the WI/IAT Workshop on Applications, Products and Services of Web-based Support Systems, pp. 1–5 (2003)

Chapter 4
Web-based Virtual Research Environments

Xiaobo Yang and Robert J. Allan

Abstract Computer-based research plays an increasingly important role around the world today. While more and more people are benefiting from research projects, they are not all well supported by advanced information and communication technologies (ICT). Information technologies have gradually been adopted by researchers, for example through digital library services, computational, and data grids, but most researchers still rely heavily on "traditional" technologies such as e-mail and telephone as their principal collaboration tools. In this chapter, we will share our recent experiences in the area of Web-based virtual research environments (VREs), which aim to provide researchers with a pervasive collaborative working environment enabling them to work together more efficiently in their daily activities to solve research "grand challenges." We first give some background and current status of VREs and then illustrate our experience in several VRE-related projects. Finally, we discuss the future of VREs and summarize the chapter.

4.1 Introduction

The UK Joint Information Systems Committee (JISC) adopted the definition that a Virtual Research Environment (VRE) is to "help researchers in all disciplines manage the increasingly complex range of tasks involved in carrying out research."[1] This statement clearly acknowledges that research today is becoming increasingly

X. Yang
STFC Daresbury Laboratory, Warrington WA4 4AD, UK, now working at IT Services, University of Birmingham, Edgbaston, Birmingham B15 2TT, UK
e-mail: x.yang.3@bham.ac.uk

R.J. Allan
STFC Daresbury Laboratory, Warrington WA4 4AD, UK
e-mail: robert.allan@stfc.ac.uk

[1] http://www.jisc.ac.uk/programme_vre.html

J.T. Yao (ed.), *Web-Based Support Systems*, Advanced Information and Knowledge Processing, DOI 10.1007/978-1-84882-628-1_4,

complex and would benefit from specialized support systems. Collaborations are now inevitable in nearly all research disciplines, particularly those addressing the so-called 'grand challenges'. One example of a distributed computing and data infrastructure is the European Enabling Grids for E-sciencE (EGEE) project,[2] which "brings together more than 140 institutions to produce a reliable and scalable computing resource available to the European and global research community..." to solve challenges "from scientific domains ranging from biomedicine to fusion science."

Today, research activities normally involve more than one individual or group, often from different institutions and potentially in different countries. This collaboration among researchers who are geographically distributed enables them to focus on tackling a shared challenge requiring multiple skills or knowledge perspectives. Using today's information technologies, in particular the Internet, such kinds of research activities can be performed online. We define these as e-Research activities – here we consider the terms *e-Science* and *e-Research* to be synonymous. While e-Learning systems are widely adopted and deployed in academic institutions for supporting education, comparable e-Research systems are still at a relatively early stage. No precise definition of VRE exists, but based on the JISC statement, a VRE system can be considered to be a flexible framework which is equipped with collaboration services for supporting research processes across all disciplines. It also needs to integrate and provide access to the full range of data management, information, computation, and potentially experimental services that are required.

We have proposed that such VRE systems can be constructed using the same frameworks which support e-Learning systems. While such frameworks are designed to support learning activities, they can support research with appropriate extensions. The key to VRE relies on efficient means of collaboration and relevant services provided to researchers. First, a VRE system should be able to act as a communication platform, which typically provides services such as instant messaging (synchronous) and online discussion (asynchronous). Furthermore, services like audio and video conferencing are also required (synchronous). This requirement is aimed at providing researchers with an efficient environment so that they can easily share data and information and exchange ideas, knowledge, work plans, etc. Second, a VRE system should at the same time act as an information provider. For example, making use of portal technology, a VRE system can collect user preferences and filter information retrieved from various sources, e.g., via RSS subscriptions or broker services, and notify end users of information they are likely to be most interested in.

In this chapter, when we talk about research, we are not limited to solely the phase of research project implementation seen in Figure 4.1, but cover the whole research lifecycle starting from call for proposals (research opportunity) to the end of a project (research project completion). Within each phase, many administrative tasks can be involved. For example, recruitment may be necessary when a project is funded. These administrative tasks are particularly important in project management, which covers a wide range of activities including risk assessment and progress

---

[2] `http://www.eu-egee.org/`

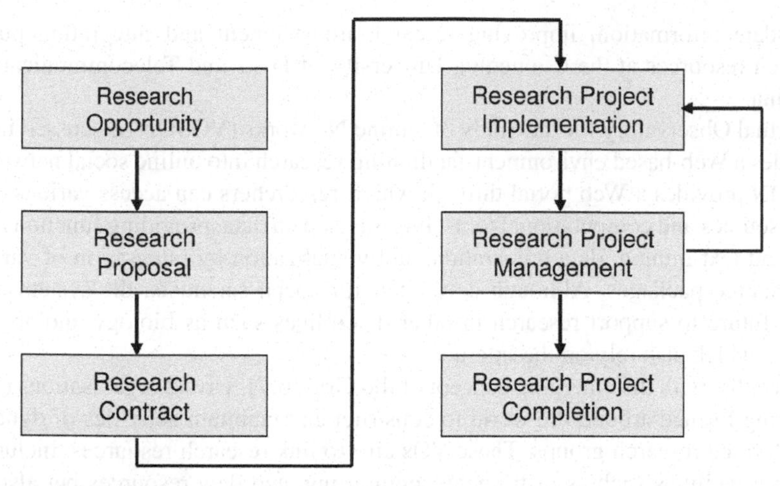

Fig. 4.1 Research lifecycle

monitoring. A VRE system should therefore provide both research and administrative support, including linking to institutional systems or shared administrative services.

A VRE system typically (but not always) has a Web-based front end which makes it available with only a Web browser required on the client side. It is thus a pervasive tool to access research-related services; this could be from a desktop or mobile device like a PDA. The development of the Web highlights ideas of e-activities including e-Research. Therefore, in this chapter, when we talk about VRE systems they are all assumed to be Web-based if not otherwise specified.

The rest of the chapter is organized as follows. First we will present a review of VRE. Then our experience on building a typical VRE system will be described. We will focus our discussion on the Sakai-based VRE framework, originally developed for the JISC-supported Sakai VRE Portal Demonstrator project, but which has been extended to serve the international materials modelling community, Psi-k, and social scientists through the ESRC-funded e-Infrastructure project. Finally, we describe our vision of the future of VRE systems.

4.2 Short Review of VREs

Before the concept of VRE emerged, Web-based research support systems (WRSS) has been studied [17] as a specific type of Web-based support systems (WSS) [16] for supporting research activities for scientists. In [17], Yao examined the principles and issues of WRSS by going through the general research process and methods. Functionalities like resource, data, or knowledge management were identified as important for designing such a WRSS system.

While there is no implementation in [17] but only guidelines, CUPTRSS, a prototype WRSS system, was discussed by Tang et al. focusing on management support [12]. CUPTRSS aims at providing researchers with accurate and up-to-date information, improving research management and integrating public research resources at the Chongqing University of Posts and Telecommunications in China.

Virtual Observatory for the Study of Online Networks (VOSON) discussed in [1] provides a Web-based environment-facilitating research into online social networks. VOSON provides a Web portal through which researchers can access various storage resources and computational tools. It is focused on data, providing functions like Web and text mining, data preparation, and visualization by integration of various open-source packages. Although developed for social scientists, the system aims in the future to support research in other disciplines such as biology and applied physics which also rely on the Internet.

Recently, with the emerging concept of the Grid [6,7], virtual organisations (VO) are being formed around the world to construct and maintain activities of dynamically formed research groups. These VOs aim to link research resources, including not only facilities such as instruments, computing, and data resources but also research staff. The key to the Grid is its ability to plug-in or remove resources dynamically if required across a multi-institutional infrastructure. This in principle brings VOs great flexibility to meet today's always changing grand challenges in research.

On the one hand, Grid and VRE are different: Grid technologies are focusing on integration of computational and data resources, while VRE is targeting collaboration among people. On the other hand, Grid and VRE both encourage collaboration. Thus, Grid is complementary to VRE; for example, a VRE system can act as a gateway for management of remote Grid resources, i.e., acting like a Grid portal. This way, VRE systems enable researchers to create and analyse scientific data via numerical simulations – one of the key everyday activities scientists do.

NEESit (Enabling Earthquake Engineering Research, Education, and Practice)[3] is such an example which links earthquake research scientists together across United States by providing them with a virtual research laboratory. NEESit makes use of Grid technology on the US TeraGrid so that geographically distributed computing resources and engineering facilities can be combined together to support collaborative research into the urban and social impact of earthquakes.

According to our understanding, WRSS systems have the same objective as VRE systems, that is, to improve research quality and productivity by providing research support systems; while underlying Grid technologies can be used to implement such a VRE system from distributed services.

At a workshop in June 2006 [3], we explained a vision of the VRE as a "one-stop-shop" for researchers to access data and global information relevant to their studies with appropriate semantic support and contextual services for discovery, location, and digital rights management. Putting the outcome of the workshop together with the outcome of other surveys which we analyzed, we identified the following 10 conclusions.

- VRE needs to provide mechanisms for discovering scientific data and linking between data, publications, and citations.

[3] `http://it.nees.org/`

- VRE discovery services need to work alongside broad-search services such as Google.
- VRE should embrace collaboration technologies to facilitate joint uses of its services.
- VRE needs to embrace Web 2.0 and semantic Web technologies.
- VRE needs to provide facilities to publish from and to make available content from personal and group information systems.
- VRE should embrace such things such as mail list servers and archives.
- VRE needs to use protocols and standards to facilitate exposing its services in a variety of user interfaces, including portals.
- VRE needs to work with and provide enhanced access for commercial sources and inter-operate with proprietary software.
- VRE interfaces need to be highly customizable and to treat users' activities "in context."
- Training for users is needed as well as increasing awareness of what is available.

In different countries, different terms are used to talk about VREs, or more widely, e-Research. In Section 4.1, we mentioned: *e-Research, e-Science, and VRE,* which are terms utilized in the United Kingdom. There are other terms with similar meaning: for example, *cyberinfrastructure, gateway* are widely utilized in the United States. Although these words are slightly different from each other, for example, VRE can be treated as an implementation of e-Research, they roughly mean the same thing – making use of the latest ICT to support research. For this reason we do not distinguish these words. If you are interested in finding out more details about how the terms are used, see the report by Lawson and Butson [9].

Lawson and Butson looked around e-Research activities in the United States, United Kingdom, Australia, and New Zealand. While work in the United States and the United Kingdom is thought to be relatively advanced, VRE-related research projects are now also commonplace in the South Pacific region, in particular, Australia. Several more recent projects are funded through VeRSI, the Victorian e-Research Strategic Initiative programme.[4] In this chapter, we focus on the UK JISC VRE programme with which we are most familiar.

The initial aim of the JISC VRE programme was "to engage the research community in:

- Building and deploying Virtual Research Environments (VREs) based on currently available tools and frameworks
- Assessing their benefits and shortcomings in supporting the research process
- Improving and extending them to meet the future needs of UK research
- Developing or integrating new tools and frameworks where appropriate solutions do not yet exist"

Phase 1 of the UK JISC VRE Programme[5] started in April 2004 and finished in March 2007. During these 3 years, 15 investigative projects were funded in various

[4] http://www.versi.edu.au/

[5] http://www.jisc.ac.uk/programme_vre.html

subjects like humanities (BVREH: Building a Virtual Research Environment for the Humanities), orthopaedic surgery (CORE: Collaborative Orthopaedic Research Environment), and biology (IBVRE: VRE to support the Integrative Biology Research Consortium). Several projects (e.g., GROWL: VRE Programming Toolkit and Applications) focused on providing general-purpose tools for supporting research. For more information about these projects, please refer to the JISC VRE programme Web site and related publications during the past few years [4, 8, 11, 14]. Information about the standards, technologies, and tools used by, or developed in, these projects is described on the eReSS Wiki.[6]

Starting from April 2007, JISC allocated funding to continue the activities of the VRE programme until March 2009. In Phase 2 four projects were funded to

- "stimulate change in research practices through the development and deployment of VRE solutions;
- continue involving and engaging the research community in building and deploying VRE solutions;
- start exploiting and extending such solutions amongst the Higher Education community;
- continue raising awareness of the benefits of VRE solutions in the research community."

The four projects are again from a range of research domains:

CREW: (Collaborative Research Events on the Web)[7] aims to "improve access to research content by capturing and publishing the scholarly communication that occurs at events like conferences and workshops."

myExperiment: "makes it really easy to find, use and share scientific workflows and other research objects, and to build communities".[8]

Oxford VRE Project: aims to "develop a VRE for the Study of Documents and Manuscripts ... to develop an integrated environment in which documents, tools and scholarly resources will be available as a complete and coherent package".[9]

VERA: (Virtual Environments for Research in Archaeology)[10] aims to "produce a fully-fledged virtual research environment for the archaeological community ... to develop utilities that help encapsulate the working practices of current research archaeologists unfamiliar with virtual research environments."

More information about these projects can be found at their own Web sites. In general, research activities are supposed to benefit from VRE systems at various levels. For instance, the CREW project is trying to solve a very basic but important issue in research by providing improved access to digital research contents. This is applicable to all research domains.

---

[6] eReSS: e-Research Standards and Specifications:
http://www.confluence.hull.ac.uk

[7] http://www.crew-vre.net/

[8] http://www.myexperiment.org/

[9] http://www.vre.ox.ac.uk/

[10] http://vera.rdg.ac.uk/

4.3 Our Experience of VRE

In this section, we present our research on VRE in several projects to support researchers from various disciplines.

4.3.1 Architecture

In Figure 4.2, we illustrate a service-oriented VRE architecture. In general, a VRE system is constructed from various services, which are divided into two groups: core and external services. Core services are designed to be pluggable, which gives flexibility – these services can be deployed or removed on demand although normally core services are unlikely to be removed. Core services are those, for example, an authentication and authorization service, or a logging service, for keeping a system up and running and ease of maintenance. In a real system, more services may be added as core services for the ease of development. For example, an abstract database service is very helpful for developers to manage database operations, but such a service may not be essential. More services called external services are outside of the core services and may even be hosted on remote servers, e.g., on a Grid. The client side of these external services is also pluggable, which makes it possible to customize a VRE system easily. An example external service may be one which provides users a

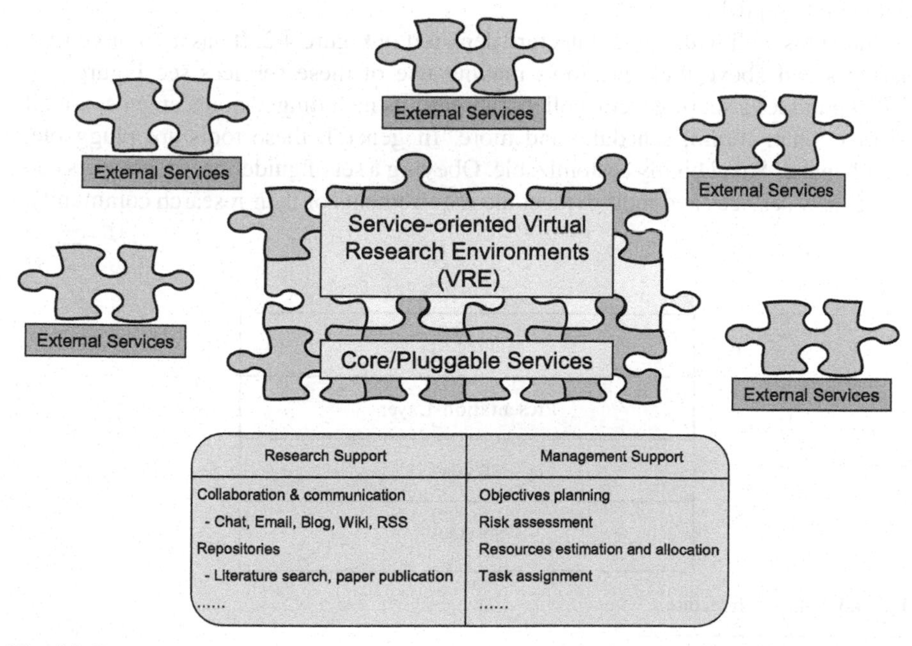

Fig. 4.2 Service-oriented VRE architecture

chance to chat using text messages. Generally speaking, with some core and external services already provided, users are encouraged to develop and share their own external services if they can not find what they want, thus creating a community of practice. This is the philosophy of the US TeraGrid Science Gateways.[11]

The core and external services should be designed to provide both research and management support to cover the whole lifecycle of a research project. More services may be added in Figure 4.2 under "Research Support" and "Management Support" accordingly.

As mentioned above, this service-based architecture gives us *flexibility*. This is key to a VRE system because we cannot design a single general-purpose system that meets requirements of researchers from all disciplines. Therefore, the only solution is to build up a flexible system using a service oriented architecture (SOA). We note that in the JISC and TeraGrid programmes there is currently no recommendation of a base framework or underlying component tool set. In our work we have therefore chosen Sakai for this purpose.

4.3.2 The Sakai Collaborative Learning Framework

Sakai[12] was designed to be an open-source Web-based collaborative learning environment. It has been deployed around the world to support teaching and learning, ad hoc group collaboration and portfolios and has been adapted for research collaboration [10].

Sakai fits well with our architecture depicted in Figure 4.2. It has a set of generic services and above them are tools making use of these services see Figure 4.3. Sakai provides a set of generic collaboration tools including announcements, e-mail archive, chat, forum, schedule, and more. In general, these tools are pluggable, which makes Sakai highly customizable. Obeying a set of guidelines, developers can create new services and tools to meet the requirements of their research community.

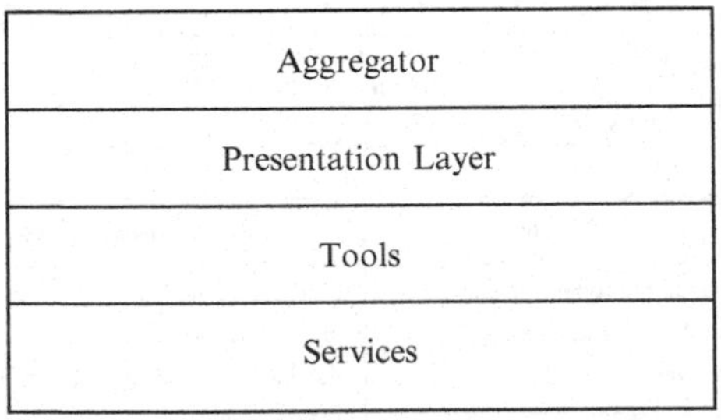

Aggregator
Presentation Layer
Tools
Services

Fig. 4.3 Sakai architecture

[11] http://www.teragrid.org/programs/sci_gateways/

[12] http://www.sakaiproject.org/

These new services and tools can then be plugged in or removed from the Sakai Java framework and configured on demand.

In the terminology of Sakai, a virtual organisation maps onto a "worksite," which appears as a tab in the Sakai portal. Through its worksite, bespoke tools are configured to be available to each VO that require them. Each of the generic tools inherits a view of the resources belonging to that VO and can be accessed only by VO members with appropriate roles and permissions within the context of the worksite. This makes Sakai a very powerful tool for "enterprise" VREs, i.e., those supporting a number of research projects, and distinguishes it from a simple project portal. One such will be described in Section 4.3.5.

The existence of a rich set of tools and services and internal role-base authorization system makes Sakai an ideal candidate for building up VRE systems and therefore it has been chosen as the basis of all the systems we have deployed. We first describe additional tool development for Sakai.

4.3.3 Prototype: Sakai VRE Portal Demonstrator

The Sakai VRE Portal Demonstrator Project [4] was one of the 15 JISC-funded projects under the VRE-1 Programme (2004–2007). The aim of the project was to create a working example of a generic VRE system based on the Sakai framework, which would engage with researchers and scientists and enable them to explore the benefits and capabilities of such a VRE in supporting their work.

This was a joint project led by Lancaster University, with three other partners, the STFC e-Science Centre (Daresbury Laboratory), and the universities of Oxford and Reading. While Sakai provides a set of generic collaboration tools, several additional tools were developed, for example, the Agora audio–video conferencing tool.[13] A requirement for this kind of tool was noted when comparing Sakai to the definition of collaborative working environment (CWE) as for instance found on Wikipedia. A screenshot of Agora is shown in Figure 4.4.

A Wiki tool was developed at University of Cambridge in the JISC VRE-1 Project for the Teaching and Learning Research Programme.[14] We used this tool extensively in our own projects.

A lot of effort was put in to solve the interoperability issues between Sakai and other Web portal frameworks based on the JSR 168 portlet standard.[15] In brief, the JSR 168 standard was proposed to enable interoperability between portal frameworks and portlets written in the Java language. With the help of this

[13] `http://agora.lancs.ac.uk/`

[14] rWiki: `http://bugs.sakaiproject.org/confluence/display/RWIKI/`

[15] `http://jcp.org/en/jsr/detail?id=168`

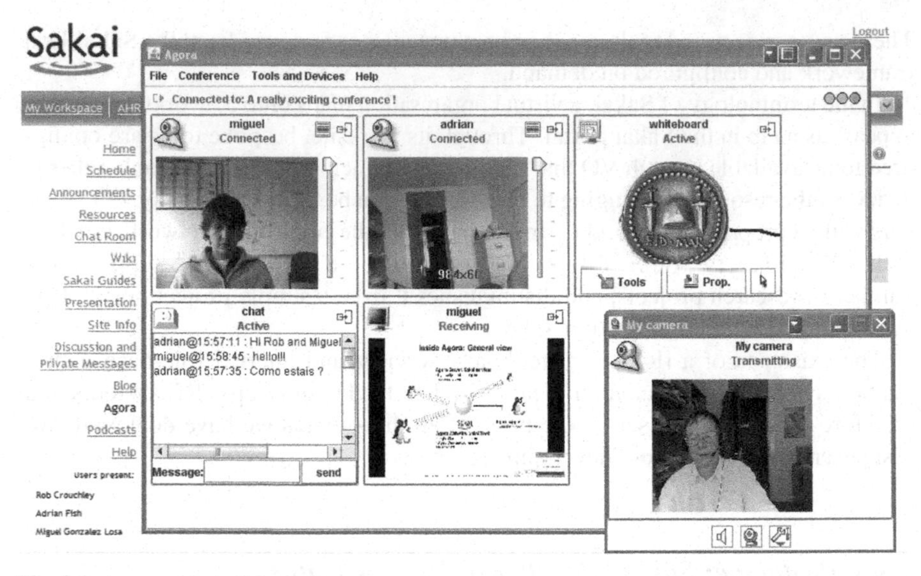

Fig. 4.4 A screenshot of Agora

specification, portal developers can use a set of standard APIs to develop portlets, which can then be deployed under any JSR 168-compliant portal framework. The main reason for wanting JSR 168 support in Sakai was to therefore to reuse research tools which had previously been developed, e.g., for Grid computing, data management, and information access [2]. Before the project was started, a Web portal [13] had been set up for researchers to seamlessly access computational and data Grid resources provided by the UK National Grid Service (NGS).[16] Release 2 of the NGS Portal described in [13] was JSR 168 compliant and had a useful set of portlets including: (1) proxy management; (2) job submission; (3) file transfer, and (4) resource discovery.

At the time, the Sakai VRE Portal Demonstrator Project was started, Sakai did not support JSR 168 portlets. While re-using those portlets developed for the NGS Portal would clearly benefit VRE users, redevelopment of the same functionality specifically for Sakai was thought to be unnecessary and time-consuming. Therefore Web Services for Remote Portlets (WSRP[17]) was studied and a Sakai WSRP consumer tool based on WSRP4J[18] was developed so that the Grid portlets became accessible to Sakai VRE users. This also demonstrated than an SOA could be achieved using this Web services standard with the actual portlets hosted and maintained elsewhere. A screenshot of the unmodified file transfer portlet used from Sakai through WSRP is shown in Figure 4.5. Within the portlet, two remote Grid resources are

[16] `http://www.grid-support.ac.uk/`

[17] `http://www.oasis-open.org/committees/download.php/3343/oasis-200304-wsrp-specification-1.0.pdf`

[18] `http://portals.apache.org/wsrp4j/`

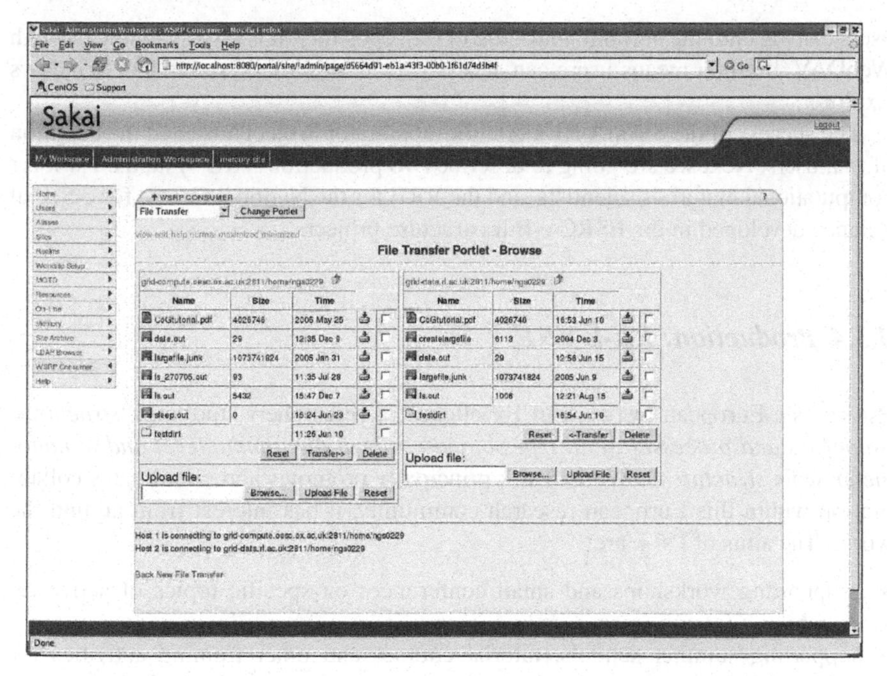

Fig. 4.5 A screenshot of the file transfer portlet accessed from Sakai using WSRP

accessed through GridFTP.[19] (*"GridFTP is a high-performance, secure, reliable data transfer protocol optimised for high-bandwidth wide-area networks."*)

At the end of the project, JSR 168 support was finally integrated in Sakai through core services. But with the WSRP consumer available, users could still consume remote portlets from any WSRP producer, for instance the library catalogue search tool developed in the JISC-funded Contextual Resource Evaluation Environment (CREE) project.[20] Developing the WSRP consumer as a Sakai tool clearly demonstrated the ease of extension of such an architecture.

Throughout the lifetime of the project, the VRE system was heavily utilized for project management, an essential part of collaborative research. For example, a repository was set up within Sakai using the shared resources tool and nearly all project-related documents added. With the search tool enabled, this central repository has proven to be very useful. It was found to be easy for project administrators use and documents, agenda, minutes, presentations, etc. could be available for meetings, including virtual meetings held over Access Grid. It was even found useful to edit the Wiki directly during meetings and workshops to record discussions. While

[19] `http://www.globus.org/grid_software/data/gridftp.php`

[20] `http://www.hull.ac.uk/cree/`

users can log onto the portal to make use of the repository, it is also available through WebDAV,[21] which means users can access it from desktop systems like Windows Explorer.

The success of the Sakai VRE Portal Demonstrator Project attracted the attention of real users. Next we are going to describe two production VRE systems: Psi-k for computational materials scientists, and the VRE for the National Centre for e-Social Science developed in the ESRC e-Infrastructure project.

4.3.4 Production: Psi-k VRE

Psi-k[22] is a European Network of Excellence of researchers studying *"structures, properties and processes in all types of materials at the atomic level and in nanometer scale structures"*. Whilst Psi-k principally promotes and coordinates collaboration within this European research community, it has interest from around the world. The aims of Psi-k are:

- "supporting workshops and small conferences on specific topics of active research;
- supporting summer schools, tutorial courses and other training activities for young and new researchers in the field;
- organising a large "Psi-k Conference" (such as those previously held in 1996, 2000, 2005) every few years;
- disseminating information about research progress, job opportunities, relevant workshops, conferences and other events, and funding opportunities via a Newsletter available freely to anyone in the world by electronic means."

Before a VRE was set up for Psi-k, it was understood that the basic user requirement was to provide an integrated platform with efficient communication and collaboration services. Other requirements included: (1) integration of Psi-k's existing Web site and mailing list; (2) workshop proposal submission support; (3) user and resource management. More requirements were added following subsequent discussions and during the development of the VRE system, as is often the case based on early user experience.

From the requirements and our experiences of VREs, a Sakai-based VRE system was set up for Psi-k in 2007 to help to realize the objectives listed above.[23] After some effort on tuning Sakai, most of the requirements were met in a straightforward way using tools from the official distribution. These tools include Schedule, Registration, Mail Archive, Profile, etc. Following this additional tools were developed to extend Sakai, for example to support the workshop proposal submission and review process. Such tools can be developed and contributed back to the official Sakai distribution.

---

[21] `http://www.webdav.org/`

[22] `http://www.psi-k.org/`

[23] `http://cselnx9.dl.ac.uk:8080/portal`

There are currently over 1,000 registered users of Psi-k and more are expected to be served, therefore scalability tests were carried out for some of the tools We already know of Sakai educational installations in the United States serving many thousands of users.

4.3.5 Production: Social Science e-Infrastructure VRE

The e-Infrastructure Project [5] is a 3 year project funded by the UK Economic and Social Research Council (ESRC) starting from January 2007. This project is hosted by NCeSS, the National Centre for e-Social Science and involves work of several of its research nodes. The aims of the project are to:

- "examine the feasibility of and begin to create a coherent e-Infrastructure for social science in the UK;
- provide a platform for disseminating the benefits of e-Social Science;
- leverage existing e-Social Science and e-Science investments;
- and define a roadmap for the sustainability of middleware, services and tools currently being developed by participants in NCeSS."

An important work package in this project was to set up a VRE to be used in the e-Infrastructure. Following a workshop (Portal Usability for e-Social Science[24]) to introduce potential users to the VRE concepts and to gather requirements, Sakai was deployed for NCeSS.[25] Besides look-and-feel customization, the major work to develop this into a VRE system has been to (1) integrate various additional services and specific requirements from research nodes; and (2) encourage the whole NCeSS community to use this VRE as the central communication and collaboration platform including project management.

The tools which NCeSS uses for collaboration include Resources, Blog, Wiki, Schedule, e-Mail, Discussion, Chat, Announcements, RSS News, Glossary and Agora. Both the Sakai VRE and Access Grid are used for project management purposes and to support the research of its nodes, each of which constitutes a VO and has one or more worksites. Quite a few of the VRE users are members of more than one of these VOs. There are also several joinable sites where all users can share information.

Agenda setting workshops have enabled the e-Infrastructure project to identify a priority area: simulation and modelling. Portlets from MoSeS and GENeSIS (Leeds and UCL) and DAMES (Stirling) are or will be included for use by their VOs and others.[26] These enable researchers discover data and run their own simulations on the NGS, visualize and analyze results and archive them for future study and reuse. This facilitates development and sharing of social simulation resources within the

[24] http://www.ncess.ac.uk/events/item/?item=164

[25] http://portal.ncess.ac.uk

[26] http://www.ncess.ac.uk/research

UK social science community, encourages cooperation between model developers and researchers, and helps foster the adoption of simulation as a research method in the social sciences. Shibboleth-enabled authentication is currently being added to enable integration with online library catalogues and other scholarly information repositories.[27]

In addition to Sakai, several NCeSS nodes have their own Web 2.0-style applications. These include ourSpaces from Policy Grid (Aberdeen), MapTube from GeoVue (UCL), and MyExperiment used in the Obesity e-Lab (Manchester). GENeSIS has demonstrated using SecondLife as a online urban laboratory to explore issues pertaining to planning and public debate in a virtually 3D collaborative environment. It is difficult to integrate such self-contained Web applications with a portlet framework unless their services can be exposed using an appropriate API such as JSR 168 or WSRP and data-interchange standards. We nevertheless concede that there is unlikely to be one single VRE that meets all requirements and a traditional portal may not always be the first choice.

4.4 Further Discussion and Summary

In this chapter, we have discussed Web-based Virtual Research Environments. VREs, acting as an implementation of e-Research, provide researchers with a platform for communication and collaboration. Besides that, VREs, enable administrative tasks to be executed within the same platform. This benefits research activities with improved efficiency and productivity.

As we can see from the VRE systems illustrated in this chapter, a basic requirement of VREs is to provide services for communication and collaboration. Both researchers and administrators benefit from these services, particularly during project planning and management. A VRE system is however not limited to provide only these services – in [15], we described more types of service a VRE may provide. For example, computational and data Grid, as shown in the Sakai VRE Demonstrator Project, provide researchers with resources for numerical simulations and data management. Since more and more documents are either in digital format or have been digitalized, repositories are becoming increasingly important. Such repositories may contain various types of information including documents, image, audio, and video, which can be easily shared among researchers by integration of appropriate tools helping them to share information and collectively create new knowledge.

On one hand, there is always the requirement for integration of external services in VRE systems, for example, JSR 168 support through WSRP as we illustrated. On the other hand, there is a requirement to consume externally hosted services from systems outside the VRE, perhaps using Web or Grid services in an SOA. For

[27] `http://shibboleth.internet2.edu/`

both scenarios, security is a major concern and we note that there is work ongoing aiming to provide a scalable distributed and ubiquitous solutions, such as Shibboleth mentioned above.

We conclude that, over the past few years, VREs have proven to be a promising approach to research support. While there is no one technical solution, research into VRE itself is making it more and more attractive as new functionalities are made available. We foresee VRE systems in the near future being adopted as turnkey solutions for commercial researchers, particularly for those wishing to collaborate with their academic counterparts and use e-Research technology.

References

1. Ackland, R., O'Neil, M., Standish, R., Buchhorn, M.: VOSON: A Web service approach for facilitating research into online networks. In: Proc. Second International Conference on e-Social Science. Manchester, UK (2006)
2. Allan, R.J., Awre, C., Baker, M. and Fish, A.: Portals and Portlets 2003. Tech. Rep. UKeS-2004-06 National e-Science Centre. See http://www.nesc.ac.uk/technical_papers/UKeS-2004-06.pdf (2004)
3. Allan, R., Crouchley, R. and Ingram, C.: Workshop on e-Research, digital repositories and portals. Ariadne Magazine, **49** (2006)
4. Allan, R., Yang, X., Crouchley, R., Fish, A., Gonzalez, M.: Virtual Research Environments: Sakai VRE Demonstrator. In: UK e-Science All Hands Meeting 2007, 469–476, Nottingham, UK (2007)
5. Daw, M., Procter, R., Lin, Y., Hewitt, T., Jie, W., Voss, A., Baird, K., Turner, A., Birkin, M., Miller, K., Dutton, W., Jirotka, M., Schroeder, R., de la Flor, G., Edwards, P., Allan, R., Yang, X., Crouchley, R.: Developing an e-Infrastructure for social science. In: Proc. Third International Conference on e-Social Science, Michigan, USA (2007)
6. Foster, I., Kishimoto, H., Savva, A., Berry, D., Djaoui, A., Grimshaw, A., Horn, B., Maciel, F., Siebenlist, F., Subramaniam, R., Treadwell, J., Von Reich, J.: The Open Grid Services Architecture, Version 1.0. In: Global Grid Forum (GGF) (2005)
7. Foster, I., Kesselman, C., Tuecke, S.: The anatomy of the Grid: enabling scalable virtual organisations. Int. J. Supercomputer Applications, **15:(3)** (2001)
8. Hodson, S.: The use of Virtual Research Environments and eScience to enhance use of online resources: the History of Political Discourse VRE Project. In: UK e-Science All Hands Meeting 2007, 284–289, Nottingham, UK (2007)
9. Lawson, I., Butson, R.: eResearch at Otago: a report of the University of Otago eResearch steering group. See http://docushare.otago.ac.nz/docushare/dsweb/Get/Document-3584/eResearch+at+Otago+Report.pdf (2007)
10. Severance, C., Hardin, J., Golden, G., Crouchley, R., Fish, A., Finholt, T., Kirshner, B., Eng, J. and Allan, R.: Using the Sakai collaborative toolkit in e-Research applications. Concurrency and Computation: Pract. Exper., **19:(12)**, 1643–1652. (2007)
11. Stanley, T.: Developing a Virtual Research Environment in a portal framework: the EVIE project. Ariadne Magazine, **51** (2007)
12. Tang, H., Wu, Y., Yao, J.T., Wang, G., Yao, Y.Y.: CUPTRSS: A Web-based research support system. In: Workshop on Applications, Products and Services of Web-based Support Systems. Halifax, Canada (2003)
13. Yang, X., Chohan, D., Wang, X.D., Allan, R.: A Web portal for the National Grid Service. In: UK e-Science All Hands Meeting 2005, 1156–1162, Nottingham, UK (2005)
14. Yang, X., Allan, R.: Web-based Virtual Research Environments (VRE): support collaboration in e-Science. In: The 3rd International Workshop on Web-Based Support Systems (WSS'06), Proc. WI-IAT 2006 Workshops, 184–187, Hong Kong, China (2006)

15. Yang, X., Allan, R.: From e-Learning to e-Research: building collaborative Virtual Research Environments using Sakai. In Proc. DCABES 2007, 995–999, Yichang, Hubei, China (2007)
16. Yao, J.T., Yao, Y.Y.: Web-based support systems. In: Workshop on Applications, Products and Services of Web-based Support Systems. Halifax, Canada (2003)
17. Yao, Y.Y.: Web-based research support systems. In: The Second International Workshop on Web-based Support Systems. Beijing, China (2004)

Chapter 5
Web-Based Learning Support System

Lisa Fan

Abstract Web-based learning support system offers many benefits over traditional learning environments and has become very popular. The Web is a powerful environment for distributing information and delivering knowledge to an increasingly wide and diverse audience. Typical Web-based learning environments, such as Web-CT, Blackboard, include course content delivery tools, quiz modules, grade reporting systems, assignment submission components, etc. They are powerful integrated learning management systems (LMS) that support a number of activities performed by teachers and students during the learning process [1]. However, students who study a course on the Internet tend to be more heterogeneously distributed than those found in a traditional classroom situation. In order to achieve optimal efficiency in a learning process, an individual learner needs his or her own personalized assistance. For a web-based open and dynamic learning environment, personalized support for learners becomes more important. This chapter demonstrates how to realize personalized learning support in dynamic and heterogeneous learning environments by utilizing Adaptive Web technologies. It focuses on course personalization in terms of contents and teaching materials that is according to each student's needs and capabilities. An example of using Rough Set to analyze student personal information to assist students with effective learning and predict student performance is presented.

5.1 Introduction

Web-based support systems (WSS) are a new research area that aim to support human activities and extend human physical limitations of information processing with Web technologies [2]. With the rapid advances in World Wide Web (WWW) interactive technologies, the use of Web-based learning support tools is rapidly

L. Fan
Department of Computer Science, University of Regina, Regina, Saskatchewan, S4S 0A2, Canada
e-mail: Lisa.fan@uregina.ca

J.T. Yao (ed.), *Web-Based Support Systems*, Advanced Information and Knowledge Processing, DOI 10.1007/978-1-84882-628-1_5,

growing. In a Web-based learning environment, instructors provide online learning material such as text, multimedia, and simulations. Learners are expected to use the resources and learning support tools provided. However, it is difficult and time-consuming for instructors to keep track and assess all the activities performed by the students on these tools [3]. Moreover, due to the lack of direct interaction between the instructor and the students, it is hard to check the students' knowledge and evaluate the effectiveness of the learning process. When instructors put together the learning material (such as class notes, examples, exercises, quizzes, etc.) online, they normally follow the curriculum and pre-design a learning path for the students, and assume that all the learners would follow this path. Often this path is not the optimal learning sequence for individual learners, and does not satisfy the learner's individual learning needs. This is typically the teacher-centered "one size fits all" approach.

Not all students have the same ability and skills to learn a subject. Students may have different background knowledge for a subject, which may affect their learning. Some students need more explanations than others. Other differences among students related to personal features such as age, interests, preferences, emotions, etc. may also affect their learning [4]. Moreover, the results of each student's work during the learning session must be taken into account in order to select the next study topics for the student [5].

By utilizing adaptive Web technologies, particularly Adaptive Educational Hypermedia (AEH) systems, it is possible to deliver, to each individual learner, a course offering that is tailored to their learning requirements and learning styles [6]. These systems combine ideas from hypermedia and intelligent tutoring systems to produce applications that adapt to meet individual educational needs. An AEH system dynamically collects and processes data about student goals, preferences, and knowledge to adapt the material being delivered to the educational needs of the students [7]. Since the learning process is influenced by many factors, including prior knowledge, experience, learning styles, and preferences, it is important that the student model of an AEH system accommodate such factors in order to adapt accurately to student needs.

This chapter first provides an overview of concepts and techniques used in adaptive Web-based learning support systems. Then it discusses and examines the use of student individual differences as a basis of adaptation in Web-based learning support systems. The chapter proposes a framework of knowledge structure-based visualization tool for representing a dynamic and personalized learning process to support students' deep learning. An example of using Rough Set to predict the students' performance is demonstrated.

5.2 Learning and Learning Support Systems

A definition of learning is given by Gagne [4]:

> a process of which man and the animals are capable. It typically involves interaction with the external environment (or with a representation of this interaction, stored in the learner's

memory). Learning is inferred when a change or modification in behavior occurs that persists over relatively long periods during the life of the individual.

Clearly, learning is an interactive, dynamic, and active feedback process with imagination driving action in exploring and interacting with an external environment [5]. It is the basic cognitive activity and accumulation procedure of experience and knowledge. There are two main learning styles: group learning and individual learning. Group learning is used in the traditional classroom learning. Teacher and students communicate in a real-time manner. This is the teacher-centered form of education. Feedback is the two-way communication between teacher and students. It requires high operational cost. Individual learning is a student-centered form of education. In this learning style, learners study the material individually, and the teacher acts as a supporter, such as in Web-based education. This learning style provides personal flexibility with low cost.

The major difference between the Web-based and conventional instruction system is that students can choose their own paces for learning. They can skip those materials that they have already learned or known. They can replay the course that they had not thoroughly understood. However, most of Web-based courses are not as "flexible" as human instructors [6]. Typically, course material is a network of static hypertext pages with some media enhancement. Neither the teacher nor the delivery system can adapt the course presentation to different students. As a result, some students waste their time learning irrelevant or already known material, and some students fail to understand (or misunderstand) the material and consequently overload a distance teacher with multiple questions and requests for additional information. Therefore, the Web-based learning support system needs to overcome the deficiencies of inflexible instruction from conventional face-to-face group learning, and potential inflexibility from not having face-to-face feedback from students to instructors.

5.3 Functions of Web-based Learning Support Systems

Building a Web-based learning support system is relatively easy from the technical point of view. However, analyzing, designing, and implementing the system with consideration of learners' learning process to achieve better teaching and learning results is a difficult process. The system should consider the following features.

- Complexity of learning support

Obtaining knowledge means going through a process of learning. The learning process is complex. Many human senses interact and collaborate. Already obtained knowledge and experiences are used to prove and verify the new cognition. Discussions are used for information exchange, and examples help to strengthen and solidify skills. Different cognitive learning styles complicate the situation. Without taking these into consideration, the knowledge is often presented in a fixed manner. Neither textbooks nor online texts can actually answer questions. The student is provided with only information.

- Individuality and adaptability support

Individuality means that a WLSS must adapt itself to the ability and skill level of individual students. Adaptive methods and techniques in learning have been introduced and evaluated since the 1950s in the area of adaptive instruction and the psychology of learning [7]. Adaptive instructional methods adapted the content of the instruction, the sequencing of learning units, the difficulty of units, and other instructional parameters to the students' knowledge. These methods have been empirically evaluated and shown to increase learning speed and to help students gain a better understanding through individualized instruction.

According to Brusilovsky [3], there are several goals that can be achieved with adaptive navigation support techniques, though they are not clearly distinct. Most of the existing adaptive systems use link hiding or link annotation in order to provide adaptive navigation support. Link hiding is currently the most frequently used technique for adaptive navigation support. The idea is to restrict the navigation space by hiding links that do not lead to "relevant" pages, i.e., not related to the user's current goal or not ready to be seen. Users with different goals and knowledge may be interested in different pieces of information and may use different links for navigation. Irrelevant information and links just overload their working memories and screen [4], in other words create, cognitive overload.

De Bra [8] presented a course that uses a system developed to track student progress and, based on that, generated documents and link structures adapted to each particular student. Links to nodes that are no longer relevant or necessary or links to information that the student is not yet ready to access are either physically removed or displayed as normal text.

Da Silva et al. [9] use typed and weighted links to link concepts to documents and to other concepts. The student's knowledge of each concept is used to guide him or her toward the appropriate documents.

- Interaction support

The Web-based learning support system must be interactive. Students must be able to communicate with the system. Users should be able to add personalized annotations and notes to the prepared knowledge base. It should allow the students to ask questions and to automatically retrieve a proper answer. Web Based Information Retrieval Support System (WBIRSS) may be a useful solution to this problem [10]. Discussion group is an important feature of the support system to improve the learning efficiency.

- Activity and assessment support

One of the most difficult challenges of Web-based learning mechanism is the assessment of students' learning process. It is hard to judge the behavior of a student since the instructor is separated spatially from the students. Testing and check points are important from the point of view of evaluating and assessing the student' progress [11]. Examples and exercises are used to strengthen the students' understanding through practice. The system should provide the students the possibility not only to look at the examples, but also to be able to modify them, try them out, and get feedback on their solutions.

5.4 Designs and Implementation of a WLSS

The design of online courses involves different actors. These actors have different requirements. The teacher needs a tool easy to use in order to create the educational material. The student needs something more than a mere transfer of a book in electronic format. They need some sort of guidance, a high level of interactivity, and a tool for accessing the learning process. The site administrator, finally, needs an easy way to maintain the system for updating the content and the information about the users.

Web Course Tools (WebCT) is a Web-based instructional delivery tool which facilitates the construction of adaptive learning environments for the Web. It enables instructors to create and customize their courses for distance education [12]. It provides a set of educational tools to facilitate learning, communication, and collaboration.

The system can facilitate active learning and handle diverse learning styles. It allows the instructor to present his or her course materials using a variety of mediums to make the course dynamic and interactive. By creating a module of selected material and having a quiz at the end, the instructor can selectively release course materials based on the quiz score from the previous module. The student tracking within the Content Module allows the instructor to observe individual student activity. This can be used for online grading based on participation.

The WebCT system uses a client–server architecture. This model is based on the distribution of functions between two types of independent and autonomous processes: client and server. The system consists of WebCT software and a number of hierarchically organized files for each course. The user accesses the data on the server through a Web browser. All the WebCT software resides on, and runs off, a server, which means any changes made to courses are accessible to students immediately after the change is made. The system provides an environment to cover all aspects of a course such as tools for creating the course materials, lectures, assignments, quizzes, and discussion groups.

Using WebCT, a core computer science course "data structure and algorithm analysis" has been designed. In order to provide an in-depth understanding of the fundamental data structures and to motivate the students, a Web-based adaptive course with an analysis tool based on student learning styles has been proposed. The course has been designed with adaptability and it has been taken into consideration for student individual learning styles.

In the following sections, the structure of the Web course and how this structure contributes to our approach to provide adaptation will be presented.

The students' goal of taking the course is to learn all or most of the course materials. However, the learning goal can be differentiated among different students according to both their expectations with respect to the course and their knowledge about the subject being taught, the latter being the most important user feature in Web-based educational system [3].

Our approach utilizes the adaptive navigation support system in WebCT. At the current stage, the structure of the course is organized around chapters that consist of several concepts. The contents of the course are designed and implemented as a collection of modules (under Content Module). Each module has well-defined learning objectives. Figure 5.1 shows the adaptive menu of the Web course material.

At the end of each chapter, the Web course provides a quiz with automatic feedback. The quiz consists of a set of questions with predefined answers, and the result obtained by the student will have influences over the presentation of the subsequent course contents. The Web course system allows the students to test his or her level of knowledge and understanding of the concepts and also permits dynamical paths among the contents of the course.

For example, as showed in Figure 5.2, if the student completes Chapter One, he or she needs to take the quiz, which will determine whether or not he or she acquired the minimum level of knowledge required to move forward to the next stage of learning. This process will identify the next chapter to be presented. Therefore the following subsequent chapters are set up as conditional.

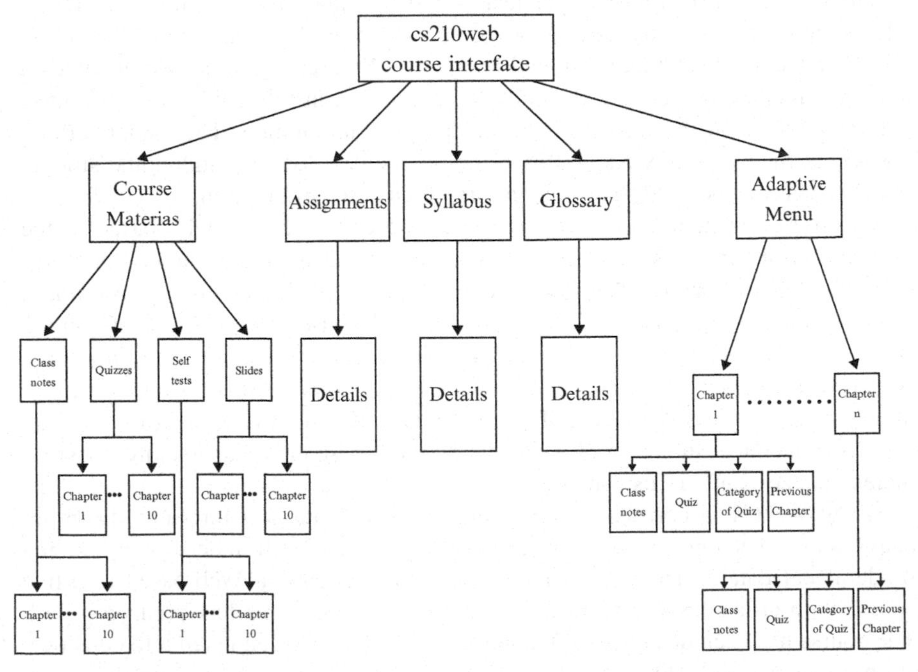

Fig. 5.1 The design structure for the online course

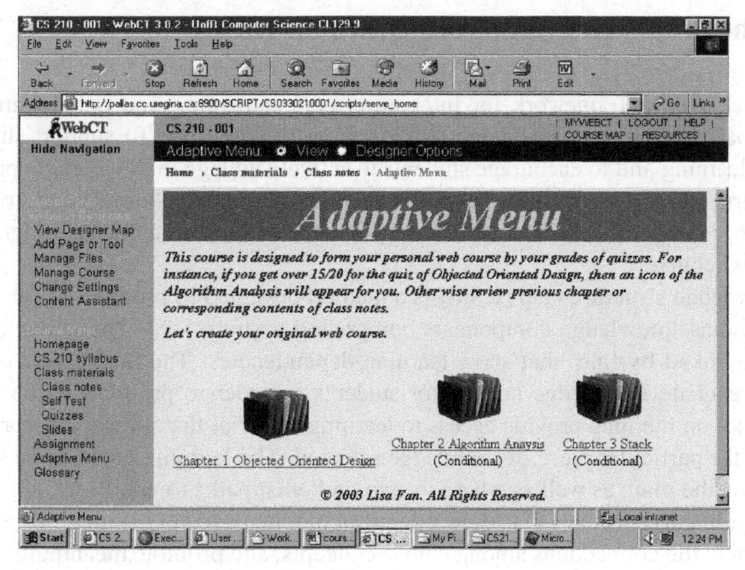

Fig. 5.2 Screen shot of the demonstration of the adaptive course material

These chapters are only presented when the student satisfies the basic requirement. In the case the student has failed the quiz, the student will be recommended to go back and study the same content one more time. The questions contained in the quiz are designed to determine whether or not the student has acquired the required level of knowledge for the chapter. The difficulty levels of the individual question associated to the content have been taken into account. Hence, each question has different weighting toward the total grade of the quiz.

The notion central to the implementation of the Web-based support system is checkpoint analysis [13]. The learning process is divided into many stages. The success of a student is checked at many points.

Several software systems have been designed with online testing [14, 15]. Most of these systems provide automatic grading for quizzes with multiple choice and true–false questions. WebCT has been designed to perform online testing of answers whose correct syntax can be specified as regular expressions. The online grading system provides limited feedback toward the students' learning problems with the course materials. As mentioned earlier, if the student has failed the quiz, he or she has to go back and study the material again. However, these systems do not provide the specifics on which section he needs to spend more time studying and which concepts he or she has difficulty understanding.

A primary concern for a Web-based learning support system is the design of an appropriate structure so that a student can easily and naturally find the most relevant information depending on his or her needs [13]. We present a model of personalization that attempts to assist learners in their cognitive learning based on their assessment results of the learning materials. It provides feedback on their performance and suggests the most suitable learning content to be learned next.

5.5 The Proposed Framework Based on KSM

In this chapter, a framework for integrating personalization and collaboration in a web-based learning management environment is proposed. To support student-centric learning and to encourage students to actively engage in the learning process to construct their own learning and define their learning goal, a knowledge structure map [16] is used as an effective learning tool for "Data Structure and Algorithm Analysis" course.

Knowledge structure map (KSM) is a method designed to produce a visual map of individual knowledge components comprising a study [16]. The nodes on the map are linked by lines that show learning dependencies. The map shows clearly the prerequisite knowledge needed for students in order to progress in the study. The nodes on the map provide access to learning material that allows the learner to acquire the particular piece of knowledge selected. The learning goal can be shown clearly on the map, as well as where to start and what paths to take.

According to Ausubel [17], knowledge structure maps foster meaningful learning by teaching the connections among course concepts, and promote meaningful learning by encouraging students to generate their own connections between concepts.

In the knowledge structure map, each node represents a piece of knowledge and the label in the box is the name of this knowledge. Links or arcs between nodes are unidirectional. A link shows that it is necessary for a student to know a particular piece of knowledge before it is possible for that student to have a full and detailed understanding of another piece of knowledge.

In order to provide an in-depth understanding of the fundamental data structures and to motivate the students, a Web-based adaptive course with an analysis tool based on student learning styles has been proposed. The idea here is to create a tool for students and teachers to visualize and reflect the learning process. The main objective is to involve the students to take active role in his or her studies. The system should encourage the students to go deeper into learning, and find efficient learning paths.

When the KSM is connected to the student database, the system reflects the study process. All the concepts that a student has learned can be marked with a color. Some comments can be added to the map. The teacher can easily see the students' progress, strengths and weakness and can help the students with their studies. Whenever a student adds a new concept to his or her personal knowledge map, the system suggests and recommends other related concepts. With this feature, a student can build a KSM quickly. The students' knowledge maps of the course can also be implemented and presented as animations. With an animation, the student can see how the learning process is proceeding and which concepts are recognized to be similar. Personalized presentation based on each student knowledge level can be presented visually. The maps can be compared. This feature makes it a reasonable platform for supporting students' collaboration. Students can compare their knowledge maps with the instructors' (expert), peer's, or mentor's. Through the process, students can view their learning process, and the dynamic changes of the student knowledge map will indicate how the student knowledge grows.

We can also construct dynamic and clickable KSMs by utilizing Web technologies. For example, if a student clicks on the node indicating "Stack," it would give options for simulation of the algorithm; examples and demos of how to use it; or a simple text file of the formal definition of stack. When a suitable knowledge structure is designed, the system can be used for effective learning, tutoring, problem solving, or diagnosing misconceptions.

5.6 Rough set-based Learning Support to Predict Academic Performance

Rough sets have been applied to many areas where multi-attribute data is required to be analyzed to acquire knowledge for decision making. WSSs are a new research area that aim to support human activities and extend human physical limitations of information processing with Web technologies. In this section, we discuss how to use rough sets to analyze student personal information to assist students with effective learning. Decision rules are obtained using rough set-based learning to predict academic performance, and it is used to determine a path for course delivery.

It is difficult to accurately measure the achievement of students. However, such measurement is essential to any web-based learning support system. Liang et al. [18] proposed a rough set-based distance-learning algorithm to understand the students' ability to learn the material as presented. They analyzed students' grade information and formed an information table to find rules behind the information table and used a table of results from course work to determine the rules associated with failure on the final exam. Then they can inform students in subsequent courses of the core sections of the course and provide guidance for online students. Magagula et al. [19] did a comparative analysis of the academic performance of distance and on-campus learners. The variables used in their study are similar to ours. Lavin [20] gave a detailed theoretical analysis of the prediction of academic performance. Vasilakos et al. [21] proposed a framework of applied computational intelligence – fuzzy systems, granular computing (including rough sets), and evolutionary computing to Web-based educational systems. Challenging issues such as knowledge representation, adaptive properties, and learning abilities and structural developments must be dealt with.

In this study, 12 variables were considered for the analysis based on an assumption that they seem to be related to academic performance. All of them are personal, such as individual background and study environment. It is not necessary to use very fine decimal values for finding out the likelihood of performance. Some of the values should be simple, descriptive, and approximated so that analysis of the data and creation of the rules would be successful. Discretization of variables is needed, so that they are used in conjunction with other variables. Some of them might be replaced with binary value. In the rules to predict academic performance, discretization of the variables has a very important role showing what they imply without any extra explanation.

Table 5.1 shows the discretization of variables:

Table 5.1 The discretization of variables

Variables	Attributes
Age	Young, post-young, pre-middle, post-middle, pre-old, old
Gender	Male, female
Financial	Support, not support
Job	Yes, no
Origin	English country, other country
Environment for study	Good, bad
Martial	Married, not married
Dependents	Yes, no
Time for study	Fair, not enough
Motivation	Yes, no
Health	Good, no
Major	Science, arts
Performance	Over and equal 80, less 80

5.6.1 Survey and Data Collection

A survey was made to collect data from the students on campus with the variables constructed previously. The sample size came to 28. In general, this size is not as big as other studies. However, this size of data is good to start with and there are varieties of age classes available for this study. As the process of constructing the survey proceeded, we realized that not only computer science knowledge, but also psychological and educational knowledge have important roles in order to collect accurate data for analysis. There are 12 variables prepared and considered for use in the analysis. Each variable has two or more choices that participants can select. A description of each choice should be understandable and concise. The discretization of variables was not performed on the variables for the survey. Simplicity is considered to be more important for the survey. For instance, instead of saying HAS_SUPPORT or DOES_NOT_SUPPORT, which is used in the database, the choice would be YES or NO corresponding to a questionnaire so that participants require less comprehension. Positive choices, for instance, YES or GOOD, should be prior to the negative description. This is based on the assumption that the majority of answers should be a positive choice rather than a negative one. If negative responses would be considered to be the majority, a questionnaire should be negated in order to have a majority of positive responses. This gives participants greater ease to answer the survey. An explanation of each choice should only appear in the corresponding question. It reduces the probability that the contents of the survey will be redundant. Considering these points, questionnaires should not give participants any difficulty to comprehend the questions, because confusing them might affect the quality of the data, which might lead to constructing undependable rules.

For constructing data, discretization of variables has to be performed on some data in the survey. For instance, for AGE, the groups in the survey are described with numerical values. These are converted to descriptive groups, such as YOUNG, POST-YOUNG, PRE-MIDDLE, MIDDLE, POST-MIDDLE, PRE-OLD, and OLD.

ID	age	gender	finance	job	origin	houseEnvi	martial	dependents	time	moti	health	major	grade
1	young	male	support	yes	english	good	no	no	fair	no	good	arts	less80
2	young	male	support	no	english	good	no	no	fair	no	good	arts	less80
3	postmiddle	male	support	yes	english	good	no	no	fair	yes	good	arts	less80
4	young	male	support	yes	english	bad	no	no	fair	no	good	science	over&equal80
5	young	female	support	yes	english	good	no	no	fair	yes	good	arts	less80
6	young	male	support	yes	english	good	no	no	fair	no	good	arts	over&equal80
7	young	male	support	yes	english	good	no	no	fair	no	good	arts	less80
8	postyoung	female	support	yes	other	good	no	no	fair	yes	good	arts	over&equal80
9	young	male	support	no	english	bad	no	no	fair	yes	good	arts	over&equal80
10	young	male	no support	yes	english	good	no	no	fair	yes	good	science	over&equal80
11	postyoung	female	support	yes	english	good	no	no	fair	yes	good	science	over&equal80
12	young	female	support	yes	english	bad	no	no	fair	no	good	arts	over&equal80
13	young	male	support	yes	english	good	no	no	fair	no	good	science/arts	less80
14	young	male	no support	no	english	good	no	no	fair	yes	good	science	over&equal80
15	postmiddle	male	no support	yes	english	good	yes	yes	fair	no	good	science	less80
16	middle	male	no support	yes	other	bad	yes	yes	not enou	yes	good	science	over&equal80
17	premiddle	male	support	yes	english	good	yes	yes	fair	yes	good	science	over&equal80

レコード: 1 / 28

Fig. 5.3 Data for analysis

The following table in Fig. 5.3 contains the data where discritization is performed. The total size of the data is 28 and the number of variables is 13.

Academic performance is a target attribute. The purpose of the analysis is to predict the degree of the students' performance, more than, equal to, or less than 80% . This degree could be changeable depending on what the purpose of the analysis is. For instance, if this outcome is used for regular prediction for future courses, all it wants to do is let the students know whether or not they will have a learning curve or issues before it is too late. In this case, the scale might be around 70%. If the system would predict probability of a good performance, 80% might be high enough.

These values form 2,048 classes that fall into either OVERANDEQUAL80 or LESS80. This does not mean that all of them affect the target attribute. Removing some of them does not change dependency. Instead of calculating probability and dependency manually, DQUEST was used to estimate dependency. The following are the reducts that were used for analysis.

1. Gender, Job, Origin, Marital, Time, Motivation, Subject (16)
2. Age, Job, Environ, Time, Motivation, Subject (14)
3. Age, Gender, Job, Time, Motivation, Subject (19)
4. Gender, Financial, Job, Origin, Environ, Marital, Motivation, Subject (20)
5. Age, Financial, Job, Environ, Motivation, Subject (16)
6. Age, Gender, Financial, Job, Motivation, Subject (17)
7. Age, Gender, Financial, Job, Environ, Subject (17)

5.6.2 Results and Discussion

The numbers in parentheses show the different combinations of these attributes appearing in the knowledge table. Low values represent strong data patterns. Since reduct #2: Age, Job, Environ, Time, Motivation, Subject (14) has the lowest value, we focus on the reduct for analysis.

Dependency is a measure of the relationship between condition attributes and the decision attribute. Significance is decreased in dependency caused by eliminating the attribute from the set. The following are rules by target attribute value. Cno implies the number of cases in the knowledge table matching the rule. Dno implies the total number of cases in the knowledge table matching. Rules representing strong data patterns will have high values for both Cno and Dno as shown in Tables 5.2 and 5.3.

The rules are 2 and 4 for ≥ 80% and 7 for < 80%, since they have the highest value among the above rules. (Age = premiddle) ⋀ (Time = fair) → (PERFORMANCE = OVERANDEQUAL80)(Time = fair) ⋀ (Motivation = yes) ⋀ (Subjects = science) → (PERFORMANCE = OVERANDEQUAL80) (Time = no) → (PERFORMANCE = LESS80)

The rules imply the following points:

1. If one is approximately in the range of 30–35 years and has enough time to study, then academic performance most likely will be greater than or equal to 80%.
2. If one has enough time, has motivation to study, and majors in science, then academic performance most likely will be greater than or equal to 80%.
3. If one does not have enough time, then performance is less than 80%.

So, many factors influence student academic performance. Rough set is used to analyze these factors in order to gain some knowledge. Using the knowledge and analysis of past grades allows us to predict academic performance.

In this chapter, we have presented the rough set approach to analysis of student academic performance. The results will assist students in their learning effectively. It also guides the instructor to improve course contents.

Based on the results, the students can be divided into guided and non-guided groups. For the guided group, the course material would be sequentially pre-

Table 5.2 Rules generated by target attribute value (≥80%)

Greater than or equal to 80%	Cno	Dno
1. (Environ = bad) ⋀ (Time = fair) ⋀ (Subject = arts)	2	11
2. (Age = premiddle) ⋀ (Time = fair)	5	11
3. (Time = not) ⋀ (Subject = arts)	1	11
4. (Time = fair) ⋀ (Motivation = yes) ⋀ (Subjects = science)	5	11

Table 5.3 Rules generated by target attribute value (<80%)

Less than 80%	Cno	Dno
1. (Age = young)⋀(Motivation = no)⋀(Subject = science)	2	14
2. (Job = yes) ⋀ (Motivation = no)	1	14
3. (Age = young) ⋀ (Job = yes) ⋀ (Motivation = yes)⋀(Subject = arts)	3	14
4. (Age = postmiddle)	2	14
5. (Job = no) ⋀ (Time = no)	1	14
6. (Age = premiddle) ⋀ (Subject = arts)	1	14
7. (Time = no)	5	14

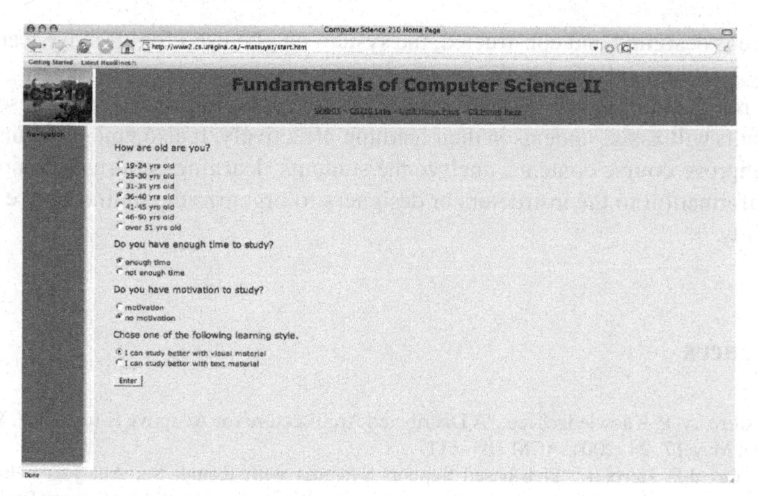

Fig. 5.4 Interface of the Web-based online course

sented with more teacher-centered contents, whereas for the non-guided group, the course material is presented non-linearly. Students have more flexibility to explore the course material that is more student-centered. Figure 5.4 is the interface of the online course. Based on students, learning styles, knowledge background, ability to learn, and personal features, Web-based learning support system provides a personalized learning environment to meet individual learners' need.

5.7 Conclusion

The primary goal is to provide an adaptive learning support environment that will effectively accommodate a wide variety of students with different skills, background, and cognitive learning styles. The web offers a dynamic and open learning environment. Based on the student-centered philosophy, personalization and adaptation are important features for Web-based learning support systems.

It is a challenging task to develop a fine-tuned WBLSS owing to the uncertainty and complexity of the student model, particularly the students' cognitive attributes. The basic functions and ideas of WLSS are discussed in this chapter. To illustrate those ideas, we implemented a learning support system for helping the students' study of the data structure and algorithm analysis course.

In order to support the students' deep learning and understanding of the difficult concepts algorithms and data structures, we proposed a feasible framework by dynamically constructing a knowledge structure map during the learning process. It can be visualized and clickable. Students can use his or her knowledge structure map to compare with the instructor's or peer's KSMs. When a suitable knowledge structure is designed and constructed, the system can be used for effective learning, tutoring, problem solving, or diagnosing misconceptions.

The rough set approach to analyze student academic performance is presented. The results will assist students in their learning effectively. It also guides the instructor to improve course contents, analyze the students' learning patterns, and provide more information to the instructors or designers to organize the online course more effectively.

References

1. Brusilovsky, P. KnowledgeTree: "A Distributed Architecture for Adaptive E-learning", WWW 2004, May 17–24, 2004, ACM 104–111.
2. J.T. Yao, J.P. Herbert: "Web-based Support Systems with Rough Set Analysis", Proceedings of International Conference on Rough Sets and Emerging Intelligent System Paradigms (RSEISP'07), June 28–30, 2007, Warsaw, Poland, LNAI 4585, pp. 360–370.
3. Brusilovsky, P., Anderson, J., "An adaptive System for Learning Cognitive Psychology on the Web", WebNet 98 World Conference of the WWW, Internet & Intranet, Orlando, Florida, November 7–12, 1998, pp. 92–97.
4. Gagne, R.M., Driscoll, M. P., "Essentials of Learning for Instruction", New Jersey, Prentice Hall, 1998.
5. Papandreou, C.A., Adamopoulos, D.X., "Modelling a multimedia communication system for education and training", Computer Communications 21, 1998, pp. 584–589.
6. Brusilovsky, P., "Methods and Techniques of Adaptive Hypermedia", User Modelling and User Adapted Interaction, Vol. 6, N2–3, 1996, pp. 87–129.
7. Tennyson, R.D., Rothen, W., "Pre-task and On-task adaptive design strategies for selecting number of instances in concept acquisition". Journal of Educational Psychology, Volume 69, 1977, pp. 586–592.
8. De Bra, P., "Teaching Through Adaptive Hypertext on the WWW." International Journal of Educational Telecommunications. August 1997, pp. 163–179.
9. Da Silva, D.P., Van Durm, R., Duval, E. & Olivi, H., "Concepts and Documents for Adaptive Educational Hypermedia: a Model and a Prototype", Second workshop on Adaptive Hypertext and Hypermedia, Ninth ACM Conference on Hypertext and Hypermedia, Pittsburgh, USA, June 20–24, 1998, pp. 35–43.
10. Yao, J.T., Yao, Y.Y., "Web-based Information Retrieval Support Systems: building research tools for scientists in the new information age", Proceedings of the IEEE/WIC International Conference on Web Intelligence, 2003.
11. Wade., V. P., Power, C., "Evaluating the Design and Delivery of WWW Based Educational Environments and Courseware", ITICSE'98, Dublin, Ireland, 1998.
12. Getting Started Tutorial for WebCT. http://www.webct.com/
13. Fan, L., Yao, Y. Y., "Web-based Learning Support System", WI/IAT 2003 Workshop on Applications, Products and Services of Web-based Support Systems, October 2003, Halifax, Canada.
14. Mason, D., Woit, D., "Integrating Technology into Computer Science Examinations." Proceeding of 29th SIGCSE., 1998, pp. 140–144.
15. Mason, D., Woit, D., "Providing Mark-up and Feedback to Students with Online Marking", Proceeding of 30th SIGCSE, 1999, pp. 3–6.
16. Novak, J. D., "The Theory Underlying Concept Maps and How to Construct Them". http://cmap.coginst.uwf.edu/info/
17. Ausubel, D. P., "Educational Psychology", A Cognitive View, Holt, Rinehart & Winston, 1968.

18. Liang, A.H., Ziarko, W., and Maguire, B.,"The Application of a Distance Learning Algorithm in Web-Based Course Delivery.", Rough Sets and Current Trends in Computing 2000, 338–345, Banff, Canada.
19. Magagula, C.M., Ngwenya, A.P., "A Comparative Analysis of the Academic Performance of Distance and On-campus Learners." Turkish Online Journal of Distance Education, October 2004 , Volume: 5 Number: 4.
20. Lavin, D.,"The Prediction of Academic Performance: A Theoretical Analysis and Review of Research." New York: Russell Sage Foundation, 1965. pp. 182.
21. Vasilakos, T., Devedzic, V., Kinshuk, Pedrycz, W., "Computational Intelligence in Web-Based Education: A Tutorial", JILR, Vol.15,No.4, 2004, pp. 299–318.

Chapter 6
A Cybernetic Design Methodology for 'Intelligent' Online Learning Support

Stephen R. Quinton

Abstract The World Wide Web (WWW) provides learners and knowledge workers convenient access to vast stores of information, so much that present methods for refinement of a query or search result are inadequate – there is far too much potentially useful material. The problem often encountered is that users usually do not recognise what may be useful until they have progressed some way through the discovery, learning, and knowledge acquisition process. Additional support is needed to structure and identify potentially relevant information, and to provide constructive feedback. In short, support for learning is needed. The learning envisioned here is not simply the capacity to recall facts or to recognise objects. The focus is on learning that results in the construction of knowledge. Although most online learning platforms are efficient at delivering information, most do not provide tools that support learning as envisaged in this chapter. It is conceivable that Web-based learning environments can incorporate software systems that assist learners to form new associations between concepts and synthesise information to create new knowledge. This chapter details the rationale and theory behind a research study that aims to evolve Web-based learning environments into 'intelligent thinking' systems that respond to natural language human input. Rather than functioning simply as a means of delivering information, it is argued that online learning solutions will 1 day interact directly with students to support their conceptual thinking and cognitive development.

6.1 Introduction

In broad terms, this chapter describes a research study that aims to evolve Web-based teaching environments into 'intelligent' learning systems modelled

S.R. Quinton
Technology Enhanced Learning and Teaching (TELT) Portfolio Leader,
Learning & Teaching @ UNSW, Division of the Deputy Vice-Chancellor (Academic), Level 4, Mathews Building, The University of New South Wales, Sydney, NSW 2052, Australia
e-mail: s.quinton@unsw.edu.au

J.T. Yao (ed.), *Web-Based Support Systems*, Advanced Information and Knowledge Processing, DOI 10.1007/978-1-84882-628-1_6,

on cybernetic, systems theory principles. For the purposes of this study, such support tools are identified as 'intelligent cognitive support systems'. The cognitive empowerment alluded to here is not intended to manage, calculate, search, order, and arrange search results as is the case with most cognitive support tools. Instead, the study targets the construction of an intelligent, interactive response system that augments the human intellect to achieve greater capacity to conceptualise and, ultimately, construct new knowledge. The primary goal is to establish design and evaluation methodologies that inform the development of educationally effective technologies to amplify and extend the human abilities to perceive, reason, comprehend, and share knowledge.

The discussions and arguments presented build in part on the seminal work of Liber [25], specifically in relation to the directions provided for the application of technology to the delivery of educationally effective learning solutions. What follows is an account of the principles and insights identified to date that hold promise for fulfilling the goals of the aforementioned study.

The central argument extends the view that what learners need is 'intelligent' technology-enhanced assistance to locate suitable information, to structure and refine potentially relevant information within the current found set, and to formulate and reformulate queries. While there are developments in search, find, and pattern-matching tools that provide partial assistance for improving human intellect, their application is limited in terms of immersive, fully interactive cognition support. In essence, learners need assistance with the query formulation → find → generate knowledge → re-formulation/integration phases of the knowledge construction activity. The synthesis of knowledge according to a technology-augmented process is referred to as the Generative Conceptualisation and Cognition Support System (GCCS) [13].

By situating the GCCS model in a learning context, it is feasible to speculate on how existing learning delivery systems can be enhanced or re-engineered to provide the 'assistance' online learners need to construct new knowledge and integrate it with their existing knowledge framework. In learning theory terms, the objective is to facilitate the learning process as students access useful information, refine it to suit the intended learning goals, structure it to match their individual learning preferences, test goal attainment and progression, and then pursue the next goal. If the preceding model is situated in an online delivery context, it is possible to speculate how existing learning delivery systems and search systems can be re-engineered and enhanced to provide the 'cognitive support' the learner needs to create new knowledge. As a general rule, the research study has revealed that existing cognitive support systems and query tools are characterised by the following limitations:

- When undertaking a specific enquiry for information, it is often necessary to search the entire Web and manually abstract information that may hold some conceptual relevance.
- Large quantities of information held in separate databases around the world are not structured. No generative infrastructure is available to assist in managing, searching, translating, categorising, indexing, and accessing the exact information required.

- At present, search results return large, unwieldy lists of potentially relevant documents. This problem presents an opportunity to provide 'generative cognitive support' to facilitate an overarching and exhaustive analysis of multiple relevant factors in the search for knowledge.
- To be effective, the learner/researcher must possess a high degree of domain-related knowledge. For example, knowledge pertaining to the 'who', 'what', 'where', and 'current developments' factors can be readily dealt with by technology, and yet this task is still carried out by the learner/researcher.
- While portions of the information or data on the Internet may be related in some way, other information may overlap or may be semi-complementary. No concept-based support tool is available to manage these needs.

The concern for this study is to provide the level of 'intelligent' interaction required for deep learning. Although there are tools that indirectly assist to enhance conceptual understanding such as concept maps, topic maps, the Compendium project, and the semantic Web, their usefulness is limited in that they are incapable of the complex levels of interactivity that guide learners to identify key relationships and derive new meanings. In other words, they do not interact with learners to augment and empower their cognitive processes.

This brief critique points to the primary aim of the study which is to determine how technology can support learners working in the online environment to gain deep understanding (or metacognition). Presently, there is no cognitive support tool which lends itself to empirical investigation from which to derive a theoretical foundation for technologically augmenting the learning process.

The effective and creative analysis of an exponentially increasing information and knowledge base requires analysis of the interconnectedness of systems that give rise to meaning. For the individual, rational, sustainable refinements to their existing knowledge schema are derived through the strategic application of the skills and processes of archiving, searching, and sharing information. Thus, for learning to be effective, something much more complex than competence in information management/processing is vital to the attainment of deep understanding. That is, online learning environments should provide tools that not only equip learners to quickly locate and sort through large quantities of data and information, but also to apply higher-order analysis techniques and thinking strategies to the construction of knowledge, all of which play a crucial role in the development of metacognition. As explained in the pages to follow, the solution to addressing these needs may be found in 'intelligent' learning systems that utilise the power of learning object technology.

6.2 Rationale

Whenever a search is conducted on the Web, it is often the case that the results obtained are little more than a collection of unconnected, unstructured information (and data) that has been removed from its original context. Having acquired a collection of such material, the problem for learners is what to do with the quantities presented, and then, how to do it. To make sense of the displayed results, the human 'inquirer' must identify meaningful relationships, and then cognitively 'process' the given results to derive potentially useful interpretations or solutions to a known problem.

An 'intelligent' support system is needed that interprets a targeted assembly of information and presents the results to the human inquirer in such a way that it elicits answers to follow-on questions, each of which are analysed and acted upon in an iterative fashion until a meaningful learning outcome is achieved. This type of cognitive 'immersion' in online information repositories is within the realms of possibility, but requires 'intelligent' assistance from advanced technology support systems that facilitate the refinement of meaning through communal discussion and consensus.

This chapter proposes that it is possible to design software systems and tools that assist learners to locate required information to structure and identify potentially relevant meaning within the current found set, and to formulate and re-formulate queries in response to human input. In short, support can be provided for learning. Most search engines locate information very efficiently, but these solutions do not support the learning process as envisaged in the study. If the goal of delivering electronic teaching materials is to assist learners to derive new knowledge and understanding, then the applied pedagogies and learning strategies must also permit complete integration of the presented learning content with the processes of information retrieval, interpretation of meaning, and construction of knowledge.

The research study is premised on the notion that for learning to be truly effective, something much more complex than competence in information management skills is vital for ensuring learners achieve deep understanding. That is, the learning environment should not only equip learners with the tools to quickly locate and sort through data and information, but also support them to develop and apply complex analysis techniques and higher-order (metacognitive) thinking strategies to the construction of knowledge.

The educational problem to be addressed is that current information search systems and cognitive support systems do not fully support cognition and metacognition development. Compounding this problem further is that most software design methods which focus on human to computer interaction and learning development processes are not suitable for the construction of educationally effective intelligent learning systems. The principal outcome of the study, therefore, is to formulate a design evaluation methodology for constructing intelligent enhancement solutions.

6.3 The Need for 'Intelligent' Cognition Support Systems

The notion of intelligent software systems is presented as a relatively recent computing-inspired abstraction that has the potential to partially overcome and resolve the complexities of learning and knowledge construction in the online environment. However, developing software solutions that satisfy today's complex learning needs poses a major challenge for developers and educationalists alike. Learners, for example, present a wide diversity of interests, backgrounds, and skills, and therefore the design of intelligent software systems that can adapt to individual differences and simultaneously enhance cognitive understanding is not a task that should be approached without a thorough examination of the learning process. In the development of intelligent learning technologies, for example, the goal is to ensure that they are human-centred, usable, useful, and understandable. Problem-solving activities, for example, generally involve recognising and analysing tacit assumptions using heuristics. Such realisations can often arise from seeing things in new perspectives. Acknowledging the human-centred perspective offers unrealised potential for enriching software design methodologies with ontological and conceptual analysis approaches to devising new and improved development approaches.

To date, few educators have grasped the enormous potential of intelligent interaction with online cognitive support systems as a tool for enhancing the learning process. This potential is further undermined by the marked absence of a theoretical foundation and evaluation methodology for the design of software solutions that serve to elicit metacognitive thinking skills. To ensure relevance to today's information management and knowledge creation needs is not lost, a new theory of technology-augmented learning must focus on individualised learning strategies that encourage learner empowerment and responsibility, initiative, diversity of understanding, and active collaboration. It is important, therefore, to devise learning theories that are learner-focused and specific to online solutions that support cognitive processing, enhance understanding, build higher-order thinking skills, and foster metacognitive thinking.

The art of teaching understanding is a matter of cultivating the learner's relationship with information and the construction of knowledge, and extending this relationship to a level where the capacity for intelligent action is demonstrated. Deep insight, leading onto metacognition, means acquiring a thorough understanding of the attributes that describe an object or event, in turn requiring extensive involvement in exploring and identifying the relationships that explain how such attributes function both individually and as a whole. These skills, in part, can be achieved for example, by designing problem-based, interdisciplinary or thematic learning environments and tailoring instructional guidance to individual preferences and subject-area idiosyncrasies.

From a learning perspective, it is no longer that acceptable practice to inform students that their knowledge of a topic is incorrect. To ensure they understand what has been misinterpreted, students should be encouraged to learn through interaction and engagement with a human tutor or an intelligent software system. The fact that every learner possesses different learning styles, cognitive abilities, preferences, motivations, and needs further complicates the latter as it clearly demands a constructivist approach to online learning design.

Educators interested in applying current learning theory and evaluation methods to new models of educational design that utilise emerging technologies inevitably discover that few theories and methods provide for new or emerging technological capabilities [29]. Moreover, research on the application of technology to educational practice brings with it new demands for developing learning theories and evaluation methods that focus on human to computer interaction in addition to human to human interaction.

The drive behind human to computer interaction is to improve electronic learning to the level where intelligent interaction with the 'machine' will support the learning process in ways traditionally thought possible only through human to human interaction. Learner-centred software tools should support accurate organisation of information, efficient analysis of information to identify meaningful relationships and concepts, and interactive feedback on derived meanings, as well as assist to evaluate and reflect on the choices made during the sequence of activities. It is to this end that the study is concerned with devising a design evaluation methodology for the construction of intelligent cognitive support systems. The outcomes will serve as 'proof of concept' for deriving a valid theoretical foundation from which to devise educationally effective interactive learning systems.

As a glimpse of the potential configuration, consider a technology solution that on receipt of human input generates conceptual categories, highlights areas of relevance by displaying visual representations of identified associations, dynamically forges new links, stores derived links and access trails, and juxtaposes what previously had not been associated with new content. In addition, users' notes about the information presented or the links that were made to other sources are stored and incorporated in the process.

The capabilities of such a solution afford the design of the ideal learning environment that assists learners to derive answers to high-level 'meta-questions', such as: How do I know what I need to learn? How do I get there? How am I progressing? Are my goals still relevant? What are the best learning models for me? What is the effect of social change, cultural differences, and economic changes on my personal learning goals? The level of complexity and interaction indicated here points to a cogent need for 'intelligent' interactive cognition support.

In effect, the proposed learning system 'makes sense' of information in a dynamic yet comprehensive, efficient fashion. All the while, the ephemeral (often tedious) task of remembering is managed by the system's recording and storage capabilities. It is in this regard that appreciating the human-centred perspective may inform and enrich educational technology design and thereby enhance the learning experience for students. In principle, this system is inspired by the MEMEX device of Bush [8] in that it supports what the human mind does by dynamically assisting learners to make sense of information. Essays in this department have introduced such notions as the Sacagawea Principle [18]:

> Human-centred computational tools need to support active organisation of information, active search for information, active exploration of information, reflection on the meaning of information, and evaluation and choice among action sequence alternatives.

Whilst from an educational standpoint this observation suggests a worthwhile goal for research design, it also raises the question of how to progress from such complex statements of intent to design solutions that serve to accomplish the desired goals? This study is an attempt to establish such a solution by resolving these issues through examining current developments in intelligent learning support systems to derive new theoretical models for prototyping the design, protocols, and strategies required to achieve educational effectiveness (in comparison to conventional learning modes). The main themes that structure this chapter are summarised to this point:

- The state of knowledge is changing at an ever increasing pace.
- Rules of evidence, discourse, and active involvement are crucial to the success of learning (constructivist and communal learning).
- Knowledge is created every time students identify relationships and construct new meanings.
- What matters most is not what is studied (content), but how students learn (the process of inquiry).
- The process of discovering knowledge is not always efficient – it is necessarily complicated and chaotic.
- Students need to make mistakes and experience cognitive conflict – answers should not be privileged over research and intellectual inquiry.
- Opportunities for reflection and exercising metacognitive thinking are essential for attaining higher order analysis skills.
- The absence of relevant design evaluation methodologies points to the need to devise new approaches in light of the specialised nature of the technology examined in this study.

In short, the research on the design and purpose of intelligent interactive learning support systems is focused not just on what to learn, but also on the methods that enhance the student's capacity to learn. The research seeks to develop technologies that enhance and extend the human ability to acquire, understand, and create knowledge. The preferred goal is to shift online learning beyond the traditional transmissionist model where it is assumed that the recipient will passively absorb and understand the displayed information, to incorporate the notion of learning environments that act as a catalyst for transforming the learner's awareness of how knowledge is constructed and absorbed. A shift in the focus of learning from the accumulation of information and knowledge to learning as a life-changing experience that is augmented by active, cognitive engagement in the learning process provides direction as to how the complex tasks of knowledge construction and concept attainment can be taught online.

6.4 Metacognition as the Primary Learning Goal

Although it is not practical to devise software design models that describe the entire human learning process, some progress can be made in relation to specific aspects. This study targets metacognition where the focus is on the development of technologies that amplify and extend the human ability to search, identify meaningful relationships, learn, understand, and create knowledge. The study therefore extends to the enhancement of learners' knowledge acquisition and awareness skills premised on the notion that effective online learning environments should incorporate cognition support tools designed to facilitate metacognitive thinking through direct 'intelligent' interaction with learners.

Basic to the concept of metacognition is the notion of thinking about one's own thoughts. Those thoughts can be of what one knows (metacognitive knowledge), what one is currently doing (metacognitive skill), or what one's current cognitive or affective state is (metacognitive experience). To differentiate metacognitive thinking from other kinds of thinking, it is necessary to consider the source of metacognitive thoughts: metacognitive thoughts do not spring from the learner's immediate external reality; rather, the source is tied to their internal mental representations of that reality, which can include what they know about that internal representation, how it works, and how they feel about it. Therefore, metacognition can be simply defined as thinking about thinking or cognition of cognition.

Where this study is concerned, metacognition is described in terms of the capacity of individuals to be aware of their own cognitive processes coupled with the strategies needed to transform a concept, a notion, or an idea into knowledge. Concepts are meanings attributed to objects that operate structurally within a broader conceptual system. A simple conceptual system is illustrated in the example of an apple. Although an object in itself, the term 'apple' is also a concept because it can be related to the ascending category of fruits, the descending category of apple types, and the horizontal category of other fruits such as bananas, grapes, and oranges.

Osman and Hannafin [32] observed that 'although existing research and theory suggest that metacognition is integral to successful learning, existing instructional design (ID) models do not typically emphasise metacognitive strategies such as planning, monitoring, revising, and other self-regulating activities'. Conceptual change requires an intentional and reflective cognitive process, which in turn leads to higher-order learning [10]. Then, whenever an individual engages in some form of discourse and interaction with one or more people, new knowledge is derived. These cognitive actions are consistent with constructivist theory, which emphasises the primacy of the learner's intentions, experience, and metacognitive strategies.

Rieber [36] described the constructivist view of learning as involving 'individual constructions of knowledge.' Learners attain a state of cognitive equilibrium through the reconstruction of concepts, schema, mental models, and other cognitive structures when exposed to new information and experiences that conflict with earlier constructions. Hannafin [21] notes that the methods consistent with constructivist foundations and assumptions typically emphasise teacher to student or student to student interactions to model or scaffold understanding and performance. Such interactions motivate teachers and students to construct learning strategies within contexts that support the learning of progressively more complex concepts. Similarly, in the same way that technology provides alternatives to teacher to student or student to student interactions, problem-based learning activities encourage learners to draw upon technological, cognitive, and social resources to solve complex, open ended problems. In effect, a constructivist approach that incorporates technology utilises the technology as a means of interpretation as understanding is derived and cognitive skills are refined. A constructivist approach therefore provides genuine opportunities for technological implementation in the learning process.

Cognitive flexibility theory (CFT) extends the accepted tenets of constructivist theory in that understandings are constructed using prior knowledge to look beyond the information given. The prior knowledge that is brought to bear is itself constructed, rather than retrieved intact from memory [37]. CFT is concerned with the transfer of knowledge and skills once students have learned the basic facts, concepts, and theories of a subject area in a linear context. CFT contends that the act of revisiting the same material, at different times, in rearranged contexts, for different purposes, and from different conceptual perspectives is essential for attaining mastery of the complexities of understanding and knowledge acquisition, and knowledge transfer. When the intended outcome is advanced knowledge skill development, a non-linear approach is needed to afford the presentation of information using multiple perspectives and diverse examples to stimulate identification of fundamental concepts.

An acknowledged challenge for this study is to circumvent the assumption that learning automatically results in a productive outcome – that is, knowledge. Simply making large quantities of information and resources available online may in fact result in disordered clutter. What is required is akin to that which occurs in nature: divergent bits of information that self-organise into coherent clusters of related phenomena. However, the relationship between self-organisation in nature and the organisation of thoughts by individuals is not identical. In the natural world organisation is causal, whereas in humans self-organisation is more complicated as it requires active cognitive recognition and formation of connections in the available data and information. Although nature appears (on the surface) to display properties of organisation, the key difference is that it is inherently chaotic, out of which order naturally forms.

Self-organisation in learning occurs whenever patterns of connection (or organisation) are either identified or created and then communicated. The process of creation is at one moment the internalisation of recognised order, and in the next, of passing the identified order onto the next person as an experience. It is in this sense that the connections between disparate ideas and knowledge give rise to the emergence of creative thinking and innovation. In a similar way, cognitive self-organisation and communication are effective means of evaluating (i.e., community judged) the depth and quality of the expertise that is distributed to the learning community.

An emphasis on community interaction and exchange (learning communities) extends the learner's involvement with the construction of knowledge to encompass multi-level, interconnected, interactive learning systems that expose them to a diverse array of perspectives, practices, interests, and knowledge domain interpretations. In this model, the learner is encouraged to negotiate pathways (either preset or user-defined) through a multiplicity of contexts whilst being simultaneously 'monitored' by community members who analyse and provide feedback on the strategies employed during the learning process. Learners engage with the displayed learning materials at varying levels of complexity that are dynamically refined and adjusted to match or accommodate their individual preferences and goals. This model of 'intelligent, interactive learning support' provides direction for further reflection on determining a theoretical foundation for designing advanced online learning solutions. In summary, active cognitive 'immersion' in complex information repositories and communities of learning is now within the realms of possibility, but requires 'intelligent' assistance from advanced technological support systems that permit iterative refinement through communal discussion and consensus. This process is situated in the context of a learner as a researcher and constructor of knowledge where the goal is to enhance learning capability both individually and collectively. The NWV technology currently under development (to be described later) provides a foundation that is conducive to the automation of intelligent interaction with the learner as it is designed to do one task well – identify and analyse concepts from typed input.

Teaching proficiency in the application of advanced cognitive competencies to the acquisition and creation of knowledge extends well beyond the transmission of knowledge and cultivation of advanced information-processing skills. Research and learning can only progress in the light of new and expanded conceptual understandings. Resolving such needs requires software systems dedicated to enabling learners to identify the key properties and relationships that serve to model the underlying conceptual structure of the targeted knowledge domain; manage and facilitate the transference of tacit knowledge to new tasks; strategically manage learning resources using dynamic assembly and re-contextualisation strategies; and extend the learning experience beyond accepted epistemological boundaries as delineated by established disciplines and specialist subject areas by assisting the student to connect and reconnect insights to multiple contexts and domains.

All this in turn raises the many latent and complex problems of how to structure and model knowledge and how to predetermine the relationships that connect existing knowledge and knowledge structures to selected teaching content-taking into account their contextual relevance and innate cultural contexts. For these reasons, it is argued, current information technology systems and cognition support tools offer little assistance for empowering human cognition.

6.5 A Brief History of Cognition Support Systems

If we look at the past decades for clues on how the problem of information overload can be managed, educational theorists and cognitive psychologists Ausubel [1], Ausubel et al. [2] advocated the use of 'advance organisers' to foster meaningful learning and assist in managing information. Advance organisers are global overviews of the material to be learned that prompt the learner to identify the pre-existing superordinate concepts formed within their own cognitive schema and thereby provide a tangible context in which to identify the most general concepts and incorporate progressively inconsistent facts and concepts. In theory, advance organisers are most effective if they make explicit the relationships amongst the concepts that learners already know and provide a structure or context into which new concepts can be integrated.

Since Ausubel's time, many knowledge (or memory) management tools and visualisation (or browsing) software tools have been developed, many of which can be grouped into several generalised categories: concept mapping tools [11,31], semantic networking tools [19], decision-making tools (also called group organisers), cognitive maps [16], topic maps [30, 34], mind maps [9, 12], and 'intelligent' personal assistants [7]. Canas et al. [11] noted that there is a considerable overlap among these categories as most tools support the visual representation of knowledge as expressed in terms of concepts, ideas, or 'thoughts'; the associative creation of links and nodes; and, the capacity to organise related concepts for specific purposes. Typically, most tools comprise a number of basic elements in that they are described as idea mapping tools. These 'mappings' are always based on causal relations; allow for the representation of a large number of concepts/ideas; use short phrases to express ideas, including active verbs; allow for single or bipolar relationships between concepts or ideas, which are usually directional, and, encourage hierarchical structuring. The brief synopsis to follow describes a range of developments accomplished to date and assist to establish a basis on which to expand on the issues and problems identified in previous pages.

The Semantic Web: The Semantic Web project intends to create a universal medium for information exchange by giving meaning (semantics) in a manner that is understandable by machines to the content of documents accessible on the Web. At present, the World Wide Web is based primarily on documents written in HTML, a language that is useful for describing (with an emphasis on visual presentation) a body of structured text interspersed with multimedia objects such as images and interactive forms. Apart from the capacity to organise a typical document and format desired visual layouts, HTML has limited ability to classify sections of text.

The Semantic Web addresses these shortcomings by using the descriptive technologies resource description framework (RDF), Web ontology language (OWL), and the data-centric, customisable extensible markup language (XML). These technologies combine to provide descriptions that supplement or replace the content of Web documents. Thus, content may manifest as descriptive data stored in Web-accessible databases, or as markup within documents. The machine-readable descriptions allow content managers to add meaning to the content, thereby facilitating automated information gathering and research-using computers. The Semantic Web aims to provide new ways to define and link data on the Web, not just for display purposes, but also for automation, integration, and reuse across various applications. The project is a mixture of theoretical and applied research work, which takes advantage of various fields not connected until now. Eventually, the project aims to provide standard solutions, independent of any proprietary implementation.

Concept Mapping: The idea of concept mapping is not new. Joseph D. Novak is often cited as having developed the idea of using a concept map as a teaching strategy in the 1960s. Novak's work was influenced by the learning theories advanced by Ausubel who emphasised the importance of a student's prior learning and knowledge (a priori knowledge) as a determinant factor in the successful learning of new concepts. Meaningful learning occurs when the student consciously and explicitly links new knowledge to concepts he or she already knows. There are many software tools that support the diagrammatic representation of conceptual thinking processes and decision making. It is not practical to list these tools out, but typically most tools comprise a number of basic elements as they

- Allow for the representation of a large number of concepts/ideas
- Are described as idea mapping tools
- Use 'mappings' that are always based on causal relations.
- Use short phrases to express ideas, including active verbs
- Allow for single or bipolar relationships that are usually directional, between concepts or ideas
- Encourage hierarchical structuring

In comparison with other cognition support solutions, concept maps are suited to identifying relationships among concepts. Understanding concepts and their underlying relationships is widely held to be necessary to the acquisition of flexible, generalisable knowledge. To date, a number of graphing tools and systems have been developed for use in decision-making activities. These tools may support a structured argument or an evidence diagramming approach such as that used in the Belvedere cognitive/metacognitive support system [33,38]. Alternatively, a number of less structured graphing systems may be used, such as causal cognitive mapping, or issue-based mapping.

In general, however, software systems developed using a 'concept mapping' approach and utilise concept maps (as interfaces) do not support or encourage overt awareness of the learning process. When compared with cognitive activities such as outlining or concept definition that inspire learners to take a thoughtful, systematic approach to engaging subject matter, the benefit of concept mapping is reduced. Evidence of effective concept map knowledge elucidation or concept map-enhanced intelligent system usefulness, usability, performance enhancement, or organisational effectiveness is not as yet conclusive [11].

Cognitive Maps: Cognitive maps, mental maps, mind maps, cognitive models, or mental models refer to a type of mental processing or cognition composed of a series of psychological transformations in which information about the relative locations and attributes of facts and events are acquired, coded, stored, recalled, and decoded. In other words, cognitive maps are a method to structure and store spatial knowledge, enabling the 'visualisation' of images to reduce cognitive load and to enhance the recall and learning of information.

To Eden [17], a cognitive map is not 'a map of cognition' but 'a map designed to aid cognition'. That is, cognitive mapping is not an assessment or measuring process, but a reflective process in which the elicitation strategies are designed to change thinking in productive ways. In this context, the term 'cognition' refers to the mental models, or belief systems, that people use to perceive, contextualise, simplify, and make sense of complex problems. These mental models are variously referred to as cognitive maps, scripts, schemata, and frames of reference. Cognitive maps can be represented and assessed on paper or screen through various methods such as concept maps, sketch maps, spider diagrams, or other suitable forms of spatial representation.

Kosko [24] extended the function of cognitive maps to incorporate strategies for learning based on what he terms a fuzzy cognitive map (FCM) that in effect draws a causal picture using concept nodes and causal edges. FCMs tie facts, ideas, and processes to values, policies, and objectives. Their purpose is to predict how complex events interact and assist to discern the big picture. The standard cognitive map does not allow for feedback as the connecting arrows do not form a closed loop as occurs in practice. The strength of the FCM approach is the inclusion of feedback mechanisms that provide bidirectional flow of information and meaning. As will become evident, the attributes outlined here contribute to the design of intelligent learning systems described in this chapter.

Topic Maps: A topic map is a document structured to improve information retrieval and navigation using topics as hubs in an information network. Topic maps can readily locate information and are formatted to provide a wide variety of search tools such as printed indexes, glossaries, and many types of high-performance online information retrieval aids [30]. In effect, they constitute an enabling technology for knowledge management. Originally designed to handle the construction of indexes, glossaries, thesauri, and tables of contents, topic maps also provide a foundation for the Semantic Web Pepper [34]. As a result, considerable effort has been directed at understanding the relationship between topic maps and RDF, the metadata framework developed by the W3C [34].

Topic maps serve to represent information currently stored as database schemas (relational and object-oriented). Where databases only capture the relations between information objects, topic maps allow these objects to be connected to the various places where they are held. Knowledge bases can be designed to not only relate concepts, but also point to the resources relevant to each concept. It is possible, for example, to represent immensely complex structures using topic maps as they support the creation and management of taxonomies, thesauri, and ontologies, and provide an encompassing knowledge model for enabling the transition from one structure to the next.

Compendium: Compendium [12] assists groups to collectively elicit, organise, and validate information. Compendium comprises three key elements: a shared visual space where ideas can be generated and analysed; a methodology that allows the exploration of different points of view; and a set of tools for quickly and easily sharing data both within and beyond the boundaries of the group. These processes enable participants to negotiate collective understanding, capture discussions, and share representations of their knowledge across communities of practice. Compendium has multiple uses, but the predominant use is for issue/argumentation/discussion support. The domain independence of Compendium's mapping technique combined with its interoperability with domain-specific applications provides an effective medium for facilitating discussions, knowledge construction, and consensus.

Smart Internet Technology: The Smart Internet's mission statement is 'to capitalise the outcomes of world class Internet research and development for

Australia' [7]. From its inception, the focus of the Smart Internet project was constructed around five major research and development programs: natural adaptive user interfaces; smart personal assistants; intelligent environments; smart networks; and user environments. The Smart Internet project integrates human factors with key technology frameworks for application where the focus is to examine how end users might interact with the many potential Internet innovations to emerge over the next few years. Eventually the Internet of the future may offer any technology, any application, any service, anywhere, anytime – for any user. The key research questions to be addressed by the project include:

- What might the Internet be like in 2010?
- What positions are taken by different people and institutional interests about the future of the Internet?
- What are the possible outcomes for end users towards 2010?

In summing up, the great strength of computers is that they can reliably manipulate vast amounts of data and information very quickly. Their weakness is that they have no awareness of what that data and information actually means. The absence of computer 'intelligence' as an aide to constructing knowledge compounds the problem in that no one knows all that is known. There is far too much knowledge for such a possibility.

6.6 Enabling Effective Cognition and Metacognition Development

Dreher et al. [15] present a vision and expectation of what a learning environment might deliver over the coming years by outlining the extended functions that could be incorporated into existing services to enhance and support the needs of learners. They argue that the basic functions and services provided by most delivery systems today do not match the design promises and implementation functionalities offered over past decades. In their view, learning solutions must be extended to include tools and services that provide access to content regardless of location. For example, learning resources may be stored in different locations throughout the world and automatically assembled and distributed in accordance with the learning needs of individuals and groups. To this end, the number of 'extended functions' range from: interactive documents that automatically generate links from annotations and comments made by learners or groups to other suitable resources, users, or groups to provide additional information, comments, or answers; expand teaching and learning support structures by delivering more than just content using advanced software systems that assist learners to structure, reflect, analyse, and synthesise new knowledge; to the implementation of intelligent search mechanisms, agents, interfaces, and portals designed to accommodate the diverse mental models or cognitive maps that learners draw on in their search for understanding and the eventual construction of knowledge.

Wobcke [45] canvassed a number of software agent applications in which he notes that the ultimate implementation of the notion of a 'smart system' will be demonstrated through its capacity to reason and think about itself and its environment. From the early days of artificial intelligence (AI) research, the focus of has been directed towards a determination of how computer technology can be harnessed to achieve machine intelligence. Minsky [28], for example, advanced the view that knowledge is an artefact that can be stored in a machine and it is the application of that stored knowledge to the real world that constitutes intelligence. This 'realist' viewpoint underpins much of the research undertaken in AI to date and has given rise to the notions of semantic networks and rule-based expert systems. However, where learning is concerned, an intelligent system must focus on the needs of students and in effect assume (in part) the role of cognition support. As Wobcke [45] observed: 'More complex personal assistants aim to address more than one of these aspects within a single system'.

More recent, Larreamendy-Joerns and Leinhardt [26] foresee the emergence of three major mutually interdependent educational visions: the presentational view that applies to the increasing use of the visualisation and presentation capabilities of online multimedia; the performance-tutoring view that heralds the promise of intelligent tutoring systems, which provide students a rich variety of informational resources, timely feedback, and suggested courses of actions, accompanied by a capacity to control their learning processes and goals in line with individual needs and preferences; and the epistemic-engagement view that emphasises the importance of multiple learning perspectives and adaptive instructional environments to provide a broad range of opportunities for intellectual engagement and interaction. As they see it, the advantages of the performance-tutoring approach to online learning can be found in an increase in student to content interaction, the individualisation of learning, and an eventual re-examination and transformation in teaching practices.

Harnad [22] argued there have been three major revolutions in human communication and cognition: language, writing, and print. All three exerted a dramatic effect on how humans think and the mediums through which those thoughts were expressed. A fourth revolution is now at the threshold, which as Harnard sees it, is 'scholarly skywriting', or Internet-based communications and publishing. Whilst the notion of a fourth revolution in human thought holds some credence, Harnard's view did not account for the full extent of technological innovation that has occurred over the past 15 years, and the extent to which a fourth revolution will be far more profound than anyone has imagined. Perhaps the fourth evolution might appear as Mittelholtz [29] claimed:

> It will soon become possible to have a program that can draw out of the learner their own 'learning schema' as they try to understand new concepts or processes and then begin developing learning modules to fit within that schema.

The preceding synopsis highlights the inadequacy of existing cognitive support tools to interact with learners to augment and empower the human intellect. Nor are there suitable evaluation methodologies to provide direction on how software systems can support learners to gain deep cognitive understanding in the online environment. If learners are to be skilled in deriving new meanings and understandings in an era marked by information overload where the focus of economic production is on the creation of new knowledge, then timely, individualised assistance with the investigative/learning process is crucial. The fundamental issue to be resolved is that no cognitive support tool lends itself to empirical investigation for deriving a theoretical foundation on which to construct software systems that effectively and efficiently augment the human intellect.

6.7 Relationships and Connectedness: Pathways to Meaning

At present, Mittelholtz's dream is yet to be realised. The task of recreating the entire spectrum of human intelligence within a computer is still some way off. Until such time as the processes of human learning can be fully described and modelled, there will always be reason to doubt that machines can mimic such capabilities. An intelligent support system of the configuration proposed in this chapter will need to be 'metacognitively aware' of all its components, capable of interacting with learners, while analysing their myriad of needs and learning and adapting as the learner also grows, learns, and adapts.

Although progress has been slow, there are nevertheless aspects of artificial intelligence and computer-aided learning that show promise for realising such goals. The key is to think small – to be less ambitious and to achieve one or two very useful outcomes very well. Give that the concern for this chapter is to determine the level of 'intelligent' technology-enhanced interaction required to facilitate deep learning, it is now argued that systems theory and the principles of cybernetics provide further direction.

Cybernetics and systems science (known also as systems theory) is derived from many traditional disciplines that include mathematics, technology, biology to philosophy, and the social sciences. It is related to the complexity sciences in that it encompasses the fields of artificial intelligence (AI), neural networks, dynamical systems, chaos, and complex adaptive systems. Whereas systems theory is focused on studying the relationships (structure) of systems as a whole (and their models), cybernetics is concerned with how systems function, how their actions are controlled, and how they communicate with other systems or with their own components. Systems science, systemics, cybernetics, and complex systems are therefore closely related fields. Given that the structure and function of a system cannot be understood in isolation, cybernetics and systems theory provide useful insights on the design of cognitive support tools that can assist the user to identify and explain the epistemological characteristics of a given subject area from multiple perspectives.

In contrast to the rule-based reductionist principles espoused by AI researchers over the past decades, the field of cybernetics evolved from a 'constructivist' view of the world [39], where objectivity is derived from group consensus on meaning. From a cybernetics perspective, information (and by implication intelligence) is an attribute of an interaction rather than a commodity stored in a computer [44]. In broad terms, cybernetics applies to the epistemological level (the limits to how humans know what they know) to explain the limitations of all mediums (technological, biological, or social) given that all human knowing is constrained by the perceptions and beliefs of individuals, and is therefore subjective. In other words, a cybernetics-based learning tool can be considered 'intelligent'. The principles of connectedness and relationships provide the core contribution of systems theory and cybernetics to this study.

A systems view provides a framework within which to examine and characterise a learning environment, its components, and all its subparts. By integrating systems concepts and principles into the design process and by learning to apply such factors, new ways of thinking, experiencing, exploring, understanding, and describing the environment at hand become apparent. Thus, the design of an intelligent cognition support system based on a systems perspective should account for a range of factors [3–5]:

- Properties of wholeness and the characteristics that emerge at various systems levels as a result of interaction and synthesis.
- Identifying the relationships, interactions, and mutual interdependencies of systems.
- The dynamics of the interactions, relationships, and patterns of connectedness among the components of systems and the changes that manifest over time.
- Interpreting and tracking interconnectedness and interdependencies in complex information systems.
- Identifying obscure relationships within information systems using complex analysis techniques to derive new insights and meanings.
- Managing the difficulties of deriving meaning and the loss of meaning from a surfeit of information.

To accommodate these factors, cognitive support systems devised to cultivate abstract thinking and conceptual understanding require a combination of 'intelligent' response mechanisms or agents, human computer interfaces (HCI), and tailored displays that supplement the diverse mental models or cognitive maps that individuals draw on in their search for understanding and construction of knowledge. The application of a systems perspective to learning design assists to determine how the cognitive processes of knowing, thinking, and reasoning can be exploited to facilitate inquiry and generate knowledge [6]. As signalled throughout the preceding account of systems concepts and principles, the identification of relationships within information using enhanced interaction methods provide direction on how learners can be empowered to organise, reflect, analyse, and synthesise information to construct knowledge.

Given such advantage, the notion of a learning model that incorporates cybernetic (and systemic) principles introduces the prospect of deriving an advanced design framework for structuring customisable, electronic learning environments assisted by 'intelligent' cognitive support systems. However, while most knowledge/memory management tools assist to make ideas explicit (therefore shareable), and permit exploration of the relationships between ideas to create new ideas, the problem, as noted, is the absence of direct interaction with the learner. The normalised word vector technology (explained in the section to follow) is fundamental to realising these goals.

In keeping with cybernetic principles, the design of learning environments of the future should incorporate principles of self-organisation, properties of emergence, and principles of connectedness, all of which inform the construction of intelligent cognition support systems and dynamic learning object assembly strategies (as outlined in the following section) to deliver more comprehensive and rewarding learning experiences. Then, given that most information and content available on the Web can be characterised as chaotic (not ordered), it is technically feasible to reshape this information and content in a multitude of ways to accommodate individual cognitive schemas and thereby encourage wider engagement in the learning process. For example, the application of selected teaching resources to varying contexts affords an effective pedagogical strategy for influencing learner understanding. Although in one context a resource may appear meaningless, the act of embedding the same resource in an alternative context exposes the learner to new insights that may further accentuate otherwise unknown or unfamiliar aspects of the broader subject/discipline area. This means in effect that the learning experience can be extended beyond current epistemological boundaries.

What should be apparent by now is that an open (multiple) systems model can be applied to educational design in which there are levels within levels, all interconnected through an intricate network of relationships. The structure is thus adaptable enough to accommodate linear, hierarchical, or even heterarchical (hyperlinked) connections to form multidimensional learning opportunities throughout all system and sub-system levels. The contextual permutations are endless, each representing additional strategies for teaching different aspects of an infinite array of knowledge. The notion of a learning model that combines systemic principles with learning object technology (described in Section 6.8) introduces the prospect of an advanced design framework for structuring customisable, 'intelligent' electronic learning environments.

Despite the long-standing vision and the ensuing advances in ICT, learners are by and large not directly supported by technology. For this reason, a core objective of the study has been to extend current understanding of how computer technologies can enhance the interactive, 'intelligent' analysis and interrogation of learners' responses and input. The challenge now is to develop a suitable theoretical foundation for devising such solutions. At this stage, it is useful to distil the design principles to be observed in the construction of intelligent learning support systems:

- Concept identification and analysis of learner input to provide immediate feedback
- Adaptive content selection and display using learning object resources
- Individualised learning solutions based on learner preferences and generational characteristics
- Learning theory principles tuned to technology-supported environments founded on systemic/cybernetic design principles
- Interactive 'cognitive immersion' in learning resources designed to cultivate metacognitive awareness
- Constructivist based, community-supported learning activities

6.8 A Model for Constructing "Intelligent" Cognition Support Systems

As emphasised in the previous section, the risk in proposing a learning model that is without reference to systemic principles, contextual flexibility, and the adaptability afforded by learning object technology is to ignore the potential of a multidimensional approach to learning environment design. Moreover, if the needs, preferences, technological propensities, and learning styles of current and future graduates are not taken into account, the design may not provide for the diversity of learning theory, pedagogical strategies, and design methodologies that guide effective learning. Resolving the issues indicated here require the construction of intelligent learning environments that draw on systems theory and cybernetic principles underpinned by learning object assembly and delivery techniques. There are several distinct advantages to this strategy.

The first advantage applies to the dynamic properties of learning objects and their capacity to be retrieved and assembled in any order at any stage during the delivery process. The dynamic attributes of learning objects are such that the structure and type of teaching content can be configured to interact with learner input. As an example, it is feasible to automatically select learning content to provide feedback in direct response to students' individual learning needs and goals. To illustrate, a number of strategies can be applied: generate activities, quizzes, or questions matched to the student's progress and competence levels; alert the student to the need for revision and iteratively present alternative content until an understanding of the required concept is indicated; analyse student input to determine comprehension levels and provide immediate feedback on progress; and, challenge students' knowledge by presenting more complex or conflicting (out of context) learning materials [35].

To expand this example further, software systems can be designed to support learners to make informed decisions about how their learning goals are met by guiding their interactions with 'intelligently' selected teaching resources. There are several ways to achieve this effect. In one scenario, learners could make decisions (with varying degrees of guidance) about both content (what to learn) and strategy (how to learn it). In another, learners could engage cognitive support tools to prompt and direct refinement to their understanding and knowledge of a subject area.

The advantages of a learning environment modelled on dynamic learning object assembly and 'intelligent' interactive responses also extend to the capacity to restructure the learning process at any stage. In effect, learning objects can be selected to automatically modify and adjust how learning occurs without restriction to the number of theoretical constructs and choice of pedagogical strategies. For example, by monitoring student progress it is feasible to deliver learning activities that range from a highly structured, preplanned sequence of tasks based on behaviourist principles through to unstructured, constructivist activities where the individual is permitted full ownership over the learning process.

Alternatively, any combination of delivery approaches can be applied to accord with changing learning needs. In place of static, inflexible teaching content, the learning process is managed in part by environments populated with software systems configured to continually monitor and accommodate the student's learning preferences and goal requirements. In effect, there are no fixed rules, only the needs of the learner and a rich diversity of learning object combinations that potentially fulfil those needs. In essence, learning object content can be designed with one overarching goal in mind – to facilitate learning in a way that caters to the preferences, values, and interests of each individual learner. As Winn [43] instructs:

> ...It follows that the only viable way to make decisions about instructional strategies that meshes with cognitive theory is to do so during instruction using a system that is in constant dialogue with the student and is capable of continuously updating information about the student's progress, attitude, expectations, and so on.

What is absent in the preceding scenarios is that there are no well-defined, universally accepted rules on how leaning objects are selected and assembled for display that will make 'instructional' sense to the user [23]. They propose two methods for automatic course sequencing: adaptive courseware generation where the entire course is adaptively generated according to defined learning outcomes and learners' knowledge levels before being presented, and dynamic courseware Generation in which the system monitors the student's progress and dynamically modifies the displayed content in response to individual needs, taking into account cognitive styles and learner preferences. A pre-filtering mechanism is employed to identify and generate a virtual pool of learning objects using filtering elements based on the IEEE P1484.12.1 Learning Object Metadata standard.

Provided the ontology matches the same IEEE standard, filtering can be achieved in combination with alternative ontologies to limit the process to a single knowledge domain. Learning object characteristics are derived from the IEEE standard, and the learner characteristics are drawn from the IMS Learner Information Package (LIP) specification. Typical elements of learning object characteristics include: general structure and aggregation level, and educational interactivity level, age range, difficulty level, context, and resource type. Learner characteristics elements refer to accessibility preference, eligibility, and disability type, qualifications, and activity attempts and results.

The methodology advanced by Karampiperis and Sampson has the capacity to correlate the dependencies between learning objects and learner characteristics that are normally made by the instructional designer when selecting teaching materials. However, this approach is not ideal when it comes to the practical realities of producing educationally effective online teaching solutions. As indicated beforehand, the ideal strategy for achieving the complex levels of interconnectivity advocated in the preceding pages is to apply a systems design approach to selecting and assembling learning objects. Concepts contained within the learner's input and responses are identified and interpreted by the NWV concept analysis system developed by Williams [40], which locates, tags, and assembles suitable learning object content to inform/teach/report in accordance with learners' needs.

All phases of the GCCS knowledge construction model apply to this approach. From the outset of a learning activity, the first step is to identify the 'information space' from which learners are expected to derive new knowledge and, represents the unordered collection of information (e.g., journal articles) that is to be 'generatively processed'. The information is scanned by the NWV analysis engine to generate a broad categorisation of concepts. Second, the 'attention gaining' aspect is initiated by 'highlighting' the key concepts to be learned (using model answers and/or predefined learning outcomes criteria), and associatively linking them to the concepts identified within the information space. Cognitive activity progresses in an interactive, recursive manner until the learner generates a response that aligns with the intended learning goals. For example, a bar graph that compares the 'concepts' contained in a model answer with the student's answer would in the case of a close match indicate that the student has understood the required concept. Where the student's response does not match the model answer, either an error has occurred, the concept has not been understood correctly, or irrelevant ideas have been introduced.

The NWV engine is currently being modified to 'interpret' input provided by the learner and present the results in an iterative process that compares their responses with a preferred model answer. Identification of the concepts to be acquired by the learner is accomplished using the NWV technology. Visual feedback is then displayed by way of graphs, concept maps, and statistical data. This interaction continues by inviting learners to pose follow-on questions, which are analysed and acted upon until a satisfactory learning outcome is achieved. Cognitive dissonance is introduced by 'highlighting' the concepts that match the concepts to be learned and visually linking them to related concepts to encourage further learning and understanding. Third, the displayed learning material is regenerated and refined or expanded on as the learner forms new ideas and understandings. Graphical representations of a concept and its relationships to other concepts acts as a visual stimulus that in effect forms a feedback bridge and provides an elaboration/explication of information found within the given information space. Feedback of this type is immediate and highly informative to both student and lecturer [20]. What have been described to this point are the key principles that underpin the GCCS model touched on beforehand.

In the preceding steps, concepts are the embodiment of the knowledge being created or synthesised by the user. They are superimposed onto the learning space and employed as temporary metatags to facilitate dynamic linking to related information and documents. The project expands on current thinking in cognition support by proposing that both cognitive (and metacognitive) tools are needed to support the immediate learning needs of individuals, groups, and organisations. The identified concepts are presented as 'knowledge about knowledge' or meta-knowledge, which contrasts with the traditional approach where the identification of related concepts requires the use of an editor or authoring tool to manually form associations with other concepts.

The unique strength of the NWV technology is to 'normalise' the words contained within documents and collections of information using an electronic thesaurus to build vector representations of all identified concepts. In brief, vector algebra techniques are used to represent similarities in the content contained within electronic documents. The advantage of using NWV as opposed to latent semantic analysis is that it is computationally more efficient and therefore faster. A digitised thesaurus is employed to 'normalise' the words contained in the scanned documents, thereby reducing all words to a predefined thesaurus root concept. In this way, the words contained with the scanned documents are aligned with the most appropriate 'within context' root concept. Vector representations are then constructed from the counts of these root concepts, where each is allocated a dimension number. The number of dimensions used to generate the vector representations is thus directly dependent upon the number of root concepts provided by the electronic thesaurus.

At present, the Macquarie Thesaurus Macquarie Library [27] is used to derive the normalised concepts. There are 812 root concepts in this thesaurus. The vectors are therefore constructed in an 812 dimensional space where the principles of vector theory carry over in exactly the same manner as for the three dimensions of space familiar to most people. In effect, each dimension consists of a concept number, which is then used as the basis for comparison with the vector representations calculated for the model answer. The angle between each vector indicates the level of accuracy. If the angle is small, the document contains content similar to the model answer. Conversely, the greater the angle, the less accurate is the match between the corresponding vectors [40]. Any electronic thesaurus is suitable for use by this technology, thus extending its functionality and application. Current applications of NWV technology are described in [14, 20, 41, 42].

The task of generating dynamic object selection is accomplished in two ways. The first refers to the capacity to dynamically select resources (in the form of learning objects) from existing repositories according to a range of criteria (e.g., learner profiles based on cognitive styles, learning preferences, learning outcomes, user input and interaction, changing performance levels and learning needs analysis, and, learning theory as applied to delivery sequencing and presentation format). The second method applies to the automatic creation of learning objects (albeit temporary) from standard (non-object formatted) electronic materials (documents, reports, papers), thus removing the need to convert such materials to a permanent XML-compliant format. Dynamic metatags are derived and generated 'on-the-fly' using 'search criteria' (i.e., the concepts) determined by the NWV engine to be the most accurate 'conceptual' match for the selection of learning object material. Thus, dynamic metatag generation, dynamic object selection, and NWV technology function as mutually dependent processes in direct compliance with systems theory/cybernetic principles.

A useful prospect to emerge from the preceding processes is that it is possible to work on the same information space in distinctly separate sessions, permitting a separation of views, concepts, and annotations – thus generating divergent sets of thoughts, ideas, and knowledge. Multiple sessions may be 'owned' by one researcher or assigned to various groups of researchers. This facility permits unlimited comparison of the derived concepts, which is of considerable value when contemplating the possibility of comparing the concepts formed by independent researchers working on the same information set. This option, which is in effect 'brainstorming', permits a range of experiments to be conducted using cognitive mapping techniques. The log of cognitive activity and subsequent analyses further inspires a range of psychological experiments perhaps capturing learners' thinking patterns in devices such as concept maps, topic maps, mind maps, Compendium, or similar artefacts. Once a GCCS system is deployed, mining the captured activity logs holds genuine promise for follow-on research studies.)

6.9 Conclusion

The recent advances in computer and communications technologies have made it feasible to not only access and manage information in productive and efficient ways, but also reveal an emerging potential to assist learners to derive conceptual understanding and generate new knowledge. It is conceivable that humans are entering into hitherto unrealised realms of thought and idea conception that may bring about a revolution in the role of technology for enhancing cognitive processing. It is important, therefore, to extend current understanding of how computer technologies can enhance 'intelligent', interactive interrogation of learners' input, thus providing new insights into how the generation of new knowledge may be supported. The nature of this exercise is such that it raises the potential to apply technology to resolving 'real-world' problems.

To re-emphasise, the scope of the study that is the subject of this chapter, is to conduct a series of investigations aimed at devising learning strategies, pedagogical approaches, quality assurance standards, assessment techniques, and evaluation methodologies to establish educational benchmarks appropriate to the development of software solutions that address current deficiencies in technology-supported learning systems. The examination of the research literature has revealed a number of issues and deficiencies that provided the impetus to conduct this study:

- There is a dearth of research literature on educationally focused evaluation design methodologies for intelligent cognitive support systems. The majority of research publications relate to technologies that are now several 'generations' old and therefore do not adequately reflect the power of current technologies, considerably improved software designs, and the increased expertise required to use it.
- Many software evaluation approaches are intended for computer-assisted instruction, integrated learning delivery systems, and online course designs – not intelligent cognitive support systems.
- Most surveys of existing evaluation studies use very small samples, and often focus on a single educational system – little is published on intelligent support systems.
- Only a handful of studies provide longitudinal data for at least 1 year or longer. Many do not include data on how the technology is actually used for instruction or teacher involvement. The majority of these studies are case studies and ad hoc in how the cases are selected rather than offer a theoretical basis for case study selection.

In short, research and evaluation in the context of improving student achievement and educational delivery is needed to guide policy and practice in the design and development of technology support systems, as well as to provide direction on areas where additional research and/or evaluation is required. While there is cause for concern over the extant ability of present-day information search and delivery systems to provide for the complex cognitive skill demands of a technologically driven future, this chapter has argued that certain pivotal factors present grounds for believing there is merit in constructing advanced cognitive support system design models that advantage the growing sophistication of computer and communication technologies. The challenge and the significance of the project is to harness emerging technological developments in ways that deliver high-quality learning outcomes relevant to the changing needs of individuals and society. This acknowledgement of course points to the need to devise innovative deployment techniques informed by reliable research studies all of which must be augmented by ongoing, open feedback measures.

An important aspect of this study to note is that it presents a unique opportunity to identify and establish the ground rules from which the benefits of applying 'intelligent' technologies to enhance conceptual thinking can be realised. The successful completion of the project, coupled with related research underway at present, will lead to a robust development methodology capable of producing cognitive support systems for applications where currently they have not been made available, for example, the Internet. The research underpinning this study is crucial, particularly given its broad application and timely relevance to improving the delivery of learning in the online environment. For the present, the main contribution to knowledge is to increase current understanding on how learning may or may not be supported and improved using technology.

Where the future is concerned, it is argued that to manage the vast quantities of information that will be generated, as well as to interpret, understand, synthesise that information to derive knowledge, requires new solutions that connect human cognitive processes to highly flexible and adaptable technologies. While it is difficult to determine the full extent of change in the short term, there is little doubt that current definitions of thinking and analysis skills will prove inadequate for addressing the complexities of an information society. In order to deal with the emerging challenges, new ways of enhancing human cognitive processing are required that support an increasing need for innovative information processing and knowledge creation skills.

The strategic use of digital technologies has enhanced the capacity of individuals and organisations to explore the inner workings of natural and human phenomena in ways that until now have not been possible. As a result, not only has there been a noticeable shift in what humans learn, but also how they learn. To keep pace with these changes, the task of devising advanced learning support models must embody the development of highly innovative technological solutions, design and evaluation methodologies, and advanced delivery strategies.

Two final assertions bring this chapter to a close: the first is that information and course delivery systems designed to cultivate abstract thinking and conceptual understanding should be supported by 'intelligent' response systems or agents, interfaces, and portals that accommodate the diverse mental models or cognitive maps that learners draw on in their search for understanding in the construction of knowledge; and second, rather than segment data and information into bounded disciplines and subject areas, the focus of learning design should be directed towards enabling identification and exploration of the rich and undiscovered connections that thread throughout all knowledge domains.

6.10 Research Questions for Further Study

In light of what has been identified and examined to this point, it has been argued that the higher-order thinking skills of problem solving and critical analysis require the ability to see parts/wholes in relationship to each other; balance the processes of both analysis and synthesis; abstract and manage complex issues; adapt to real-world change; and command multiple strategies for solving problems. It is further argued that without a thorough examination of the relationships between technology, communication, media, human interactions, and cognitive development, the full extent of the power of electronic support environments cannot be fully realised.

For every successive level of understanding aspired to new and increasingly more complex strategies are required to ensure continued progression towards higher levels of cognitive development and smart information use. For these reasons, the following research questions provide a platform from which to begin the investigation process:

- How to construct 'intelligent' learning guidance and response mechanisms that assists users to derive deep cognitive understanding and solve immediate problems?
- How to manage the dynamic generation of data and information so that it aligns with the individual's needs and interactively provides constructive responses?
- How to impart the higher-order thinking skills that foster conceptual understanding with a strong emphasis on enhancing knowledge construction and tacit knowledge skills?
- How to identify and provide automated support for the needs of networked 'communities of learning' within any given population of learners?
- How to design delivery systems and services that support interactive interrogation of information on demand in a way that affords flexible mobility and seamlessly integrates with the personal circumstances and needs of individuals?
- How to demonstrate and evaluate the benefits of advanced ICT solutions (and their advantages over conventional cognition support systems)?

Resolving the issues that arise from the preceding questions should be guided by three broad themes of inquiry:

- Is it feasible to design computer systems that support the way individuals acquire, generate, and apply knowledge, as opposed to the generation and application of information?

- Is it possible to mimic or emulate the way the human mind works using current and emerging technologies?
- Can human understanding be advanced by developments in computer-mediated synthetic reasoning that are designed to support conceptual association, problem solving, and knowledge construction?

The strategies proposed thus far in this chapter provide partial direction, but might be further strengthened by adapting games theory principles to enhance interaction and motivation to engage in information processing that place less emphasis on memorising facts and more on utilising cognitive strategies for discovering knowledge; and extending the nature of digitised repositories to deliver more than just content and instead assist the user to organise, reflect, analyse, and synthesise new knowledge.

The design and development of information delivery solutions that assist users to cultivate effective knowledge construction skills requires several interrelated areas of research: an analysis of the properties, modelling structures and representation of knowledge domains; methods for managing and transferring tacit and cognitive knowledge; contextualisation of information and knowledge through identification of the interrelationships that elicit metacognitive thinking; and, the strategic structuring of information and knowledge that permit navigation using multiple learning pathways.

To have any noticeable effect on the quality of outcomes, attention should also be directed towards profiling learner behaviours, values, attitudes, communication skills, and display preferences, all of which require continuous monitoring, refinement, and updating. Just as important, the tacit knowledge skills of predictive analysis, creative thinking, entrepreneurial acumen, and the ability to move from problem solving to opportunity identification and acquisition, have not as yet been comprehensively examined.

Acknowledgements I am highly indebted to Dong Won Kim who willingly and patiently provided considerable support and editorial assistance in completing this book chapter. He also undertook the difficult task of converting the original Word document to LaTeX on my behalf for which I am very grateful.

References

1. Ausubel, D. P. (1968). Educational Psychology: A Cognitive View. New York: Holt, Rinehart and Winston.
2. Ausubel, D. P. Novak, J. D. and Hanesian, H. (1978). Educational Psychology: A Cognitive View (2nd ed.). New York: Holt, Rinehart and Winston.
3. Banathy, B.H. (1988a). Systems inquiry in education. Systems Practice, Vol. 1, No. 2. pp. 193-211.
4. Banathy, B.H. (1991). Systems design of education. Englewood Cliffs, NJ: Educational Technology.
5. Banathy, B.H. (1996). Designing social systems in a changing world. New York: Plenum Press.

6. Banathy, Bela H. and Jenlink, Patrick M. (2004). Systems Inquiry and Its Application in Education. In David H. Jonassen, ed., Handbook of Research on Educational Communications and Technology, 2nd ed. Mahwah, N.J.: Lawrence Erlbaum Associates, Inc. p. 49.
7. Barr, Trevor. Burns, Alex. and Sharp, Darren. (2005). Smart Internet 2010. Faculty of Life and Social Sciences, Swinburne University of Technology. Report produced for the Smart Internet Technology CRC Pty Ltd, Australian Technology Park, Eveleigh, NSW, Australia. p. 4. WWW Ref: http://www.smartinternet.com.au/SITWEB/publication/publications.jsp
8. Bush, V. (1945), As We May Think. Atlantic Monthly, Vol. 176, No. 1. July. pp. 101-108.
9. Buzan, T. and Buzan, B. (2000). The Mind Map Book. London: BBC Worldwide.
10. Campos, Milton. (2004). A Constructivist Method for the Analysis of Networked Cognitive Communication and the Assessment of Collaborative Learning and Knowledge-Building. JALN. Vol. 8, No. 2. April. pp. 9–10.
11. Canas, Albert. Coffey, John. W. Carnot, Mary Jo. Feltovich, Paul. Hoffman, Robert R. Feltovich, Joan. and Novak, Joseph D. (2003). A Summary of Literature Pertaining to the Use of Concept Mapping Techniques and Technologies for Education and Performance Support. Report prepared by The Institute for Human and Machine Cognition and The University of West Florida for The Chief of Naval Education and Training, Pensacola FL 32500. pp. 54–60. WWW Ref: http://cmap.ihmc.us/Publications/
12. Compendium. (2008). WWW Ref: http://www.compendiuminstitute.org/
13. Dreher, H. (1997) Empowering Human Cognitive Activity through Hypertext Technology. PhD thesis. WWW Ref: http://adt.curtin.edu.au/theses/available/adt-WCU2000046.121219/
14. Dreher, H. (2006). Interactive Online Formative Evaluation of Student Assignments. Presented at InSITE 2006, June 25-28, Greater Manchester, England. WWW Ref: http://2006.informingscience.org/
15. Dreher, Heinz. Krottmaier, Harold. and Maurer, Hermann. (2004). What We Expect form Digital Libraries. Journal of Universal Computer Science. September. pp. 1–7. WWW Ref: http://www.jucs.org
16. Eden, C. (1988), Cognitive Mapping. European Journal of Operational Research, Vol. 36, pp. 1–13. Elsevier Science Publishers B V (North Holland).
17. Eden, C. (1992). On the Nature of Cognitive Maps. Journal of Management Studies. Vol. 29, No. 3. pp. 261–265.
18. Endsley, M. and Hoffman, R.R. (2002). The Sacagawea Principle. Intelligent Systems. Vol. 17, Issue 6. Nov/Dec. pp. 80–85.
19. Fisher, K.M. (2000). SemNet Software as an assessment tool. Assessing Science Understanding. pp. 197–221.
20. Guetl, C., Dreher, H. and Williams, R. (2005) E-TESTER: a Computer-based Tool for Auto-generated Question and Answer Assessment. In Proceedings of World Conference on eLearning in Corporate, Government, Healthcare, and Higher Education, pp. 2929–2936. October. E-Learn 2005. WWW Ref: http://www.editlib.org/index.cfm
21. Hannafin, Michael J. (1997). The Case for Grounded Learning Systems Design: What the Literature Suggests About Effective Teaching, Learning, and Technology, Educational Technology Research & Development, Vol. 45, No. 3, pp. 101–117.
22. Harnad, S. (1991). Post-Gutenberg Galaxy: The Fourth Revolution in the Means of Production of Knowledge. Public Access Computer Systems Review, Vol. 2, No. 1. pp. 39–53.
23. Karampiperis, Pythagoras and Sampson, Demetrios. (2004). Adaptive Learning Selection in Intelligent Learning Systems. Journal of Interactive Learning Research, Vol. 15, No. 4. pp. 390–395.
24. Kosko, Bart. (1994). Fuzzy Thinking – A New Science of Fuzzy Logic. Hammersmith, London, UK: Harper Collins.
25. Liber, Oleg. (2004). Cybernetics. e-Learning and the Educations System. International Journal of Learning Technology, Vol. 1, No. 1.
26. Larreamendy-Joerns, J. and Leinhardt, Gaea. (2006). Going the distance with online education. Review of Educational Research, Vol. 76, No. 4. Winter. pp. 580–589.
27. Macquarie Library. (2008). WWW Ref: http://www.macquariedictionary.com.au

28. Minsky, Marvin. Ed. (1968). Semantic Information Processing. Cambridge, Massachusetts: The MIT Press.
29. Mittelholtz, Daniel. (1997). Metacognitive Cybernetics: The Chess Master is No Longer Human! p. 4. WWW Ref: http://www.usask.ca/education/coursework/802papers/Mittelholtz/MC.pdf
30. Newcomb, Steven R. and Biezunski, Michel. (2000). Topic Maps go XML. WWW Ref: http://www.gca.org/papers/xmleurope2000/pdf/s11-02.pdf
31. Novak, J. (circa 1999). The Theory Underlying Concept Maps and how to Construct Them. WWW Ref: http://cmap.coginst.uwf.edu/info/
32. Osman, M.E. and Hannafin, M.J. (1992). Metacognition research and theory: Analysis and implications for instructional design. Educational Technology Research and Development, Vol. 40, No. 2. pp. 83–99.
33. Paolucci, M. Suthers, D. and Weiner, A. (1995). Belvedere: Stimulating Students' Critical Discussion. WWW Ref: http://lilt.ics.hawaii.edu/lilt/papers/1995/paolucci-et-al-chi95.pdf
34. Pepper, S. (2000). The TAO of Topic Maps: Finding the Way in the Age of Infoglut. WWW Ref: http://www.ontopia.net/topicmaps/materials/tao.html
35. Quinton, S. (2004). Towards Dynamically Generated, Individualised Learning. Refereed paper published in e-University: International Conference on ICT and Higher Education. Dr. Wichian Premchaiswadi (Ed.). Siam University, Bangkok: Thailand.
36. Rieber, L.P. (1992). Computer-based microworlds: A bridge between constructivism and direct instruction. Educational Technology Research and Development, Vol. 40, No. 1. pp. 93–106.
37. Spiro, Rand J. Feltovich, Paul J. Jacobson, Michael J. Coulson, Richard L. (1991). Knowledge representation, content specification, and the development of skill in situation specific knowledge assembly: some constructivist issues as they relate to cognitive flexibility theory and hypertext. Educational Technology, Vol. XXXI, No. 9. Sept. p. 5.
38. Suthers, D.D. and Hundhausen, C. (2001). Learning by constructing collaborative representations: An empirical comparison of three alternatives. Paper presented at the European Perspectives on Computer Supported Collaborative Learning, Universiteit Masstricht, Maastrict, The Netherlands.
39. von Glasersfeld, E. (1987). The Construction of Knowledge, Contributions to Conceptual Semantics. Seaside, California: Intersystems Publications.
40. Williams, R. (2006). The Power of Normalised Word Vectors for Automatically Grading Essays. Presented at InSITE 2006, June 25-28, Greater Manchester, England. pp. 3–4. WWW Ref: http://2006.informingscience.org/
41. Williams, R. and Dreher, H. (2004) Automatically Grading Essays with Markit©. Issues in Informing Science and Information Technology, Vol. 1, pp. 693–700. WWW Ref.: http://articles.iisit.org/092willi.pdf
42. Williams, R. and Dreher, H. (2005) Formative Assessment Visual Feedback in Computer Graded Essays. The Journal of Issues in Informing Science and Information Technology, 2005. Vol. 2. pp. 23–32. WWW Ref. http://2005papers.iisit.org/I03f95Will.pdf
43. Winn, W. (1989). Toward a rational and theoretical basis for educational technology. Educational Technology Research & Development, Vol. 37, No. 1. pp. 39–41.
44. Winograd, Terry and Fernando Flores. (1986). Understanding Computers and Cognition: A New Foundation for Design. Norwood, New Jersey: Ablex Publishing Corporation.
45. Wobcke, Wayne. (2004). Intelligent Agents Technology Review. Smart Internet Technology CRC, Sydney. October.

Chapter 7
A Web-Based Learning Support System for Inquiry-Based Learning

Dong Won Kim and JingTao Yao

Abstract The emergence of the Internet and Web technology makes it possible to implement the ideals of inquiry-based learning, in which students seek truth, information, or knowledge by questioning. Web-based learning support systems can provide a good framework for inquiry-based learning. This article presents a study on a Web-based learning support system called *Online Treasure Hunt*. The Web-based learning support system mainly consists of a teaching support subsystem, a learning support subsystem, and a treasure hunt game. The teaching support subsystem allows instructors to design their own inquiry-based learning environments. The learning support subsystem supports students' inquiry activities. The treasure hunt game enables students to investigate new knowledge, develop ideas, and review their findings. *Online Treasure Hunt* complies with a treasure hunt model. The treasure hunt model formalizes a general treasure hunt game to contain the learning strategies of inquiry-based learning. This Web-based learning support system empowered with the online-learning game and founded on the sound learning strategies furnishes students with the interactive and collaborative student-centered learning environment.

7.1 Introduction

The impact of advances in World Wide Web technology has shifted the paradigm of education from the traditional teacher-centered form of education to the collaborative student-centered form of education [10]. Web-based learning support systems support teaching and learning by taking full advantage of Web technology [31]. Web-based learning support systems have no barriers regarding time and place to

D.W. Kim and J.T. Yao
Department of Computer Science, University of Regina, 3737 Wascana Parkway, Regina, Saskatchewan, Canada
e-mail: kim263@cs.uregina.ca; jtyao@cs.uregina.ca

J.T. Yao (ed.), *Web-Based Support Systems*, Advanced Information and Knowledge Processing, DOI 10.1007/978-1-84882-628-1_7,

learn, and provide learning environments that are accessible to all [31]. Nowadays, many Web-based learning support systems are available to support instructors and students, and improve their works in education [2, 25, 26].

Inquiry-based learning is a student-centered educational approach that is driven more by student questions than by the instructor's lessons [8, 10]. The educational philosophy of inquiry-based learning is founded on the ideals and principles of constructivism [17]. In constructivism, knowledge is defined as a cognitive structure of a person [14]. Learning is an active process of constructing knowledge rather than the process of knowledge acquisition. Teaching is supporting the student's constructive processing of understanding rather than delivering the information to the student [14].

Student motivation is a key to the success of learning [20]. Modern digital games provide young people with learning opportunities at every moment, and game-based learning stimulates students' motivation for learning with challenges and entertainment [20]. Recently, game-based learning has been adopted for adult education as well as children's learning [6]. Online-learning games are more effective than standalone learning applications by encouraging collaborative learning [5, 13].

In this chapter, we will present a Web-based learning support system called *Online Treasure Hunt*. We combine Web technology with online game technology to support education, and provide an efficient model for inquiry-based learning through this study. The design of *Online Treasure Hunt* depends entirely on a treasure hunt model. The treasure hunt model takes inspiration from a treasure hunt in order to integrate the idea of inquiry-based learning into Web-based learning support systems. *Online Treasure Hunt* helps both instructors and students work toward common educational goals by providing an interactive and collaborative student-centered learning environment.

7.2 Web-Based Learning Support Systems and Inquiry-Based Learning

7.2.1 Web-Based Learning Support Systems

The Web is a new medium for storing, presenting, gathering, sharing, processing, and using information [29]. The Web offers the following benefits for learning: (1) it provides a distributed infrastructure for educational content; (2) it can be used as a channel in which students and instructors collaborate; (3) it delivers in a timely manner various forms of learning content and secure information; (4) it provides a student-friendly interface; (5) there are no time-frame or geographic restrictions on attending online classes; and (6) it can remotely and instantly manage and retrieve knowledge.

Web technology refers to all the technologies that implement, maintain, and use the Web. Web-based learning support systems (WLSS) are redesigned or modified systems of the traditional computerized learning support systems, which support teaching and learning, by using Web technology. WLSS not only support students in learning, but also provide student-centered learning environments that are accessible to all [31].

Well-designed WLSS can provide a student-centered form of education. In the student-centered education, students choose and study learning content based on their background knowledge, and the instructor just plays a supporter's role [9]. The main considerations of WLSS are as follows: encouraging users to communicate with each other, delivering adapted content based on students' knowledge, providing an interactive interface, evaluating students' learning process, and stimulating student motivation for learning [31].

7.2.2 Web Services

Web services are considered one of the emerging Web technologies. They are self-contained, self-describing, modular applications that operate over the Internet. In the development of Web services, the same technical standards are applied to self-description, publication, location, communication, invocation, and data-exchange capabilities. It increases the interoperability and reusability of Web-based applications, and greatly reduces the time and effort spent on implementing the applications as well. The field of online learning may benefit from the advantages of Web services [13]. Especially the Web services can be combined together to implement a Web-based learning support system as necessary.

Google Calendar is a Web service provided by Google and allows applications to read and update calendar events using Google Calendar Data API. A client application can use the Google Calendar Data API to create new events, edit or delete existing events, and query for events that match particular criteria [12]. The Google Calendar is a useful coordination tool for students to make an appointment with other colleagues and manage their schedules to work together [19].

MetaWeblog API is an application programming interface that enables the implementation of blog-based Web services by allowing applications to publish and update the content and attributes of weblog posts [15]. A *weblog* (usually shortened to blog) furnishes an excellent new channel for discussion, communication and collaboration not only in research [30], but also in learning. Furthermore, *blogs* enable students to develop a deep understanding of and to take more responsibility for their own knowledge [27].

7.2.3 Online-Learning Games

Online games are highly graphical two- or three-dimensional video games played over the Internet [24]. Players, through their self-created digital characters or avatars, are allowed to interact not only with non-player characters, which are computer-controlled characters, but with other players' avatars as well. *Online-learning games* are derived from an attempt to not only bring the concept of the online games to interactive learning tools, but also provide students with learning experience through the online games.

Dziabenko et al. [6] implemented an online-learning game called "UNIGAME: Social Skills and Knowledge Training." The game is accessible through its Web site, which furnishes the following features: training and help, community area, user registration, game introduction, virtual conference, and so on. They believed the game would be used as a complement to lectures or even as a stand-alone learning process, and enable students to understand better the theory as they practically apply it.

Global Goonzu [11] is a massively multiplayer online role-playing game simulating the real-world society. A player can experience this world as a warrior, merchant, and politician. Its free market economic system enables players to experience real-time trading of stocks, real estate, and other goods. Furthermore, players can take part in the shareholder election to elect or to become a town chief. Although the game has been developed for fun, it is very useful for economics and politics education.

Childress and Braswell [5] described current uses of multiplayer online games in learning. They derived the process of using multiplayer online games for cooperative learning activities from their experiences in *Second Life* [22], a massively multiplayer online role-playing game. They also emphasized that multiplayer online games can provide students with many meaningful and enriching learning experiences, and the virtual environments of the games enable instructors to easily and efficiently design highly social cooperative learning activities.

7.2.4 Inquiry-Based Learning

Inquiry-based learning (IBL) is an educational approach in which students seek truth, information, or knowledge by questioning [4]. Inquiry, in the *Merriam-Webster* dictionary,[1] is defined as "examination into facts or principles, research, a request for information, and a systematic investigation often of a matter of public interest." In the literature, Russell [21] presents inquiry as an investigation of certain problems. Fleissner [10] considers inquiry "as a seeking for truth, information, or knowledge; that is, seeking information by questioning."

The process of IBL commonly described in the literature [7, 8, 16] involves the following four phases: (1) presenting phase: present information and background of an inquiry; (2) retrieving phase: explore and query data and information for the inquiry; (3) developing phase: develop and generalize ideas or concepts by the evaluation and interpretation of data and information collected; and (4) evaluating phase: evaluate the developed ideas and inquiry process, and give constructive feedback. Throughout the whole process of IBL, students may revise their ideas or decide to go forward in the intended direction by looking back at the questions, process and direction of the inquiry. Students also communicate with others to fully understand the given information, and clarify or solidify their findings.

---

[1] http://www.merriam-webster.com/

7.2.5 *Web-Based Learning Support Systems for Inquiry-Based Learning*

WLSS can provide ideal solutions with which to implement IBL environments. Edelson et al. [8] describe the computing technologies necessary to properly implement IBL as follows: (1) storing and manipulating large quantities of information; (2) presenting and permitting interaction with information in a variety of visual and audio formats; (3) performing complex computations; (4) supporting communication and expression; and (5) responding rapidly and individually to users.

Web-based learning support system architecture generally follows a three-layer framework proposed by Yao and Yao [29] for Web-based support systems. The three-layer architecture includes: an interface and presentation layer, a data layer, and a management layer. The interface and presentation layer presents and permits interaction with information in a variety of visual and audio formats, and supports communication and expression through a Web browser. It also responds individually to the request of users. The data layer is responsible for storing and manipulating large quantities of information. The management layer performs complex computations.

In addition to its architecture satisfying the computing requirements for IBL, the Web provides vast resources for inquiry, and many of the resources are free for anyone to access and use through the Internet [16]. Therefore, WLSS having strong support from Web technology can become a good framework or systems with which to accomplish IBL. In the next section, we present a treasure hunt model which allows IBL to be integrated into WLSS by taking advantage of the Web and online game technologies.

7.3 Modeling *Online Treasure Hunt*

7.3.1 *Treasure Hunts*

A treasure hunt is originally an outdoor activity and a game played by children and occasionally by adults. For a treasure hunt, an adult prepares a list of hidden objects which children will try to find. Each team of children receives a duplicate list of the hidden objects. The winner is the first team to find all the items on the list.

Mechling [18] laid out a treasure hunt game played by a Boy Scout troop. The game requires the game organizers to spend a few days preparing for the hunt. Their tasks include: (1) deciding game boundaries, determining the location for hiding the treasure and the location of the stations along the route from the camp to the treasure site; (2) creating a chain of clues that leads the game teams from one station to another; and (3) allocating the tasks and necessary equipment to the stations. He also describes the important elements of the modern treasure hunt: rhymed clues, riddles, clues written in code, performances at stations of the hunt, and foodstuffs as the treasure. In the treasure hunt, the scouts are requested to follow all the rules and use their physical skills and strategies to overcome challenges.

Skelly [23] defines the treasure hunt as an online activity "in which one searches for clues and answers." Blas et al. [3] introduce an online treasure hunt game in the SEE project that provides students with a virtual learning environment. Students, represented by avatars, have online meetings together and discuss previously studied themes under the active supervision of a guide, also represented by an avatar, in a virtual museum. A treasure hunt in the SEE project helps students find a solution to cultural riddles provided by the museum, and enables them to review their learning in exciting ways.

A treasure hunt can be well matched to IBL. *Treasure* is considered as information, truth, or knowledge, and *hunt* implies inquiry, which is a systematic investigation. A *treasure hunt* is an inquiry activity in which one systematically seeks knowledge with questions.

7.3.2 Treasure Hunt Model for Inquiry-Based Learning

The treasure hunt model formalizes the treasure hunt in order to efficiently integrate the learning strategies of IBL into WLSS. It also provides an effective way to combine the Web and online game technologies in the designing of WLSS for IBL.

The treasure hunt model consists of seven components: *agents*, *treasures*, *information*, *events*, *actions*, *stations*, and *logbooks*. Its structure is defined as follows:

$$S = (AGT, TRE, INF, EVT, ACT, STA, LOG). \tag{7.1}$$

- AGT is a finite nonempty set of *agents* that represent students or guides in the system, and expressed as $AGT = SD \cup GI$, where SD is the set of all students and GI is the set of all guides. Guides help students find treasures by providing useful information about the treasures.
- TRE is a finite nonempty set of *treasures*. A *treasure* is a conclusion of a topic and a reward for learning the topic. The *reward* is a useful thing to help students overcome obstacles in finding other treasures.
- INF is a finite nonempty set of *information* about the treasures. This is defined as $INF = \{TP, RS, QS, CL\}$, where TP is a finite nonempty set of learning topics, RS is a finite nonempty set of resources, QS is a finite nonempty set of questions, and CL is a family of nonempty set of clues. A *learning topic* is a finite nonempty set of *learning objects* [31] necessary to master a topic. The *resource* is a collection of information sources related to a learning object. The *question* is what students need to investigate while learning a topic. The *clue* is a rhymed hint about the next learning object to find and learn.
- EVT is a finite nonempty set of *events* which request student participation. There are four different types of events: orientation events, help events, obstacle events, and evaluation events.

- *ACT* is a finite nonempty set of *actions* that students can take in order to find the *treasure*. Some major actions are as follows: getting information, analyzing information, retrieving resource, writing findings, and determining direction.
- *STA* is a finite nonempty set of *stations* where the event occurs and students take the actions. In accordance with the event, the stations can be grouped into the following types: orientation stations, help stations, obstacle stations, and evaluation stations.
- *LOG* is a finite nonempty set of *logbooks* that contain all the records of the students' important achievements in the treasure hunt. Students write their findings into the logbooks, the contents of which help them construct new knowledge.

Let n be the number of topics for a course and let m be the number of sub-topics of each topic. TP of INF in Equation 7.1 can be expressed as $TP = \{L_1, L_2, \ldots, L_n\}$, where L_i (for $1 \leq i \leq n$) is the learning topic. A learning topic L is described as $L = \{k_0, k_1, k_2, \ldots, k_m, k_{m+1}\}$, where k_0 is the *introduction* of the topic, k_i (for $1 \leq i \leq m$) is a *sub-topic* of the topic, and k_{m+1} is the *conclusion* of the topic. In the treasure hunt model, a set T of all learning objects which students need to understand in order to complete *a course* through the treasure hunt is defined as

$$T = \bigcup_{L \in TP} L. \tag{7.2}$$

Let $IT \subset T$ be a set of all the *introductions* of topics for a course, let $SU \subset T$ be a set of all the *sub-topics*, and let $CO \subset T$ be a set of all the *conclusions* of topics. Thus, the set T of all the objects of the course is also defined as:

$$T = IT \cup SU \cup CO. \tag{7.3}$$

The learning objects are allocated to the stations and guides by using a learning object allocation function f_{OA} given by

$$f_{OA} : T \rightarrow STA \times GI. \tag{7.4}$$

If TN is a finite nonempty set of names of topics for a course, an orientation function f_{OT} representing the orientation event of EVT in Equation 7.1 is described as

$$f_{OT} : STA \times GI \times TN \rightarrow IT \times QS \times CL. \tag{7.5}$$

The orientation function f_{OT} allows a guide to provide an introduction, a question, and a clue for a given topic name at a station. A help function f_{HL} for the help event of EVT is expressed as

$$f_{HL} : STA \times GI \times QS \rightarrow SU \times CL. \tag{7.6}$$

The definition of the help function f_{HL} describes that a guide furnishes a sub-topic description and a clue for a given question at a station.

Let $f_{SM}(q, t, t^{'})$ be a similarity function to measure the similarity of a student answer $t^{'}$ to a sample answer t, provided by the instructor, for a given question $q \in QS$. If S_q is a set of significant terms appeared in the question q, S_t is a set of significant terms appeared in the sample answer t, and $S_{t'}$ is a set of significant terms appeared in the student answer t', then the similarity can be calculated by

$$\frac{|S_{t''}|}{|S_{qt}|}, \tag{7.7}$$

where $S_{qt} = S_q \cup S_t$ and $S_{t''} = S_{t'} \cap S_{qt}$.

If RW is a set of rewards that each student can receive after passing the test at an evaluation event, the set of treasures TRE in Equation 7.1 can be described by $TRE = CO \times RW$, where CO is a set of all the *conclusions* of topics as described before. Therefore, an evaluation function f_{EV} for the evaluation event of EVT is defined as

$$f_{EV} : STA \times R(f_{SM}) \rightarrow TRE, \tag{7.8}$$

where $R(f_{SM})$ is the range of f_{SM}. The evaluation function f_{EV} expresses that a student can obtain a treasure at a particular station based on the similarity of an answer to a sample answer.

The obstacle event gives students enjoyment and hinders them from finding the treasures. If OB is the finite nonempty set of obstacles, an obstacle function f_{OB} for the obstacle event of EVT is described as

$$f_{OB} : STA \times OB \times 2^{RW} \rightarrow \mathbb{N}, \tag{7.9}$$

which represents how well students overcome an obstacle using rewards, which they have obtained at a station. It shows how many points students acquire in struggling with the obstacle at a station using the things that they have received as their reward.

7.4 Implementation of *Online Treasure Hunt*

7.4.1 Architecture of *Online Treasure Hunt*

Online Treasure Hunt is a Web-based learning support system based on the treasure hunt model. This is designed to support IBL under the context of the treasure hunt. The Web-based learning support system is enhanced by employing a treasure hunt, online-learning game. The architecture of the system is depicted in Figure 7.1. *Online Treasure Hunt* has the following components such as a teaching support subsystem, a learning support subsystem, and a treasure hunt game.

The teaching support subsystem helps instructors design their courses as IBL is asserted. The system provides a graphical user interface to instructors defining course outlines, topics, questions, clues, and related information. The system stores the information into the course knowledge base and generates treasure hunt environments for the treasure hunt games. From this generation, treasure boxes are placed at certain stations containing questions, conclusions, and rewards. In addition, non-

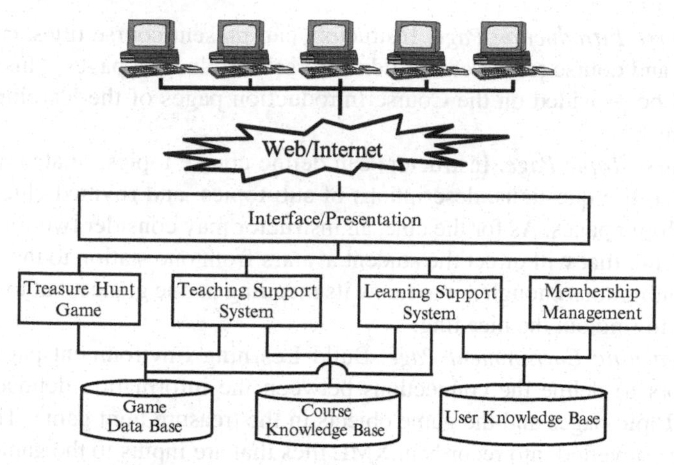

Fig. 7.1 An architecture of online treasure hunt

player characters can be located at certain stations and have their own words to guide students as the *guides* of the treasure hunt model.

The learning support subsystem provides students with resources related to each topic and information about the treasure hunt. Students can search for more information about the topics using a Web-based information retrieval support system [28]. In addition, they can post their findings to *blogs*, whose data is stored into the user knowledge base. The system automatically posts student progress to the *blogs*. The *blogs*, which are the logbooks of the treasure hunt model, must encourage students to discuss their progress with peers and help students develop their ideas by tracing the findings.

The treasure hunt game provides a playground for the treasure hunt. All of the information that the treasure hunt game requires is in the game database. Students can be motivated and experience the treasure hunt described in Figure 7.2 by playing the game. The learning support subsystem provides students with virtual places, such as their homes, where students complete homework and prepare for their next classes, while the treasure hunt game provides different virtual places, such as classrooms, where students learn from teachers.

7.4.2 Teaching Support Subsystem

The teaching support subsystem allows instructors to lay out their courses and build their own IBL environments in the treasure hunt game. The teaching support subsystem has five major components as follows.

1. *Teaching Guide Page*. Teaching Guide pages provide the information about how to use the system for IBL and how to build their own IBL environments on the system.

2. *Edit Course Introduction Page*. Instructors can present course titles, course objectives, and course references on Edit Course Introduction pages. This information will be provided on the Course Introduction pages of the learning support subsystem.
3. *Edit Course Topic Page*. Instructors can define course topics, related questions, answers to the questions, descriptions of sub-topics, and rhymed clues on Edit Course Topic pages. As for the clue, an instructor may consider two- or four-line rhymed clue that will direct the student avatars from one station to the next. The clue is ambiguous enough not to know its meaning at one glance, but to stimulate students to find out the meaning.
4. *Build Learning Environment Page*. Build Learning Environment pages enable instructors to define the connections between the information defined on Edit Course Topic pages and the game objects in the treasure hunt game. The definitions are converted into records in XML files that are inputs to the game.
5. *Teaching Forum*. Instructors can share their ideas for teaching with other instructors and propose any improvements on the system for IBL through the teaching forum pages.

7.4.3 Learning Support Subsystem

The learning support subsystem supports the treasure hunt of the student. The learning support subsystem provides the resources of the topics to students and helps them manage the complex, extended activities in the treasure hunt. The learning support subsystem consists of the following seven components.

1. *Learning Guide Page*. Learning guide pages provide helpful information about how to effectively use the system for IBL.
2. *Course Introduction Page*. Course Introduction pages provide course titles, course objectives, and course references.
3. *Course Topic Page*. Course Topic pages furnish the information about the topic, the question, the descriptions of sub-topics, and the clues.
4. *Learning State*. Learning state shows each student's learning progress with the student's current *knowledge state* [1].
5. *Weblog*. Weblog works as the logbook of the treasure hunt model containing the information about the student's important findings in the treasure hunt. Especially, this not only allows students to write what they have learned, but also allows the system to post related information onto the Weblog by using MetaWeblog API described in Section 7.2.2.
6. *Calendar*. Calendar enables students to make an appointment with others to collaborate like Google Calendar described in Section 7.2.2.
7. *Web-Based Information Retrieval Support System*. Web-based information retrieval support system provides students with the resources *RS* mentioned in *INF* of Equation 7.1.

The knowledge state is the subset of learning objects that a student needs to learn for a course [1]. From Equation 7.2, a knowledge state K is the subset of T. Let $\mathscr{K}_S$ be a set of the knowledge states. The learning state LS is defined as a tuple:

$$LS = (SD, t_0, \lambda, LE, f_{LE}), \tag{7.10}$$

where SD is a finite nonempty set of students, t_0 is a course introduction, λ is the sequence of stations, LE is a set of learning experiences defined as, from Equations 7.1 and 7.2, $LE = STA \times \mathscr{K}_S \times T \times 2^{ACT}$, and f_{LE} is an experience interpretation function. If there is a function $f(\lambda) : \mathbb{N} \rightarrow LE$, f_{LE} can be defined as

$$f_{LE} : t_0 \times \prod_{\lambda} f(\lambda) \rightarrow K_n, \tag{7.11}$$

where $K_n \in \mathscr{K}_S$ is the new knowledge state of the student.

7.4.4 Treasure Hunt Game

Treasure hunt game is a two-dimensional online learning game developed with Java. Students can access the treasure hunt game through an authentication process. Basically, students try to find hidden treasures based on quests with given clues, exploring the game world and overcoming challenges. If students acquire enough experience points, they can level up, become more strong and faster, and are allowed to equip with higher-level weapons and armors. Completing the quests, overcoming challenges, and finding treasures give students rewards and experience points. Some equipment, tools, and aids are needed to find the treasures and defeat monsters considered as the *obstacles* in the treasure hunt model. Students can buy these useful things with gold acquired as their reward.

The hidden treasure is the proper answer to the question, represented by a quest, about a topic. The quests are requests that students explore and find the answers by visiting stations and constructing the ideas. Non-player characters (NPC) work as the guides and personify one of the sub-topics of the topic. All the names of NPC and places in the game world may consist of the words relevant to the sub-topics. If a student meets an NPC and the help event occurs, the NPC tells the student the description of a sub-topic of the topic and a clue for the next station. It offers an opportunity to further investigate the given question and learn the topic in-depth. Every finding of the treasure causes the evaluation event, in which students should take a test and give the answer developed through playing the game. The testing scores show students' understanding of the topic. Lack of scores results in a failed quest.

7.4.5 Treasure Hunt Process

Figure 7.2 shows the process of the treasure hunt involving the following four phases:

1. *Presenting Phase.* An NPC in the treasure hunt game provides students with a question about a topic, which they investigate through the treasure hunt, as well as the outlines of the topic in the *orientation*. The learning support subsystem also furnishes the outlines of the topic and the related basic information to help students easily understand the objectives of the topic and the question.
2. *Retrieving Phase.* Once students have the orientation in the presenting phase, they are required to find every station, where an NPC offers a related sub-topic description for solving the question, using the clues in the treasure hunt game. Students need to understand the sub-topic by retrieving its available resource using the learning support subsystem in order to construct an answer to the question. While exploring stations, students may meet and defeat some monsters, and find treasures.
3. *Developing Phase.* Students develop their ideas through the "Construct knowledge" shown in Figure 7.2 by posting their findings and some meaningful

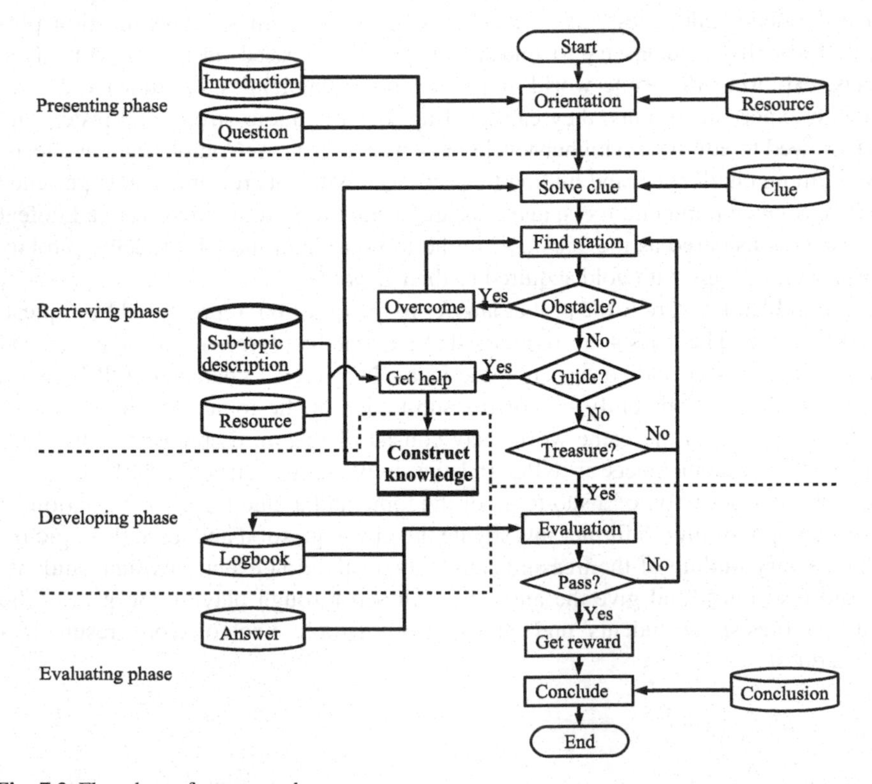

Fig. 7.2 Flowchart of a treasure hunt

achievements onto their *blogs* using the learning support subsystem. The learning support subsystem also automatically writes some messages on the *blogs* whenever students reach milestones in completing quests in the game.

4. *Evaluating Phase*. Students are required to take a test when they find a treasure at an evaluation station. In the *evaluation*, students need to answer the question that they got in the presenting phase, referring to the ideas on their *blogs*. The answer is compared with the sample answer provided by the instructor. Using the similarity function described in Equation 7.7, the system computes marks from how similar the student's answer is to the sample answer, and decides whether the student passes or not.

If IS is the idea structure describing the idea to be posted onto the *blog*, IS can be defined as a tuple:

$$IS = (QS, \lambda, SBJ, CMT, REL), \tag{7.12}$$

where QS is a finite nonempty set of questions from INF of Equation 7.1, λ is the sequence of stations, SBJ is a set of the *subjects*, CMT is a set of *comments*, and REL is a set of *relationships*. The subject is the title of the comment and the relationship represents the relationship between the question and the comment. A *writing idea* function f_{WI} posting the findings of the student onto the *blog* in the developing phase is defined as

$$f_{WI} : QS \times \mathcal{K}_S \times T \times RS \rightarrow IS, \tag{7.13}$$

where RS is a finite set of resources from INF of Equation 7.1. The function f_{WI} specifies an idea developed from the given learning object and related resource with respect to the current knowledge state of the student for a question.

7.5 A Demonstrative Example of the System

Suppose an instructor prepares for a lecture about the *data link layer* in a Data Communications and Network course using the teaching support subsystem. The topic must be the data link layer (DLL). According to the instructor's lecture notes, the instructor may have the following sub-topics: DLL introduction, DLL duties, error detection, flow control, parity checking, and so on.

Figures 7.3 and 7.4 show examples of Edit Course Topic pages: where to input the information about the topic, the question, the descriptions of sub-topics, and the clues. The instructor enters the topic title, the introduction, and the question for the orientation. As for the evaluation, the instructor provides a sample answer to the question presented on the orientation page, and a conclusion for the lecture. The instructor also defines the description of the sub-topic related to the DLL on the sub-topic page. Defining the description of sub-topic "DLL duties," and the clue to reach this sub-topic, is shown in Figure 7.4. Other sub-topic definitions can be performed in a similar manner.

Edit Course Topic - Orientation
Course: CS335 Data Communications and Networks
Topic Title
Data Link Layer
Introduction
The data link layer supervises the flow of information between adjacent network nodes. It uses error detection or correction techniques to ensure that a transmission contains no errors.
Question
Find what DLL is, and describe what DLL can do for networking in detail.
Save Cancel

Edit Course Topic - Evaluation
Course: CS335 Data Communications and Networks
Topic: Data Link Layer
Question
Find what DLL is, and describe what DLL can do for networking in detail.
Sample answer
DLL stands for data link layer which is layer two of the seven-layer OSI model as well as of the five-layer TCP/IP reference model. It receives services from physical layer and provides services to network layer. DLL is responsible for carrying a packet from one node to another and has only a local responsibility...
Conclusion
You understand the data link layer and its functions. The next topic can be the network layer.
Save Cancel

Fig. 7.3 Examples of edit course topic pages for orientation and evaluation

Edit Course Topic - Sub-Topic
Course: CS335 Data Communications and Networks
Topic: Data Link Layer
Sub-Topic Title
DLL duties
Description
The duties of the data link layer include packetizing and framing, addressing error control, and flow control.
Clue
A boy is working in a town hall.
Save Cancel

Edit Course Topic - Sub-Topic
Course: CS335 Data Communications and Networks
Topic: Data Link Layer
Sub-Topic Title
Error detection
Description
The data link layer performs the error check using the Frame Check Sequence (FCS) in the trailer and discards the frame if an error is detected. The corrupted message frame may need to be retransmitted under control of an upper layer.
Clue
A woman is waiting for you in a city.
Save Cancel

Fig. 7.4 Examples of edit course topic pages for sub-topic

Once the instructor finishes making up a topic on the Edit Course Topic pages, the instructor needs to build a learning environment for the topic in the treasure hunt game using the Build Learning Environment pages, where the connections between the information defined on the Edit Course Topic pages and the game objects in the treasure hunt game are defined. As shown in Figure 7.5, for the topic "data link layer," the instructor assigns a game map "Town Hall" as the orientation station and non-player character (NPC) "Mayor" as the guide, in accordance with the object allocation function defined in Equation 7.4. The instructor also modifies the introduction of the topic to simulate a conversation. For the evaluation station, the instructor chooses a treasure box, "Treasure Box 12," in "Network Mountain" and sets 1,000 gold and 100 game experience points as the rewards for the quest. Meanwhile, according to the building definition for "DLL Duties" shown in Figure 7.6, NPC "boy" in "Town Hall" will deliver the description of sub-topic "DLL duties" to students.

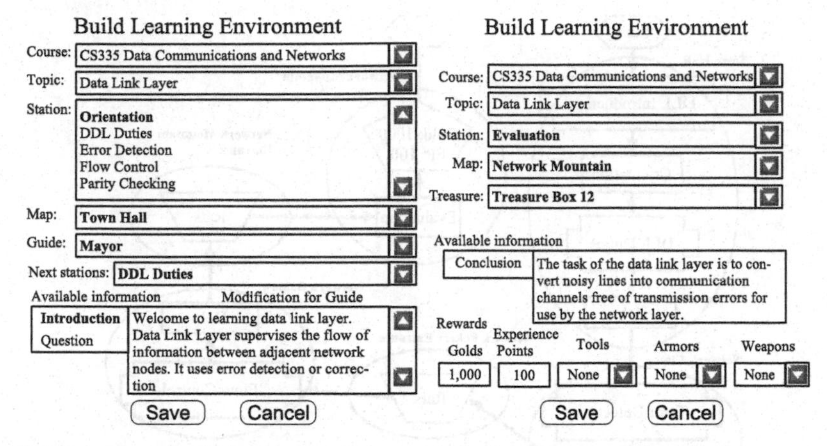

Fig. 7.5 Examples of build learning environment pages for orientation and evaluation

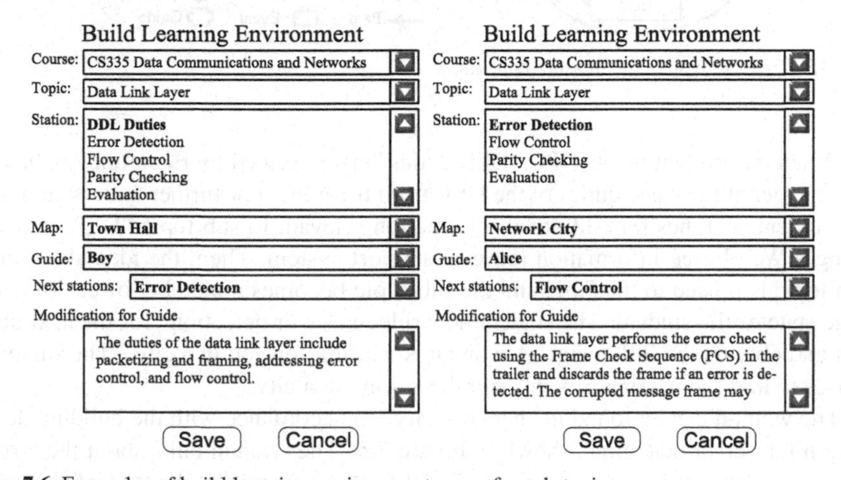

Fig. 7.6 Examples of build learning environment pages for sub-topic

Figure 7.7 shows a possible path to the treasure. In the game, the name of the learning object, the information of which is provided by an NPC, is displayed as the name of the NPC. From the orientation function defined in Equation 7.5, a student receives the quest about the DLL from NPC "DLL introduction" with the background information and the question. The student posts what he or she has understood with the orientation information to the *blog* as the writing idea function described in Equation 7.13. Then, the introduction described in Figure 7.3 is added to the knowledge state of the student according to the experience interpretation function expressed in Equation 7.11. A list of next learnable objects is displayed when the student clicks the "Next Stations" button. For this case, there is only one sub-topic available, "DLL duties," as the building definition for the orientation shown in Figure 7.5. When the student selects the object on the list, NPC "DLL introduction" tells the student the clue: "A boy is working in a town hall." The student finds the boy called "DLL duties" in the town hall.

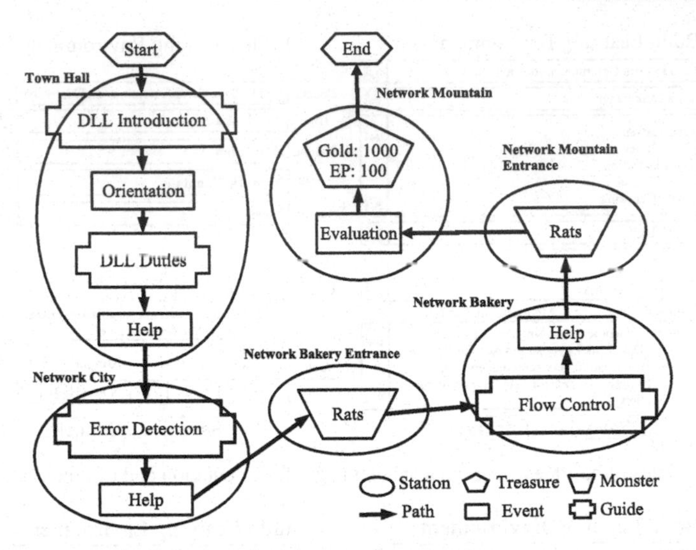

Fig. 7.7 Treasure map in the treasure hunt game

When the student meets NPC "DLL duties," as described by Equation 7.6, he or she may learn the basic duties of the DLL from the NPC. For further understanding, the student searches for additional information relevant to sub-topic "DLL duties" using a Web-based information retrieval support system. Then, the idea about the sub-topic is posted to the *blog*, and the sub-topic becomes a member of the knowledge state of the student. The student now selects "error detection" for the next object and obtains the following: "A woman is waiting for you in a city." The student moves to find the woman called "error detection" in a city.

The woman can be found in "network city" in accordance with the building definition for "error detection" shown in Figure 7.6. The woman talks about the error detection of the DLL and shows a clue for NPC "flow control" if the student chooses sub-topic "flow control" as the next object to learn. The student meets rats when he or she tries to enter "network bakery" on the way to find the NPC. The student is required to defeat them in order to enter this place and to check if NPC "Flow Control" is there. According to the obstacle function defined in Equation 7.9, some experience points are obtained as the student expels the rats using his or her weapons.

After overcoming all the challenges, the student finds a treasure box in "network mountain" in the end. When the student opens it, he or she is asked to take a test to complete this quest in accordance with the evaluation function described in Equation 7.8. For the treasure, as depicted in Figure 7.3, the student can get 1,000 gold and 100 game experience points if he or she passes. Enough experience points enable the student to "level up." The gold the student received will be spent on arms to help him or her defeat the challenges he or she will face in the next round of the treasure hunt. The game level of the student is one of the measures indicating how well he or she has performed learning.

Weblog

Course: CS335 Data Communications and Networks
Topic: Data Link Layer

Question

What is DLL? Describe what DLL can do for networking in detail.

Error Detection – Sub topic by userid

DDL has abilities to detect when a transmission has been changed, and correct the error without a second transmission.

•••

Data Link Layer – Introduction by userid

The data link layer is layer two not only of the five-layer TCP/IP reference model, but also of the seven-layer OSI model.

Data Link Layer – Introduction by System

I'm DLL Introduction. I receive services from physical layer and I provide services to network layer.

Fig. 7.8 Examples of blog posts

As previously mentioned, during the journey to the treasure hunt, the student is required to write his or her idea about each sub-topic in the *blog*, as shown in Figure 7.8. Each post on the *blog* forms the idea structure defined in Equation 7.12. For example, in the last post of Figure 7.8, "error detection" is the subject and "Sub-topic" is the relationship in the *idea structure*. The post also contains the comment below the subject part. The question is shown on top of the *blog*. This mechanism helps the student construct ideas about the question and gives the proper answer when the student finds the treasure and takes the test.

7.6 Conclusion

Web-based learning support systems support the activities of the instructors and students in education using Web technology. The Web offers global accessibility and student-friendly interfaces. Furthermore, the Web is capable of delivering various forms of learning content such as text, audio, graphics, animation, and video. Web-based learning support systems can be a good framework for implementing educational methods.

Inquiry-based learning is a student-centered educational method in which students investigate new knowledge, develop ideas, and learn topics through a series of inquiry activities. A treasure hunt model contains the learning strategies of inquiry-based learning and is applied to a Web-based learning support system that takes advantage of the Web and online game technologies in implementing the inquiry-based learning environments. *Online Treasure Hunt* is the Web-based learning support system designed with the treasure hunt model.

The Web-based learning support system consists of three major components: the teaching support subsystem, learning support subsystem, and treasure hunt game. The teaching support subsystem makes it possible for instructors to design and build their own learning environments for their students. Learning support subsystem helps students manage their learning activities and provides the resources. The treasure hunt game encourages students to investigate topics, develop their ideas, and test their knowledge through playing the game.

Online Treasure Hunt can be an effective Web-based learning support system. The system not only enables instructors to bring the ideals of inquiry-based learning to their teachings, but also promotes students' active participation in learning with the intrinsic motivation for learning.

References

1. D. Albert and C. Hockemeyer, Adaptive and dynamic hypertext tutoring systems based on knowledge space theory, In: B. du Boulay and R. Mizoguchi (eds.) Artificial Intelligence in Education: Knowledge and Media in Learning Systems, 553-555, IOS Press, Amsterdam, 1997.
2. Blackboard Inc., http://www.blackboard.com/. Accessed 21 April 2008.
3. N. D. Blas, P. Paolini and C. Poggi, Learning by Playing An Edutainment 3D Environment for Schools, World Conference on Educational Multimedia, Hypermedia and Telecommunications, Lugano, Switzerland, 1313-1320, 2004.
4. Y. Y. Chan, Teaching Queueing Theory with an Inquiry-based Learning Approach: A Case for Applying WebQuest in a Course in Simulation and Statistical Analysis, 37th ASEE/IEEE Frontiers in Education Conference, F3C-(1-6), 2007.
5. M. D. Childress and R. Braswell, Using Massively Multiplayer Online Role-Playing Games for Online Learning, Distance Education, 27(2):187-196, 2006.
6. O. Dziabenko, M. Pivec, C. Bouras, V. Igglesis, V. Kapoulas and I. Misedakis, A Web-Based Game For Supporting Game-Based Learning, GAME-ON 2003, London, England, 111-118, 2003.
7. D. C. Edelson, Matching the Design of Activities to the Affordances of Software to Support Inquiry-Based Learning, Proceedings of International Conference on the Learning, 77-83, 1998.
8. D. C. Edelson, D. Gordin and R. Pea, Addressing the Challenges of Inquiry-Based Learning through Technology and Curriculum Design, Journal of the Learning Sciences, 8:391-450, 1999.
9. L. Fan and Y. Y. Yao, Web-based Learning Support Systems, WI/IAT 2003 Workshop on Applications, Products and Services of Web-based Support Systems, Halifax, Canada, 43-48, 2003.
10. S. Fleissner, Y. Y. Chan, T. H. Yuen, and V. Ng, WebQuest Markup Language (WQML) for Sharable Inquiry-Based Learning, LNCS, 3980:383-392, 2006.
11. Global Goonzu, http://global.goonzu.com/. Accessed 21 April 2008.
12. Google Calendar, http://code.google.com/apis/calendar/. Accessed 21 April 2008.
13. G. Katsionis and M. Virvou, Personalised e-learning through an educational virtual reality game using Web services, Multimedia Tools and Applications, 39(1):47-71, 2008.
14. J. S. Kim, The Effects of a Constructivist Teaching Approach on Student Academic Achievement, Self-concept, and Learning Strategies, Asia Pacific Education Review, 6(1):7-19, 2005.
15. R. Klamma, Y. Cao and M. Spaniol, Watching the Blogosphere: Knowledge Sharing in the Web 2.0, International Conference on Weblogs and Social Media, 105-112, 2007.

16. B. R. Lim, Challenges and issues in designing inquiry on the Web, British Journal of Educational Technology, 35(5):627-643, 2004.
17. D. Llewellyn, Teaching High School Science Through Inquiry A Case Study Approach, Corwin Press, 2004.
18. J. Mechling, Patois and Paradox in a Boy Scout Treasure Hunt, The Journal of American Folklore, 97(383):24-42, 1984.
19. L. Oishi, Working together: Google Apps goes to school.(Emerging Tech), Technology & Learning, 27(9):46-47, 2007.
20. M. Prensky, Digital Game-Based Learning, ACM Computers in Entertainment, 1(1):21-21, 2003.
21. B. Russell, An Inquiry into Meaning and Truth, Penguin Books, Harmondsworth, Middlesex, UK, 1962.
22. Second Life: Official site of the 3D online virtual world, http://www.secondlife.com/. Accessed 21 April 2008.
23. L. M. Skelly, Is Your Student Project Better Than a Blackline Master?, MACUL Conference 2006, Grand Rapids, MI, USA, 2006.
24. C. A. Steinkuehler, Learning in Massively Multiplayer Online Games, Proceedings of the 6th international conference on Learning sciences, 521-528, 2004.
25. Swiss Virtual Campus, http://virtualcampus.ch/. Accessed 21 April 2008.
26. WebQuest.Org, http://www.webquest.org/. Accessed 21 April 2008.
27. K.T. Wang, Y.M. Huang, Y.L. Jeng, T.I. Wang, A blog-based dynamic learning map, Computers & Education, 51(1):262-278, 2008.
28. J. T. Yao and Y. Y. Yao, Web-based information retrieval support systems: building research tools for scientists in the new information age, Proceedings of the IEEE/WIC International Conference on Web Intelligence, Halifax, Canada, 570-573, 2003.
29. J. T. Yao and Y. Y. Yao, Web-based support systems, Proceedings of 2003 WI/IAT Workshop on Applications, Products and Services of Web-based Support System, 1-5, 2003.
30. J. T. Yao, Supporting Research with Weblogs: A Study on Web-Based Research Support Systems, Proceedings of the 3rd International Workshop on Web-Based Support Systems, 161-164, 2006.
31. J. T. Yao, D. W. Kim and J. P. Herbert, Supporting Online Learning with Games, Proceedings of SPIE Vol. 6570, Data Mining, Intrusion Detection, Information Assurance, and Data Networks Security 2007, Orlando, Florida, USA, 65700G-(1-11), 2007.

Part II
Web-Based Applications and WSS Techniques

Chapter 8
Combinatorial Fusion Analysis for Meta Search Information Retrieval

D. Frank Hsu and Isak Taksa

Abstract Leading commercial search engines are built as single event systems. In response to a particular search query, the search engine returns a single list of ranked search results. To find more relevant results the user must frequently try several other search engines. A meta search engine was developed to enhance the process of multi-engine querying. The meta search engine queries several engines at the same time and fuses individual engine results into a single search results list. The fusion of multiple search results has been shown (mostly experimentally) to be highly effective. However, the question of why and how the fusion should be done still remains largely unanswered. In this chapter, we utilize the combinatorial fusion analysis proposed by Hsu et al. to analyze combination and fusion of multiple sources of information. A rank/score function is used in the design and analysis of our framework. The framework provides a better understanding of the fusion phenomenon in information retrieval. For example, to improve the performance of the combined multiple scoring systems, it is necessary that each of the individual scoring systems has relatively high performance and the individual scoring systems are diverse. Additionally, we illustrate various applications of the framework using two examples from the information retrieval domain.

8.1 Introduction

Yao [35] defines Web-based support systems (WSS) as a multidisciplinary area of research that focuses on assisting the user involved in various Web activities. Whether shopping [40], learning [9], or searching [32], WSS is an indispensable

D.F. Hsu
Department of Computer and Information Sciences, Fordham University, Bronx, NY 10458
e-mail: hsu@cis.fordham.edu

I. Taksa
Department of Computer Information Systems, Baruch College, New York, NY 10010
e-mail: Isak.Taksa@baruch.cuny.edu

J.T. Yao (ed.), *Web-Based Support Systems*, Advanced Information and Knowledge Processing, DOI 10.1007/978-1-84882-628-1_8,

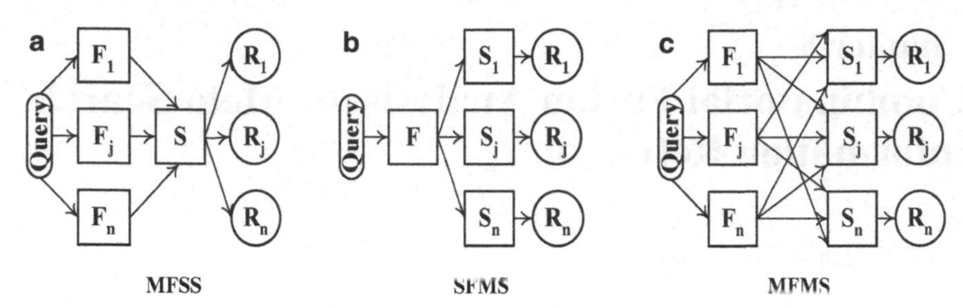

Fig. 8.1 Multiple formulations and multiple schemes. **(a)** MFSS – multiple formulations, single retrieval scheme **(b)** SFMS – single formulation, multiple retrieval schemes **(c)** MMFS – multiple formulations, multiple retrieval schemes

element of a successful and satisfactory user experience. A frequent topic in WSS research is the improvement of the information retrieval process. Yao [37] introduced the concept of Web-based Information retrieval support systems (WIRSS) and a framework for research. Later on, Yao and Yao [36] used this framework to conduct an extensive case study of WIRSS. This chapter concentrates on a specific issue of WIRSS – information fusion.

The advent of computer science and information technology has enabled us to improve information retrieval system performance. Instead of the typical *single query formulation – single retrieval scheme* approach, the user is assisted by a WSS that allows for *single or multiple formulations–single or multiple schemes*. Meta search engines that implement the latter approach assist the user with several basic functions: query formulation F_i (single or multiple) and subsequent submission of all query formulations to a single or multiple search engines S_i. The third, and most important, issue is the fusion (without redundancy) of all search results R_i into a single rank list of search results. Figure 8.1 depicts the three distinct combinations of query formulation and retrieval schemes (see [12]).

Shapiro and Taksa [29] investigated an MFSS scheme using a long-query meta search engine that accepts an original unlimited size query and replaces it with a number of shorter queries. Shorter queries are algorithmically formulated and submitted to a single search engine and search results of all submissions are combined to create a unified search result. To demonstrate an SFMS scheme, Diaz and Raghavan [6] used a meta search engine where a single query was submitted to a dynamically selected group of search engines and all individual search results were subsequently combined into a single search result. In a recent study of the MFMS scheme, Keyhanipoor et al. [17] examined a combination of individual search results based on users' preferences. In this research, a WebFusion meta search engine is used to collect users' implicit feedback to rank search results in the final joined result.

In the earlier research into combining multiple evidences in information retrieval, Belkin et al. [2] and Fox and Shaw [10] investigated the effect of combination of multiple representations of TREC topics on retrieval performance. Both projects found that the best method of combination often led to results that were better than the best performing single query. However, they indicated that choosing the best query often results in significant performance differences from combined queries. They also pointed out that in any single run there are always instances of combined queries performing better than the best; on average, combination does better. Belkin et al. [2] reported on two studies conducted at Rutgers University [3] and at Virginia Tech [10] that investigated the effect on retrieval performance of a combination of multiple representations of TREC-2 topics. When performing query combination, the rules used (CombSUM, CombANZ, and CombMNZ) were based on similarity scores between a topic and a document. On the other hand, when considering multiple evidences from different schemes (or systems), combinations (MAX, MIN, and MED) were based on rank information. Encouraged by the interesting and generally positive results of the two separate studies involving combination of evidences (using similarity scores) or data fusion (using rank information), Belkin et al. [3] performed two other experiments and observed conditions impacting the results of various combination methods (see Remark 1).

Examining the foundations for evidence combination, Lee [20] demonstrated that different runs return similar sets of relevant documents but retrieve different sets of nonrelevant documents. He also investigated the effect of using ranks instead of using similarity on retrieval effectiveness. In particular, he showed experimentally that in some circumstances, using ranks works better than using similarity score. He also investigated the effect of using ranks instead of similarity score on retrieval effectiveness and determined conditions using ranks works better than using similarity scores (see Remark 2).

In developing various combination schemes, researchers attempted to solve a problem of predicting, in advance, whether combination (or fusion) of two or more retrieval schemes will be worth doing. In their earlier research, Ng and Kantor [26] identified two variables that predict effectiveness of the combination method (see Remark 3). In a subsequent study, Ng and Kantor [27] investigated the prediction power of these two variables using symmetrical data fusion and the receiver operating characteristic (ROC) curve. Using precision at the 100th document, $P_{@100}$, to represent efficacy similarity, they usc, ratio P_l/P_h (P_l and P_h are $P_{@100}$ for the lower and higher performance schemes, respectively) as a variable to measure the similarity of performance of the two IR schemes. Although they found that most of the positive cases (the combination performs better than or equal to each individual case) have a ratio of precision P_l/P_h close to 1, they also stated that the two predictive variables do not completely determine whether simple (linear) and symmetric data fusion will be effective (see Remark 4).

Previous empirical and experimental results (including those reviewed in this section) have achieved certain statistical success in understanding the effectiveness of data fusion (with multiple formulations of queries, or multiple schemes, or in different runs) in information retrieval. However, the general questions of "why" and "how" data fusion in information retrieval can be effective still remain unanswered. All these indicate that the problem involves tremendously high complexity and dimensionality. They have become both quantitatively and qualitatively difficult to trace. In an information retrieval system, different schemes (systems or engines) can use different techniques (or algorithms) to measure the likelihood or probability of relevance of a document to a given query. Moreover, the choices of techniques (or algorithms) rely heavily on the application domain they are applied to or used in. This situation is complicated by having a variety of multiple formulations of the information need and a large and multifaceted collection of documents (see Figure 8.1(a–c)). Multiple representations (or query formulations) can occur either as a result of the interpretation of the original need by multiple experts or as disjoint or non-disjoint subsets from the partition of the original query (such as a long query). In both cases, they also involve semantic consideration. On the other hand, the document space consists of not only large and different structured database systems but also a variety of sites (such as the World Wide Web) located in different networks and different countries.

Hsu and Taksa [12] continue the study of the problem of fusing data and information in an information retrieval domain using similarity measures to search for proper (relevant) documents in the databases or on the World Wide Web when presented with an information need (a query). They include the general MFMS setting (see Figure 8.1(c)), considering the case of combining results of search in the same database or search space. They established the predictive power of the graphical behaviors of the combination function (see Remark 5).

Remarks: (Hsu and Taksa [12])

1. (a) When different systems are commensurable, combination using similarity scores is better than combination using only ranks; (b) when multiple systems have incompatible scores, a combination method based on ranked outputs rather than the scores directly is the proper method for combination; and (c) although results from the experiments for combination of results from different databases are encouraging, it is not clear that such combination is possible among systems that have different methods for computing similarity scores.
2. Data fusion using rank works better than using similarity scores if the runs in the combination have "different" rank-similarity curves.
3. Two predictive variables for the effectiveness of the combination: (a) z – a list-based measure of output dissimilarity, and (b) r – a pair-wise measure of the similarity of the performance of the two schemes.
4. The results of the NK study show that (a) the positive cases tend to lie above the diagonal line $r+z=1$, and (b) the negative cases (the combination is not positive) are more likely to scatter around the diagonal line $r+z=1$.
5. The performance of rank combination is at least as good as that of score combination under certain conditions. Graphical behaviors of the rank/score function of each individual system might be a feasible predictive variable for the effectiveness of combination.

In this introductory section of the chapter, we discussed the data fusion (DF) research in the information retrieval domain. The concept of DF has been used to study the combination of multiple evidences resulting from different query formulations or from different schemes [2,3,10,15,20,26–28]. Many empirical studies have been performed and various results have been obtained. While some of the major issues related to questions such as why and how multiple evidences should be combined remain unanswered, researchers have come to realize the advantage and benefit of combining multiple evidences. Data fusion in its broader context could also be viewed as a process (acquisition, design, and interpretation) of combining information gathered by multiple agents (sources, schemes, sensors, or systems) into a single representation (or result). As such, data fusion has been studied in various application domains such as pattern recognition [33], social choice functions [1, 38, 39], virtual screening and drug discovery [34], target tracking and surveillance [22, 30]. Recent research in data fusion is focused on dynamic online learning algorithms for efficient DF [25], on diversity (dissimilarity between scoring systems) [4, 5, 18], and on use of support vector machines and neural networks for protein structure prediction [7, 14, 21].

In Section 8.2 of this chapter, we review an information fusion approach designed and developed by Hsu et al. [13] called combinatorial fusion analysis (CFA). The concept of multiple scoring systems is introduced, and a variety of ways of combinations is discussed. The concept of diversity between two scoring systems is defined and rank/score function is computed as a composite function of the reverse rank function followed by the score function. Section 8.3 demonstrates two examples of information fusion in the information retrieval domain. They are (a) prediction of fusion results and (b) comparison of rank and score combination method. Section 8.4 concludes the chapter.

8.2 Combinatorial Fusion Analysis

8.2.1 Multiple Scoring Systems

In order to turn raw data into useful information and then into valuable knowledge successfully, it requires the application of scientific methods to the study of storage, retrieval, extraction, transmission, diffusion, fusion/combination, manipulation, and interpretation at each stage of the process. Most scientific problems are multifaceted and can be quantified in a variety of ways. Among many methodologies and approaches to solve complex scientific problems and deal with large datasets are (a) classification, (b) artificial neural nets, (c) clustering, (d) association rule, (e) statistical and stochastic rules, and (f) similarity measurement. Hybrid methods combining these methods have been used.

Large data sets can be collected from the environment of multi-sensor devices, multisource systems, or multi-information inputs; or generated by experiments, surveys, recognition, and judging systems. These are stored in a data grid $G(n, m, q)$ with $n$objects in $D = \{d_1, d_2, \ldots, d_n\}$, m features/attributes/indicators/cues in $G = \{a_1, a_2, \ldots, a_m\}$, and, possibly, q temporal traces in $T = \{t_1, t_2, \ldots, t_q\}$. We call this three-dimensional grid the data space.

Because both m and q can be very big and the size of the datasets may limit the utility of single informatics approaches, it is not a good idea to use/design a single method/system because of the following reasons (see Hsu et al. [13] for detailed description):

1. Different methods/systems are appropriate for different features/attributes/indicators/cues and different temporal traces.
2. Different features/attributes/indicators/cues may use different kinds of measurements.
3. Different methods/systems may be good for the same problem with different data sets generated from different information sources/experiments.
4. Different methods/systems may be good for the same problem with the same data sets generated or collected from different devices/sources.

The first item indicates that performance of a single system/method may be most advantageous for some features/attributes/indicators/cues, but may diminish its performance for other features/attributes/indicators/cues. The second item indicates the need to normalize the measurement employed in each a_i across all m features, attributes, indicators, and cues. The last two items indicate that each single system/method, when applied to the problem, can be improved in performance to some extent, but due to the complexity of the problem involved it is extremely difficult, if not impossible, to achieve perfect results.

Every system/method has the ability to study different classes of outcomes, e.g., class assignment in a classification problem or similarity score assignment in the similarity measurement problem. In CFA, we treat the outcome of each system/method as a scoring system A which assigns (a) an object as a class among all objects in D, (b) a score to each object in D, and (c) a rank number to each object in D. These three outcomes were described as the abstract, score, and rank level, respectively, by Xu et al. [33] . Now we are ready to construct the system grid $H(n, p, q)$ with n objects in $D = \{d_1, d_2, \ldots, d_n\}$, p systems in $H = \{A_1, A_2, \ldots, A_p\}$, and possibly q temporal traces in $T = \{t_1, t_2, \ldots, t_q\}$. This three dimensional grid is called the system space for the multiple scoring systems.

In the next section, the rank/score function for each scoring system A is defined using score function and rank function. In Section 8.2.3 rank and score combination are defined and studied.

8.2.2 Rank/Score Function and the Rank-Score Characteristics (RSC) Graph

Let $D = \{d_1, d_2, ..., d_n\}$ be a set of n objects and $N = \{1, 2, 3, \ldots, n\}$ be the set of all positive integers less than or equal to n. Let $\mathbf{R}$ be the set of real numbers. We now describe the rank/score function and the rank-score characteristics graph that were previously defined and studied by Hsu and Taksa [12], Hsu et al. [13], and Lyons and Hsu [22].

Definition 1. Let s be a score function from D to $\mathbf{R}$ in which the function s assigns a score (a real number) to each object in D (we write s: $D \rightarrow \mathbf{R}$ such that for every d_i in D, there exists a real number $s(d_i)$ in $\mathbf{R}$ corresponding to d_i), and r be a rank function from D to N such that the function r: $D \rightarrow N$ maps each element d_i in D to a natural number $r(di)$ in N (the number $r(d_i)$ stands for the rank number $r(d_i)$ assigned to the object d_i). We have:

(a) **Rank/Score Function:** Given r and s as rank and score function on the set D of objects respectively, the rank/score function f is defined to be f: $N \rightarrow \mathbf{R}$ such that $f(i) = (s \circ r^{-1}(i)) = s(r^{-1}(i))$. In other words, the score function $s = f \circ r$ is the composite function of the rank/score function and the rank function.
(b) **Rank-Score Characteristics (RSC) Graph:** The graph representation of a rank/score function f on a two-dimensional coordinate system.

We note that in several application domains, one has to normalize the score function values before any combination can be performed. Hence, it is quite natural to define the two functions s and f in a way that each of them has $[0, 1] = \{x \mid x \text{ in } \mathbf{R}, 0 \leq x \leq 1\}$ instead of $\mathbf{R}$ as their function range (see Figure 8.2).

Other intervals of real numbers can also be used depending on the situation and environment. We also note that since the rank function r' defined by Hsu and Taksa [12] is the inverse of r defined above, the rank/score function f would be such that $f = s \circ r'$ instead of $f = s \circ r^{-1}$.

In theory and practice, the Score function depends more on the variate data in the parametric domain and the scores are more popularly used in sciences, engineering, finance, and business. The rank function depends more on the ordinal data in the nonparametric fashion and is more used in social choices, ordinal data analysis, and decision science. The comparison continues of score vs. rank, data level vs. decision level, variate data vs. ordinal data, and parametric vs. nonparametric. However, in biomedical informatics, since the data collected are numerous (and of multiple dimensions) and the information we are seeking from biological and physiological

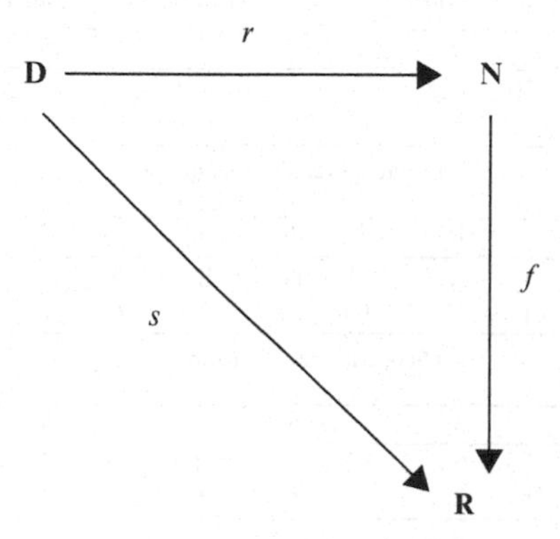

Fig. 8.2 Functions s, r, and f

systems is complex (and multivariable), the information we find (or strive to find) from the rank/score function and hence the RSC graph would become valuable in biological, physiological, and pharmaceutical studies.

The concept of a RSC graph that depicts the graph representation of a rank-score function has the following properties (see Hsu et al. [13]):

Remark 6

(a) The rank/score function f is easy to compute. When a score function s_A is assigned resulting from scoring system A by either laboratory work or field study (conducted in vivo or in vitro), treating s_A as an array and sorting the array of scores into descending order would give rise to the rank function r_A. The rank/score function can be obtained accordingly as the composite function of the inverse function of r_A and of s_A. (If there are n objects, this transformation takes $O(n\log n)$ steps.

(b) The rank/score function f is independent of the objects in D. Since the rank/score function f is defined from N to $\mathbf{R}$ or from N to [0, 1], it does not depend on the set of objects $D = \{d_1, d_2, \ldots, d_n\}$. Hence the rank/score function f_A of a scoring system A exhibits the behavior of the scoring system (scorer or ranker) A and is independent of who (or which object) has what rank or what score. The rank/score function f also fills the gap in the relationship between the three sets D, N, and $\mathbf{R}$ (see Table 8.1).

(c) The RSC graph can be easily and clearly visualized. From the graph of f_A, it is readily concluded that the function f_A is a nonincreasing monotonic function on N. This easy-to-visualize property enables us to compare two functions f_A and f_B by drawing the graph of f_A and f_B on the same coordinate system (see Figure 8.3).

Two examples of score functions, rank functions (derived from score function), rank/score functions with respect to scoring systems A and B are illustrated in

Table 8.1 Score function, rank function, rank/score function of A and B

D	d_1	d_2	d_3	d_4	d_5	d_6	d_7	d_8	d_9	d_{10}
$s_A(di)$	4	10	4.2	3	6.4	6.2	2	7	0	1
$r_A(di)$	6	1	5	7	3	4	8	2	10	9

(a) Score and rank function for A

D	d_1	d_2	d_3	d_4	d_5	d_6	d_7	d_8	d_9	d_{10}
$s_B(di)$	4	7	3	1	10	8	5	6	9	2
$r_B(di)$	7	4	8	10	1	3	6	5	2	9

(b) Score and rank function for B

N	1	2	3	4	5	6	7	8	9	10
$f_A(di)$	10	7	6.4	6.2	4.2	4	3	2	1	0
$f_A(di)$	10	9	8	7	6	5	4	3	2	1

(c) Rank/score function of A and B

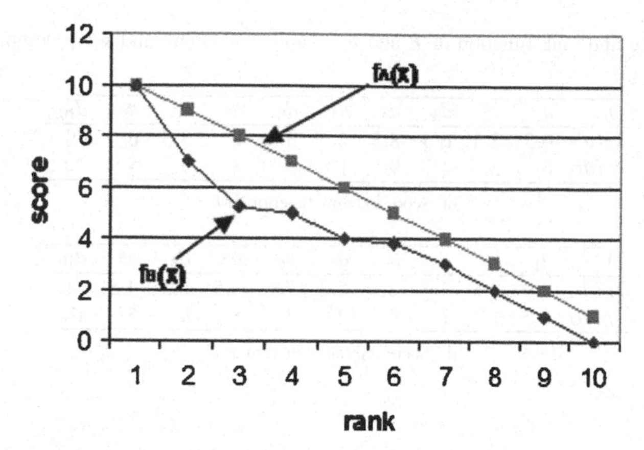

Fig. 8.3 RSC graph of f_A and f_B

Table 8.1 (see Hsu and Taksa [12]) and the RSC graphs of f_A and f_B are shown in Figure 8.3, where $D = \{d_i \mid i = 1 \text{ to } 10\}$ and $s(d_i)$ is in [0, 10].

8.2.3 Rank and Score Combination

As mentioned earlier in Section 8.1, the combination (or fusion) of multiple classifiers, multiple evidences, or multiple scoring functions in various domains has gained tremendous momentum in the past decade. In this section, we deal with combinations of two functions with respect to both score and rank combinations. The following definitions were used by Hsu and Taksa [12] and Hsu et al. [13]:

Definition 2. Let s_A, s_B, r_A, r_B be the score function and rank function as in Definition 1 for scoring systems A and B respectively. We have

(a) **Score combinations:** Given two score functions s_A and s_B, the score function of the score combined function s_F is defined as $s_F(d) = \frac{1}{2}[\mathrm{s}_A(\mathrm{d}) + \mathrm{s}_B(\mathrm{d})]$ for every object d in D.

(b) **Rank combinations:** Given two rank functions r_A and r_B, the score function of the rank combined function s_E is defined as $s_E(d) = \frac{1}{2}[\mathrm{r}_A(\mathrm{d}) + \mathrm{r}_B(\mathrm{d})]$ for every object d in D.

Note that both scoring systems A and B have their score and rank function s_A, r_A and s_B, r_B, respectively. Subsequently, each of the combined scoring systems E and F (by rank and by score combination) has s_E, r_E and s_F, r_F, respectively. Using the rank combination (as defined in Definition 2b) to obtain the score function r_E of the rank combination from r_A and r_B and then sorting s_E into ascending order gives rise to r_E. Correspondingly, using the score combination (as defined

Table 8.2 Score and rank function of E and F (E and F are rank and score combinations of A and B)

D	d_1	d_2	d_3	d_4	d_5	d_6	d_7	d_8	d_9	d_{10}
$s_E(di)$	6.5	2.5	6.5	8.5	2	3.5	7	3.5	6	9
$r_E(di)$	6	2	4	9	1	3	8	7	5	10

(a) Score and rank function of E

D	d_1	d_2	d_3	d_4	d_5	d_6	d_7	d_8	d_9	d_{10}
$s_F(di)$	4.0	8.5	3.6	2.0	8.2	7.1	3.5	6.5	4.5	1.5
$r_F(di)$	6	1	7	9	2	3	8	4	5	10

(b) Score and rank function of F

in Definition 2a) we obtain the score combination s_F from s_A and s_B. Subsequent sorting s_F into descending order gives rise to r_F. Hence, both scoring systems E and F wind up having the score function and rank function s_E, r_E and s_F, r_F respectively.

The readers are reminded of the difference in the conversion process from s_E to r_E and that from s_F to r_F. The conversion from s_E to r_E is by sorting into ascending order while that from s_F to r_F is by sorting into descending order. The difference is due to the fact that both scoring systems E and F are rank and score combination of A and B respectively. We use the framework for the fusion method established by the authors in earlier research [12]. Table 8.2(a) and (b) show the rank and score function of the rank and score combination E and F(both related to the scoring systems $A$and B in Table 8.1) respectively. We use a simple average combination of A and B to obtain the combination (rank or score). Use of weighted proportion on A and B to obtain the combination is studied in Hsu and Palumbo [11].

8.2.4 Performance Evaluation

The chief rationale for data fusion is for the fused scoring system to outperform each of the individual scoring system. To be able to say "outperform" in a meaningful way, undoubtedly depends on the concept of "performance" clearly defined for a given study (or a class of studies).

In information retrieval, a query Q is given. Each of the ranking algorithms calculates similarity between the query and the documents in the database of n documents $D = \{d_1, d_2, \ldots, d_n\}$. A score function s_A for algorithm A is assigned and a rank function r_A is obtained. The performance of the algorithm for the query, written $P_q(A)$, is defined as the precision of A at q with respect to the query Q. More specifically, the following is defined and widely used in the information retrieval community (see Hsu and Taksa [12] and Hsu et al. [13]).

Definition 3.

(a) Let *Rel(Q)* be the set of all documents that are judged to be relevant with respect to the query Q. Let $|Rel(Q)| = q$ for some q, $0 \leq q \leq n$. On the other hand, let $A_{(k)} = \{d | d \text{ in } D \text{ and } r_A(d) \leq k\}$.

(b) **Precision of** A **at** q. The performance of the scoring system A with respect to the query Q is defined to be $P_q(A) = |Rel(Q) \cap A_{(q)}|/q$, where $q = |Rel(Q)|$.

Now we come back to the fundamental issue of when the combined scoring system E or F outperforms its individual scoring system A and B. The following two central questions regarding combination and fusion were asked by Hsu and Taksa [12]:

Remark 7

Let E and F be the rank combination and score combination of A and B, respectively. We have

(a) For what scoring systems A and B and with what combination or fusion algorithm, $P(E)$ (or $P(F)$) $\geq \max\{P(A), P(B)\}$, and

(b) For what A and B, $P(E) \geq P(F)$?

Four important issues, as described in Hsu et al. [13], are central to CFA: (1) Which is the best fusion algorithm/method to use? (2) Does the performance of E or F depend very much (or how much) on the relationship between A and B? (3) Given a limited number of primary scoring systems, can we optimize the specifics of the fusion? and (4) Can we answer any or all of the previous three issues without sorting to empirical validation? The general issue of the combination algorithm/method will be discussed in this section. In Section 8.2.3, we simply use the average combination regardless of rank or score combination.

Arguably, issues (1) and (2) may be considered primary issues, while issues (3) and (4) secondary. We stipulate that issue (2) is as, if not more, important as issue (1). It has been observed and reported that the combined scoring system E or F performs better than each individual scoring system A and B when A and B are "different," "diverse," or "orthogonal." In particular, Vogt and Cottrell [31] studied the problem of predicting the performance of a linearly combined system and stated that the linear combination should only be used when the individual systems involved have high performance, a large overlap of relevant documents, and a small overlap of nonrelevant documents. Ng and Kantor [27] identified two predictive variables for the effectiveness of the combination: (a) the output dissimilarity of $A$and B, and (b) the ratio of the performance of A and B. Then, Hsu and Taksa [12] suggested using the difference between the two rank/score functions f_A and f_B as the diversity measurement to predict the effectiveness of the combination. This diversity measurement has been used in microarray gene expression analysis by Hsu et al. [13] and in virtual screening by Yang et al. [34] We briefly describe the following six methods/algorithms/approaches that aim to fuse the given p scoring functions (Hsu et al. [13]). This list is by no means complete or is exhaustive.

(a) **Voting:** Scoring function $s^*(d)$, d in D. The score of the object d_i, $s^*(d_i)$, is obtained by a voting scheme among the p values $M(i, j)$, $j = 1, 2, \ldots, p$. These include max, min, median, and majority voting.

(b) **Linear combination:** These are the cases that $s^*(d_i)$ is a weighted linear combination of the $M(i, j)$'s, i.e., $s^*(d_i) = \sum_{j=1}^{p} w_j \times M(i, j)$ for some weighted function so that $= \sum_{j=1}^{p} w_j = 1$. When $w_j = 1/p$, $s^*(d_i)$ is the average of the ranks $M(i, j)$'s, $j = 1, 2, \ldots, p$.

(c) **Probability method:** Two examples are the Bayes rule that uses the information from the given p nodes A_j, $j = 1, 2, \ldots, p$ to predict the node A^*, and the Markov Chain method that calculates a stochastic transition matrix.

(d) **Rank statistics:** Suppose that the p rank functions are obtained by the p scoring systems or observers who are ranking the n objects. We may ask: What is the true ranking of each of the n objects? Since the real (or true) ranking is difficult to come by, we may ask another question: What is the best estimate of the true ranking when we are given the p observations? Rank correlation among the p rank functions can be calculated as $W = \frac{12S}{p^2(n^3-n)}$, where $S = \sum_{j=1}^{p}(\mathrm{R}_j{}^2 - \frac{np^2(n+1)^2}{4})$ and $\mathrm{R}_j = \sum_{j=1}^{p} M(i, j)$. The significance of an observed value of W is then tested in the $(n!)^p$ possible sets of rank functions.

(e) **Combinatorial algorithm:** For each of the n objects and its set of p elements $\{M(i, j) | j = 1, 2, \ldots, p\} = C$ as the candidate set for $s^*(d_i)$, one combinatorial algorithm considers the power set 2^C and explores all the possible combinatorial combinations. Another algorithm treats the n objects as n vectors $d_i = (a_{i1}, a_{i2}, \ldots, a_{ip})$ where $a_{ij} = M(i, j)$, $i = 1, 2, \ldots, n$ and $1 \leq a_{ij} \leq n$. It then places these n vectors in the context of a partially ordered set (Poset) L consisting of all the n^p vectors $(a_{i1}, a_{i2}, \ldots, a_{ip})$, $i = 1, 2, \ldots, n$ and a_{ij} in N. The scores $s^*(d_i)$, $i = 1, 2, \ldots, n$ is then calculated based on the relative position of the vector d_i in the Poset L.

(f) **Evolutional approaches:** Genetic algorithms and other machine learning techniques such as neural networks and support vector machines can be used on the p rank functions to process a (large) number of iterations in order to produce a rank function that is closest to the node A^*.

Arrow [1], Young and Levenglick [38] and Young [39] have extensively used the voting schemes in social choice functions research. Hsu and Palumbo [11], Hsu and Taksa [12], Kendall and Gibbons [16], Kuriakose et al. [19], Marden [23], and Vogt and Cottrell [31] have extensively used the linear combination and average linear combination in many application domains. The widely used Borda count is conceptually equivalent to the average linear combination. Dwork et al. [8] used the probability method (Markov chain) to aggregate the rank functions for the Web. As described in rank statistics, the significance of S depends on the distribution of S in the $(n!)^p$ possible set of rank functions. Due to the manner that S is defined, it may be shown that the average linear combination gives a "best" estimate in the sense of Spearman's rho distance (see Kendall and Gibbons [16], Chapter 6). Hsu et al. [13] and Melnik et al. [24] used combinatorial algorithms in the Rank and Combine method and the Mixed Group Ranks research. Yang et al. [34] use linear combination and the RSC graph as a diversity measurement, instead of genetic algorithms (evolution approaches) such as GemDOCK and GOLD that are typically used to study the docking of ligands into a protein.

The above list enumerates and describes six different groups of methods/algorithms/approaches for performing combination. Here we return to the second issue raised in Remark 7. That is: What are the predictive variables/parameters/criteria for effective combination? In accordance with Remark 7, we focus on two functions $A$and B (i.e., $p = 2$) at this moment although the methods/algorithms/approaches in the list above, are able to deal with the multiple functions ($p \geq 2$). We summarize, in Definition 4, the two variables for the prediction of effective combination among two scoring systems A and B as outlined by Hsu and Taksa [12], Hsu et al. [13], Ng and Kantor [27], Vogt and Cottrell [31], and Yang et al. [34].

Definition 4. Let P_l and P_h be the lower and high performance among $P(A)$ and $P(B)$. We have:

(a) **The performance ratio,** P_l / P_h, measures the relative performance of A and B.
(b) **The bi-diversity between** A **and** B, $d_2(A, B)$, measures the "difference/dissimilarity/diversity" between the two scoring systems A and B.

8.2.5 Diversity

In order to properly use diversity $d_2(A, B)$ as a predictive parameter for effective combination of scoring functions A and B, $d_2(A, B)$ has to be defined to reflect different combination algorithms and different domain applications. Moreover, for the diversity measurement to be effective, it has to be rather "universal" among data sets in a variety of applications domain. Diversity measurement between two scoring systems A and B, $d_2(A, B)$ have been studied by Hsu et al. [13], Ng and Kantor [27], and Yang et al. [34].

Definition 5.

Diversity measure: The bi-diversity (or 2-diversity) measure $d_2(A, B)$ between two scoring systems $A$and B is defined as one of the following:

(a) $d_2(A, B) = d(s_A, s_B)$, the distance between score functions s_A and s_B. One example of $d(s_A, s_B)$ is the covariance of s_A and s_B, $\mathrm{Cov}(s_A, s_B)$, when s_A and s_B are viewed as two random variables.
(b) $d_2(A, B) = d(r_A, r_B)$, the distance between rank functions r_A and r_B. One example of $d(r_A, r_B)$ is the Kendall's tau distance.
(c) $d_2(A, B) = d(f_A, f_B)$, the distance between rank/score functions f_A and f_B.

The concept of a diversity measure for multiple classifier systems in pattern recognition and classification has been studied extensively by Kuncheva [18]. The concept of a diversity, rank/score function and its diversity RSC graph is defined in Lin et al. [21].

Definition 6. In the data space $G(n, m, q)$, $m = 2$, given a temporal step t_i in $T = \{t1, t2, \ldots, tq\}$ and the two scoring systems A and B, let $d_{ti}(A,B) = \Sigma_j |f_A(j) - f_B(j)|$ where j is in $N = \{1, 2, \ldots, n\}$ be the function value of the diversity score function $d_x(A, B)$ for t_i. If we let i vary and fix the system pair A and B, then $s(A,B)(x)$ is the diversity score function, defined as $s_{(A,B)}(t_i) = d_{ti}(AB)$, from $T = \{t_1, t_2, \ldots, tq\}$ to **R**. Sorting $s_{(A,B)}(x)$ into ascending order leads to the diversity rank function $r_{(A,B)}(x)$ from T to $\{1, 2, \ldots, q\}$. We define:

(a) the **diversity rank/score function** $f_{(A,B)}(j)$ as
$f_{(A,B)}(j) = (s_{(A,B)} \circ r^{-1}_{(A,B)})(j) = s_{(A,B)}\ r^{-1}_{(A,B)}(j))$, where j is in $\{1, 2, \ldots, q\}$;

(b) the diversity RSC graph (or diversity RSC graph) as the graph representation of the diversity rank/score function $f_{(A,B)}(j)$ from $\{1, 2, \ldots, q\}$ to **R**.

Note the difference between the rank/score function (see Definition 1a) and the diversity rank/score function (see Definition 6a). In prior definition, the set N is used as the index set for the rank function values and the set D is the set of objects (classes, documents, ligands, and classes or folding patterns). The rank/score function f_A (f_A: $N \rightarrow [0, 1]$) describes the scoring (or ranking) behavior of the scoring system A and is independent of the objects under consideration. In the later definition, the set Q is used as the index set for the diversity rank function values and the set $T = \{t_1, t_2, \ldots, t_q\}$ is the set of temporal steps under study. The diversity rank/score function $f_{(A,B)} f_{(A,B)}(j)$ is defined as a function from $Q = \{1, 2, \ldots, q\}$ to **R** (or [0, 1])) describes the diversity trend of the pair of scoring systems A and B and is independent of the specific temporal step t_i for some i under study.

8.3 Combinatorial Fusion Analysis Applications in Information Retrieval

8.3.1 Predicting Fusion Results

Our first example is the study by Ng and Kantor [26,27], also called NK-study. Their exploratory analysis considered data from TREC competition with 26 systems and 50 queries for each system on a large but fixed database consisting of about 1,000 documents. The results from these 26 systems are then fused in a paired manner. Consequently, there are $\frac{(26\times25)}{2} \times 50 = 16,250$ cases of data fusion in the training data set. In 3,623 of these cases, the performance measures as P_{100} of the combined system is better than the best of the two original systems. We refer to these as positive cases. There are 9,171 negative cases where the performance of the combined system is worse than the best of the two original systems. In order to explain these two outcomes, two predictive variables are used. The first is the ratio of P_{100}, $r = P_l/P_h$ (see Definition 4a). The second variable is the normalized dissimilarity $z = d(r_A, r_B)$ (see Definition 5b). We summarize the results of the NK study as follows:

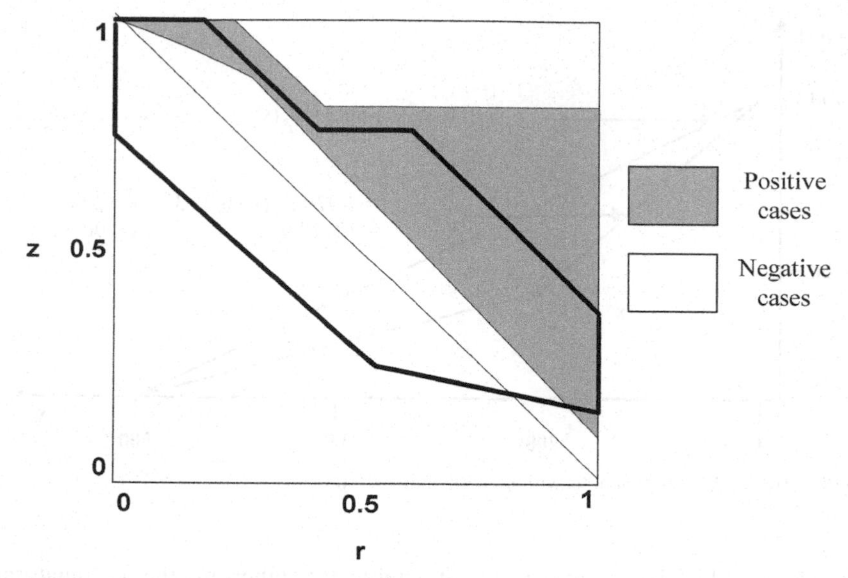

Fig. 8.4 The two predictive variables proposed in NK study – r and z, and the regions that positive and negative cases are most likely to scatter around

Remark 8 (Hsu et al. [13])

The results of the NK-study shows that (a) the positive cases tend to lie above the diagonal line $r+z=1$, and (b) the negative cases are more likely to scatter around the line $r+z=1$ (Figure 8.4).

Figure 8.4 depicts the general trend where the negative and positive cases should fall. First, negative cases hardly fall where r and z are small. Second, there are relatively few negative cases where r and z are high. Third, the majority of negative cases fall around the line $r+z=1$ (z approaches 0 as r approaches 1 and vice versa). This indicates that for negative cases with similar performance P_{100}, their rank functions are comparable to each other. For the positive cases, Figure 8.4 shows very few cases fall where r and z are small, a slightly larger number falls where r and z are higher, and the majority of cases fall above the diagonal $r+z=1$. This suggests that systems with divergent rank functions but comparable performance are estimated to lead to effective fusion.

8.3.2 Comparing Rank and Score Combination

Our next example is the comparative study of rank and score combination methods for data fusion in IR conducted by Hsu and Taksa [12]. The approach to the study consisted of a model of simulation and analysis, and an architecture. The authors use the symmetric group S_n as the sample spacc (also called rank space) with respect to n documents. Since the total number of possible rank data written as permutations of n

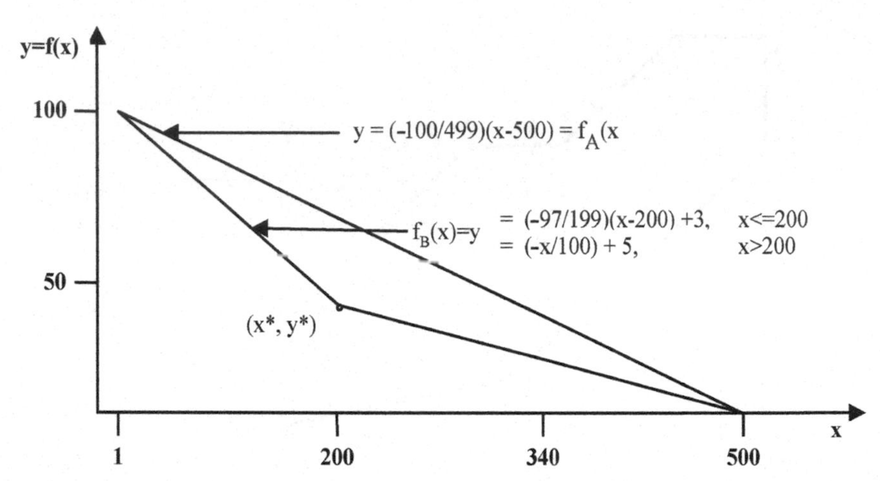

Fig. 8.5 Two rank/score functions with $n = 500$ and $s = 100$

objects is $n!$, which is computationally intractable, the authors use the combinatorial fusion analysis and algorithm to simulate the phenomena. Two basic rankers (or ranked lists) were used (called A and B). Ranker A has a rank function $r_A(x)$, a score function $s_A(d)$, and the performance P(A), $P_{@q}$(A) or P_{avg}(A). Ranker B is represented in the same fashion. Ranked lists C and D are rank combinations and score combinations of A and B respectively. By employing different variations of A and B, the authors extended and generalize their results.

Remark 9 (Hsu and Taksa [12])

Let A, B, r_A, r_B, s_A, and s_B be two scoring systems A and B and their rank functions and score functions. Let C and D be the rank combinations and score combinations of A and B respectively. Under certain conditions such as the greatest value of the diversity $d(f_A, f_B)$ as defined in Definition 5c, we have the P(C) $\geq$ P(D). Figure 8.5 depicts such an example having two RSC graphs f_A and f_B.

8.4 Conclusion and Future Work

WSSs open a new era for Web users. Combining search results from several search engines (meta searching) became an appealing way to improve the quality (relevance) of search results. There are many ways to combine (fuse) information and this chapter describes a combinatorial fusion methodology, and provides a theoretical framework, with illustrations of various applications of the framework using examples from the information retrieval domain. The advantage of the discussed methodology lies in its ability to combine information from multiple scoring systems, i.e., information coming in numerous styles and formats from multiple homogeneous or heterogeneous resources.

To generically represent a scoring system, we use the concepts of score function, rank function and rank/score function as defined by the authors in their prior work [12, 13]. Using these common definitions we describe various performance measurements with respect to different application domains. We also explore various algorithms for combining multiple scoring systems. Theoretical analysis provides insights into system selection and methods of fusion.

We observe that combining multiple scoring systems improves the performance only if each of the individual scoring systems has relatively high performance (performance criterion), and the individual scoring systems are diverse (diversity criterion). While the first criterion is relatively easy to obtain, the second is more complex. It is a bi-diversity between two scoring systems and is defined as either a distance between two score functions, between two rank functions, or between two rank/score functions.

While diversity measures have been studied extensively by Brown et al. [4], Chung et al. [5], Hsu et al. [11–13], Kuncheva [18], Ng and Kantor [26, 27], and Yang et al. [34], it would be of interest to extend the research into a higher level of diversity measurement, e.g., n-diversity. The question is how to approach this issue – direct, as a single problem, or as a series of lower-level diversity problems (2-diversity, 3-diversity, etc.).

Another interesting issue is how to select the most (domain) appropriate scoring systems for combination and then how to select the most (effective and feasible) appropriate method to combine them.

References

1. Arrow KJ (1963) Social choices and individual values. John Wiley, New York
2. Belkin NJ, Kantor PB, Cool C, Quatrain R (1994) Combining evidence for information retrieval. In: Harman D (ed.) TREC-2. Proc of the Second Text Retrieval Conf, Washington, D.C, GPO: 35–44
3. Belkin NJ, Kantor PB, Fox EA, Shaw JA (1995) Combining evidence of multiple query representation for information retrieval. Inf Proces & Manag 31(3): 431–448
4. Brown G, Wyatt J, Harris, R, Yao X (2005) Diversity creation methods: a survey and categorization. Inf Fusion 6: 5-20
5. Chung YS, Hsu DF, Tang CY (2007) On the diversity-performance relationship for majority voting in classifier ensembles. In:Haidl M, Kittler J, Roli F (ed) Multiple Classifier Systems, 7th Int. Workshop, LN in Comput Sci 4472 Springer: 407–420
6. Diaz ED, De A, Raghavan VV (2004) On Selective Result Merging in a Metasearch Environment. In: Yao JT, Raghvan VV, Wang GY (ed) WSS'04: 52–59
7. Ding CHQ, Dubchak I (2001) Multi-class protein fold recognition using support vector machines and neural networks. Bioinformatics 17(4): 349-–358
8. Dwork C, Kumar R, Naor M, Sivakumar D (2001) Rank aggregation methods for the web. In Proc of the Tenth Int World Wide Web Conf WWW 10: 613–622
9. Fan L, Yao YY (2003) Web-Based Learning Support Systems. In: Yao JT, Lingras P (ed) WSS'03: 43–48
10. Fox EA, Shaw JA (1994) Combination of multiple searches. Proc of the Second Text Retrieval Conf (TREC-2), NIST Special Publ 500-215: 243–252

11. Hsu DF, Palumbo A (2004) A study of data fusion in Cayley graphs G(Sn, Pn). In: Hsu DF et al. (ed.), Proc of the 7th Int Symp on Parallel Archit, Algorithms and Netw (I-SPAN'04): 557–562
12. Hsu DF, Taksa I (2005) Comparing Rank and Score Combination Methods for Data Fusion in Information Retrieval. Inf Retr 8(3): 449–480
13. Hsu DF, Chung YS, Kristal BS (2006) Combinatorial Fusion Aanalysis: Methods and Practices of Combining Multiple Scoring Systems. In: Hsu HH (ed) Advanced Data Mining Technologies in Bioinformatics, Idea Group Inc: 32–62
14. Huang CD, Lin CT, Pal NR (2003) Hierarchical learning architecture with automatic feature selection for multi-class protein fold classification. IEEE Trans on NanoBioscience 2(4): 503–517
15. Kantor PB (1998) Semantic dimension: On the effectiveness of naive data fusion methods in certain learning and detection problems. Fifth Int Symp on Artif Intel and Math Ft. Lauderdale, FL
16. Kendall M, Gibbons JD (1990) Rank correlation methods. Edward Arnold, London
17. Keyhanipour AH, Moshiri B, Kazemian M, Piroozmand M, Lucas C (2007) Aggregation of web search engines based on users' preferences in WebFusion. Knowl based syst 20(4): 321–328
18. Kuncheva LI (2005) Diversity in multiple classifier systems [Guest editorial]. Inf Fusion 6: 3–4
19. Kuriakose MA, Chen WT, He ZM, Sikora AG, Zhang P, Zhang ZY, Qiu WL, Hsu DF, McMunn-Coffran C, Brown SM, Elango EM, Delacure MD, Chen FA (2004) Selection and Validation of differentially expressed genes in head and neck cancer. Cell and Mo Life Sci 61: 1372–1383
20. Lee JH (1997) Analyses of Multiple Evidence Combination. Proc of the 20th Annu Int ACM SIGIR Conf on Res and Dev in Inf Retr, Philadelphia, PA: 267–276.
21. Lin KL, Lin CY, Huang CD, Chang HM, Lin CT, Yang CY, Lin CT, Tang CY, Hsu DF (2007) Feature selection and combination criteria for improving accuracy in protein structure prediction. IEEE Trans on NanoBioscience 6(2): 186–196
22. Lyons DM, Hsu DF (2008) Method of combining multiple scoring systems for target tracking using Rank-Score characteristics. Inf Fusion.
23. Marden JI (1995) Analyzing and modeling rank data. (Monographs on Statistics and Applied Probability) 64. Chapman & Hall, London
24. Melnik D, Vardi Y, Zhang CU (2004) Mixed group ranks: preference and confidence in classifier combination. IEEE Trans on Pattern Analysis and Machine Intelligence 26(8): 973–981
25. Mesterharm C, Hsu DF (2008) Combinatorial fusion with on-line learning algorithms. Proc of 11th Int Conf on Inf Fusion: 1–8
26. Ng KB, Kantor PB (1998) An investigation of the preconditions for effective data fusion in information retrieval: A pilot study. Proc of the 61st Annu Meet of the Am Soc for Inf Sci: 166–178
27. Ng KB, Kantor PB (2000) Predicting the effectiveness of naïve data fusion on the basis of system characteristics. J of the Am Soc for Inf Sci 51(13): 1177–1189
28. Pfeifer U, Poersch T, Fuhr N (1996) Retrieval effectiveness of proper name search methods. Inf Proces and Manag 32(6): 667–679
29. Shapiro J, Taksa I (2003) Constructing Web search queries from the user's information need expressed in a natural language. Proc of the Eighteenth Annu ACM Symp on Appl Computing: 1157–1162
30. Varshney PK (ed) (1997) Proc of the IEEE, Special issue on data fusion 85(1): 3–183
31. Vogt CC, Cottrell GW (1999) Fusion via a linear combination of scores. Inf Retr 1(3): 151–173
32. Wu Z, Mundluru D, Raghavan VV (2004) Automatically Detecting Boolean Operations Supported by Search Engines, towards Search Engine Query Language Discovery. In: Yao JT, Raghvan VV, Wang GY (ed) WSS'04: 171–178
33. Xu L, Krzyzak A, Suen CY (1992) Methods of combining multiple classifiers and their applications to handwriting recognition. IEEE Trans on Syst, Man, and Cybern 22(3): 418–435

34. Yang JM, Chen YF, Shen TW, Kristal BS & Hsu DF (2005) Consensus scoring for improving enrichment in virtual screening. J of Chem Inf and Model 45: 1134–1146
35. Yao JT (2005) Design of web-based support systems. Proc. 8th Int Conf on Comp Sci and Inform: 349–352
36. Yao JT, Yao YY (2003) Web-based Information Retrieval Support Systems: building research tools for scientists in the new information age. Proc IEEE/WIC Int Conf on Web Intel: 570–573
37. Yao YY (2002); Information Retrieval Support Systems. Proc FUZZ-IEEE: 773–778
38. Young HP, Levenglick A (1978) A consistent extension of Condorcet's election principle. SIAM J on Appl Math 35(2): 285–300
39. Young HP (1975) Social choice scoring functions. SIAM J on Appl Math 28(4): 824-838
40. Zhong Y, Hu D, Huang M, Chen D (2006) A Comprehensive RMS Model for P2P e-Commerce Communities. Proc of the 2006 IEEE/WIC/ACM Int Conf on Intel Agent Technol – Workshops: 320–323

Chapter 9
Automating Information Discovery Within the Invisible Web

Edwina Sweeney, Kevin Curran, and Ermai Xie

Abstract A Web crawler or spider crawls through the Web looking for pages to index, and when it locates a new page it passes the page on to an indexer. The indexer identifies links, keywords, and other content and stores these within its database. This database is searched by entering keywords through an interface and suitable Web pages are returned in a results page in the form of hyperlinks accompanied by short descriptions. The Web, however, is increasingly moving away from being a collection of documents to a multidimensional repository for sounds, images, audio, and other formats. This is leading to a situation where certain parts of the Web are invisible or hidden. The term known as the "Deep Web" has emerged to refer to the mass of information that can be accessed via the Web but cannot be indexed by conventional search engines. The concept of the Deep Web makes searches quite complex for search engines. Google states that the claim that conventional search engines cannot find such documents as PDFs, Word, PowerPoint, Excel, or any non-HTML page is not fully accurate and steps have been taken to address this problem by implementing procedures to search items such as academic publications, news, blogs, videos, books, and real-time information. However, Google still only provides access to a fraction of the Deep Web. This chapter explores the Deep Web and the current tools available in accessing it.

E. Sweeney
Computing Department, Letterkenny Institute of Technology, Letterkenny, Ireland
e-mail: Edwina.sweeney@lyit.ie

K. Curran and E. Xie
School of Computing and Intelligent Systems, University of Ulster, Londonderry, Northern Ireland, UK
e-mail: KJ.Curran@ulster.ac.uk; Xie-e@email.ulster.ac.uk

J.T. Yao (ed.), *Web-Based Support Systems*, Advanced Information and Knowledge Processing, DOI 10.1007/978-1-84882-628-1_9,

9.1 Introduction

In all 85% of Web users use search engines to find information about a specific topic. However, nearly an equal amount state their inability to find desired information as one of their biggest frustrations [7]. The fact that so much information is available on the Internet and that the inadequacies of search engines are denying us access is why search engines are an obvious focus of much investigation. With conventional search engines, indexing a document so that it can be found by a search engine means that the crawler must follow a link from some other document. Therefore, the more links to your document, the more chance it has of being indexed. This leaves a major loophole for documents that are generated dynamically. Because no links exist to these documents, they will never be located to be indexed. Also, for Web sites that host databases, subscriber information or registration information are sometime required before access is given to their resources. Typically, this type of information is never accessed because the crawler does not have the capability to submit registration or subscriber information. Millions of these documents exist on the Internet and because of this a substantial amount of valuable information is never read by Internet users. Steve Lawerence of the NEC Research Institute writes "Articles freely available online are more highly cited. For greater impact and faster scientific progress, authors and publishers should aim to make their research easy to access. *Evidence shows that usage increases when access is more convenient and maximizing the usage of the scientific record benefits all of society*" [9].

The Surface Web[1] contains an estimated 2.5 billion documents, growing at a rate of 7.5 million documents per day [7]. The Deep Web is estimated to be well in excess of 307,000 Deep Web sites, with 450,000 online databases. Furthermore, the content provided by many Deep Web sites is often of very high quality and can be extremely valuable to many users [11]. Some researchers will point out the fact that the problem is with the database owners, who need to adopt an approach which makes it easier for search engines to gain access. This research is concerned with the problem of accessing the Deep Web. There are a number of projects [6, 10, 11, 13] that have investigated differing aspects of the Deep Web and later we examine each of these in more detail and outline how we intend to utilize aspects of the research in an innovative custom Deep Web search engine prototype. Lin and Chen [10] implement a system that primarily creates a database of Deep Web search engines and subsequently submits queries to specific search engines to find the relevant results. Kabra et al. [6] examine user queries with regard to the Deep Web sources. They attempt to find the most relevant Deep Web resources from a given imprecise query by employing a co-occurrence-based attribute graph which captures the relevance of attributes. Based on this relevancy, Web pages can be ranked. The key insight underlying this study is that although autonomous heterogeneous sources are seemingly independent, the information that they provide is revealed through the query user interfaces. Ntoulas et al. [11] devised a Hidden Web Crawler that uses an algorithm to

---

[1] A term which is used to describe information that is indexed and accessible through conventional search engines.

identify relevant keywords which can be used to query Web sources that cannot be indexed by conventional crawlers. Results are downloaded and indexed at a central location so that the Internet users can access all the information at their convenience from a central location. They provide a framework to investigate the query generation problem for the Hidden Web and propose policies for generating queries automatically. The results of their experiments were notable where they claim to have downloaded more than 90% of a Deep Web site after issuing fewer than 100 queries. Raghavan and Garcia-Molina [13] propose a wrapper and alignment program that will work in conjunction with a Web crawler. It extracts data objects from multiple Deep Web sources and integrates the data together in one table. Their research contributes largely to fully automating the data annotations of Web sites. Many techniques can be adopted to preclude the Deep or invisible Web problem and a number of them concentrate on improving a user's query to obtain more relevant results. A user's query has been shown to be extremely effective when searching for relevant information. In addition, directing the user to appropriate search engines or Deep Web sites could ensure a more fruitful search.

9.2 The Deep Web

General search engines are the first place a typical user will go in order to find information. There are common approaches implemented by many indexed search engines. They all include three programs such as a crawler, an indexer, and a searcher. This architecture can be seen in systems including Google, FAST, etc. Some search engines also offer directory categories which involve human intervention in selecting appropriate Web sites for certain categories. Directories naturally complement search engines. There is now a trend developing towards the use of directories because, in addition to their classification, their content is prescreened, evaluated, and annotated by humans [8]. Search engines do not search the WWW directly. In fact, when searching for a Web page through a search engine, it is always searching a somewhat stale copy of the real Web page. It is only when that page's URL is returned via a results page that a fresh copy of the page is made available. A Web crawler or spider crawls through the Web looking for Web pages to index. Its journey is directed by links within Web pages. Through the process of following links from different Web pages, the spider's journey can become infinite. When a spider locates a new page it passes the Web page to another program called an indexer. This program identifies links, keywords, and other content and stores these within its database. This database is then searched by entering keywords through a search interface and a number of suitable Web pages will be returned in a results page in the form of hyperlinks accompanied by short descriptions (see Figure 9.1).

Chang et al. claim that although "the surface Web has linked billions of static HTML pages, an equally or even more significant amount of information is 'hid-den' on the Deep Web, behind the query forms of searchable databases" [4]. Conventional search engines create their indices by crawling static Web pages.

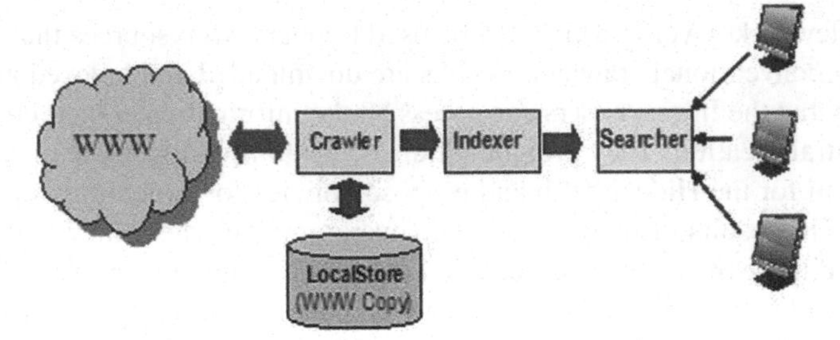

Fig. 9.1 Conventional search engine configuration

In order to be discovered, the page must be static and linked to other pages [1]. It is thought that Web crawlers cannot browse or enter certain types of files, for example, dynamic Web pages, and therefore cannot index the information within these files. However, claims have now been made that conventional search engines have the capabilities to locate dynamic Web pages, pages that are generated in an ad hoc manner [2]. Index Web crawlers cannot index pages that are in non-HTML format, for example, PDFs, spreadsheets, presentations, and some word-processing files, script-based pages which include Perl, JavaScript, CGI', pages generated dynamically by active server pages, for example, Database files and images, video and music files. The biggest problems faced by search engines to date are their inability to login, to subscribe, or to enter relevant keywords to specialized databases or catalogs. When you use a search form to query a back-end database, the information generated is just in response to your particular query. It is much more cost-effective to use these Web resources in this manner as opposed to entering all possible queries and then storing all the generated content in some repository. There are also pages that the search engine companies exclude for specific nontechnical reasons. Some pages would not provide a financial benefit to store, while other pages would not be of great interest to many people. There are thousands of public, official, and special-purpose databases containing government, financial, and other types of information that is needed to answer very specific inquiries. This information may include a stable link; however, many search engines do not warrant them important enough to store [2]. The Web is increasingly moving away from being a collection of documents and is becoming multidimensional repository for sounds, images, audio, and other formats. This has led to a situation where certain parts of the Web are invisible or hidden. The term known as the "Deep Web" has been used repeatedly by many writers to refer to the mass of information that can be accessed via the WWW but cannot be indexed by conventional search engines. Most people accessing the WWW through a search engine assume that all documents available on the Internet will be found through an efficient search engine like Google. In August 2005, Google claimed to have indexed 8.2 billion Web pages [7]. This figure sounds impressive. However, when it is compared with the estimated size of the Deep or hidden Web, it becomes apparent that a whole wealth of information is effectively hidden to Internet users. Extensive research was carried out by a corporation called BrightPlanet in 2001 [1]. Its comprehensive findings concluded that

1. Public information on the Deep Web is currently 400 to 550 times larger than the commonly defined WWW.
2. The Deep Web contains nearly 550 billion individual documents compared to the 1 billion of the surface Web.
3. More than 200,000 Deep Web sites exist.
4. Sixty of the largest Deep Web sites collectively contain about 750 TB of information, which is sufficient by themselves to exceed the size of the surface Web 40 times.
5. On an average, Deep Web sites receive 50.
6. The Deep Web is the largest growing category of new information on the Web.
7. Total quality content of the Deep Web is 1,000 to 2,000 times greater than that of the surface Web.
8. More than half of the Deep Web resides in topic-specific databases.
9. A full 95% of the Deep Web is publicly accessible information and is not subject to fees or subscription.

Table 9.1 represents the subject coverage across all 17 K Deep Web sites used in their study. The table shows uniform distribution of content across all areas, with no category lacking significant representation of content illustrating that the Deep Web content has relevance to every information need and market [1].

Table 9.2 indicates the five queries that were issued to three search engines, namely, AltaVista, Fast, and Northern Lights, and three well-known Deep Web sites. The results show that a Deep Web site is three times more likely to provide "quality"[2] information than a surface Web site. Professional content suppliers typically have the kinds of database sites that make up the Deep Web; static HTML pages that typically make up the surface Web are less likely to be from professional content suppliers [1].

Table 9.1 Distribution of Deep Web sites by subject area [1]

Deep Web coverage			
General (%)		Specific (%)	
Humanities	13	Agriculture	2.7
Lifestyles	4	Arts	6.6
News, media	12.2	Business	5.9
People companies	4.9	Computing	6.9
Recreation, sports	3.5	Education	4.3
Travel	3.4	Government	3.9
Shopping	3.2	Engineering	3.1
Employment	4.1	Health	5.5
Science, math	4	Law	3.9

[2] Quality is a metric value that can be difficult to determine or assign.

Table 9.2 "Quality" documents retrieved from deep and surface Web [1]

Query	Surface Web			Deep Web		
	Total	Quality	Yield (%)	Total	Quality	Yield (%)
Agriculture	400	20	5.0	300	42	14.0
Medicine	500	23	4.6	400	50	12.5
Finance	350	18	5.1	600	75	12.5
Science	700	30	4.3	700	80	11.4
Law	260	12	4.6	320	38	11.9
Total	2,210	103	4.7	2,320	285	12.3

The concept of the Deep Web makes searches quite complex for search engines. Even Google has attempted to integrate the Deep Web into its centralized search function. Google provides specific searches, for example, to academic publications, news, blogs, video, books, and real-time information. However, even a search engine such as Google provides access to only a fraction of the Deep Web [5]. When the Internet evolved, Web pages were structured HTML documents. Managing such static documents was quite achievable. Since then, the growth rate of the Web has been 200% annually [1]. Since 1996, three developments within the WWW have taken place. Firstly, database technology was introduced to the Internet; secondly, the Web became commercialized; and thirdly, Web servers became capable of delivering dynamic information through Web pages. It is now quite common for most organizations, public and private, to transfer, seek, and provide information through a database-driven application. Governments at all levels around the world have made commitments to make their official documents and records available on the Web through single-access portals [7]. Lately, an increasing amount of the Deep Web has become available on the Web. As publishers and libraries make agreements with commercial search engines, undoubtedly more content will be searchable through a centralized location. Also, as the amount of online information grows, the amount of dynamically generated Web pages will grow. Search engines cannot ignore and exclude these pages.

9.3 State of the Art in Searching the Deep Web

Currently many tools exist that allow Internet users to find information. However, a large number of Web pages require Internet users to submit a query form. The information generated from these query forms is not indexable by most search engines since they are dynamically generated by querying back-end databases [3]. The majority of these resources are within the Deep Web, and therefore are rarely indexed [6, 8]. When examining the contents of the Deep Web, it became very obvious that much of the information stored here is very accurate and relevant, for example, scientific, historic, medical data that would dramatically improve communication, research, and progress in these areas [9]. This section provides an overview of research addressing the Deep Web problem. Each project proposes a different method; however, there are some similarities between them.

9.3.1 Automatic Information Discovery from the Invisible Web

Lin and Chen [10] propose a process of examining search engines to select the most appropriate search engine to query. Information was discovered that is not found by conventional search engine. This system maintains information regarding specialized search engines in the Deep Web. When a query is submitted, a selection is made as to the most appropriate search engine and then a query is redirected to that search engine automatically so that the user can directly receive the appropriate query results. The system maintains a database of specialized search engines, storing information such as URL, domain, and search fields. The system utilizes the search engine database to automatically decide which search engine to use and route the query to. It also applies a data-mining technique to discover information related to the keywords specified so as to facilitate the search engine selection and the query specification. This would enable more relevant results to be returned. The system is divided into various components [10]. A crawler is used to populate a database with information regarding specialized search engines. A search engine database stores the information regarding the specialized search engines. A query preprocessor receives user input and finds phrases associated with the query keywords. These phrases form the basis of the search engine selection. A search engine selector is finally sent the keyword phrase and with access to the search engine database, it selects the specialized search engine to send the query to. It also proceeds to send the actual query to the selected databases. The search engine database is populated by a crawler. It visits every page, but only extracts the portion that corresponds to a search engine by identifying form tags. The search engine's URL and other information are also stored. This becomes essential when constructing the query string to send to the search engines.

In order to select the most appropriate search engines, a query preprocessing stage is carried out to determine the best query to send. This will supplement the keywords supplied by the user with other information, i.e., keywords and phrases returned from an episode rules discovery technique. Each keyword phrase generated from the preprocessing stage is matched with the three fields of each search engine within the database. The results are ranked and those search engines with the most matches are returned. The system now sends the user query to each of these search engines and the results are returned through a user interface (see Figure 9.2). The main objective of this system is to use specialized search engines to discover information that cannot be retrieved from a conventional search engine. To measure the effectiveness of this system, the Web pages returned from a typical query were compared against results obtained from the Google search engine [10]. Only in 3% of the cases do both systems return similar or related results, while a majority of the time, the system was able to access Web pages that are not directly accessible through Google.

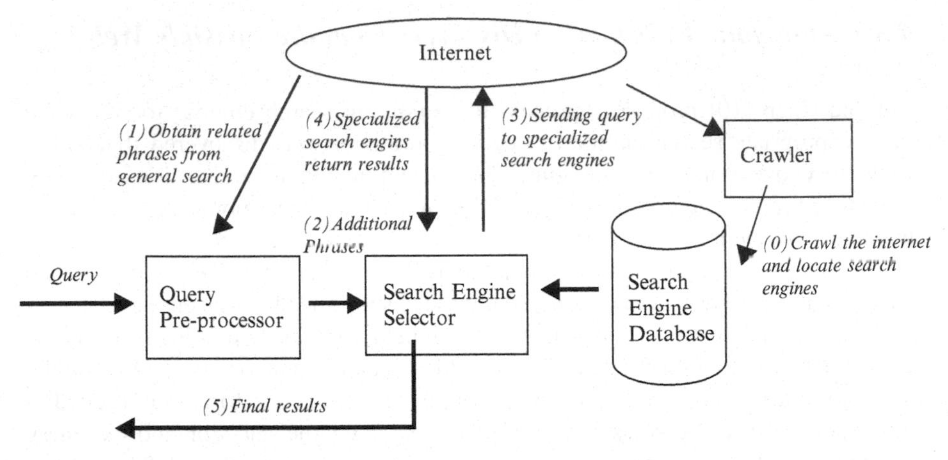

Fig. 9.2 Architecture of system

9.3.2 Query Routing: Finding Ways in the Maze of the Deep Web

Although autonomous heterogeneous sources are seemingly independent, the information that they provide is revealed through the query user interfaces. A search form within a book sales Web site like Barnes and Noble for example, contains attributes like author, title, ISBN, edition, etc. This would indicate that this site is about books. Within an airline site, attributes like from, to, departure date, return date, etc., would reveal that the site is about flights. Kabra et al. [6] have built a source selection system which examine user queries with regard to "deep" Web sources. Given the huge number of heterogeneous Deep Web sources, Internet users may not be aware of all the sources that can satisfy their information needs. Therefore, the system attempts to find the most relevant Deep Web resources from a given simple imprecise query. It designs a co-occurrence-based attribute graph for capturing the relevance of attributes. It employs this graph to rank sources in order of relevance to a user requirement. The system counteracts the impreciseness and incompleteness in user queries. It assumes that the user may not be aware of the best attributes to enter as a query and therefore limits the results that will be returned to them. The main contributions of the work by Kabra et al. are an attribute co-occurrence graph for modeling the relevance of attributes and an iterative algorithm that computes attributes relevance given just an imprecise query. A methodology was created in which the relevance of an attribute was determined using the relevance of other attributes. There may be some attributes that are not included in a user's query but which do appear in many of the data source query interfaces. These attributes that co-occur with the input query attributes will also be relevant to the search. Also, there are similar attributes that co-occur with the other attributes that are likely to have some relevance to the query in question. Therefore, any relevant attribute shall lead to increasing the relevance of its co-occurring attributes. In order to quantify this approach, a relevance score was associated with each attribute

Query 1	(from, to, departure data, return date)
Query 2	(author, title, ISBN)
Query 3	(make, model, price)
Query 4	(song, album, artist)
Query 5	(title, actor, director)
Query 6	(from, to)
Query 7	(from, to, adult, child)

Fig. 9.3 Queries used to evaluate the algorithm

which represented how likely a query interface containing this attribute will provide appropriate information to the Internet user. A graph was constructed which is similar to the graphs used by search engines to determine the ranking of pages. Each graph contains nodes and edges. Each node corresponds to an attribute and the edges connect two co-occurring nodes that have associated weighting. The weighting indicates the relevant scores, where given an attribute A that is known to be relevant and which co-occurs with attribute B, the more frequently the attribute B occurs with attribute A as compared with other attributes co-occurring with attribute A, the higher the degree to which it gets relevance induced from attribute A [6]. Consequently, Web sources that contain more attributes with higher relevancy are going to have a higher probability of being more relevant to user requirements. During its experimental evaluation it manually collected a deep set of Web sources that contained 494 Web query interfaces and a total of 370 attributes. These Web sources included a set of diverse domains including airlines, books, car rentals, jobs, music, movies, and hotels.

It constructed a number of queries which are listed in Figure 9.3 and tested it against the algorithm. The algorithm takes as input the set of attributes in the user query. The algorithm was run against all the different queries. A relevance score was associated with each of the attributes and the top five attributes were identified. Figure 9.4 shows a listing of the top 15 attributes identified from query 1. Based on the co-occurrence analysis, the algorithm is able to find other attributes that are also relevant. The algorithm then uses the top five attribute relevancies to compute a relevance score for each data source. The results indicated that the system frees users from worrying about finding the right set of query attributes. Even with a set of attributes that might not necessarily best describe the type of information required, the algorithm will direct the Internet user to the most appropriate set of Web sources.

9.3.3 Downloading the Hidden Web Content

Much of the Deep Web sources today are only searchable through query interfaces where the Internet users type certain keywords in a search form in order to access

Rank	(Index)	Name
1	(6)	departure date
2	(7)	return date
3	(18)	to
4	(4)	from
5	(9)	adult
6	(10)	child
7	(16)	trip type
8	(8)	class
9	(19)	return
10	(12)	infant
11	(1)	city
12	(14)	departure
13	(15)	senior
14	(28)	airline
15	(2)	destination

Fig. 9.4 Top 15 relevant attributes for query 1

pages from different Web sites. Ntoulas et al. [11] produced a hidden Web crawler which automatically generates meaningful queries to a Web site by using an adaptive algorithm. Results are downloaded and indexed at a central location so that the Internet users can access all the information at their convenience from a central location. They provide a framework to investigate the query generation problem for the hidden Web and propose policies for generating queries automatically. The results of their experiments were notable where they claim to have downloaded more than 90% of a Deep Web site after issuing fewer than 100 queries. The challenge undertaken by this project was that the Web crawler must generate meaningful queries so that it can discover and download the Deep Web pages. This project also investigated a number of crawling policies in order to find the best policy that could download the most pages with the fewest queries. It proposed a new policy called the adaptive policy that examines the pages returned from the previous queries and adapts its query-selection policy automatically based on them [11]. The experiments were carried out on real Web sites and the results noted for each crawling policy. The proposal focuses on textual databases that support single-attributed keyword queries. The main objectives of this crawler are to gain entry to a hidden Web site, generate a query, issue it to a Web site, download the results page, and then follow the links to download the actual pages. This process is repeated until all the resources are used up, i.e., the crawler has limited time and network resources. The most important decision that the crawler has to make is what query to issue with each iteration. If the crawler issues meaningful queries and returns many Web pages, then the crawlers will expose much of the Hidden Web in the least amount of time. However, if the crawler issues inappropriate queries then very few pages will be returned, which will result in a minimum amount of the hidden Web pages being exposed. Therefore, how the crawler selects its next query greatly affects its effectiveness. There are three different methods to select meaningful queries.

- Random: Selecting random keywords from an English dictionary and then issuing them a database.

- Generic frequency: Analyzing a generic document body collected elsewhere and obtaining the frequency distribution of each keyword. Based on this selection, the most frequent keyword was selected first and issued to the database and then the next most frequent keyword and so on. The hope was that the list of generic keywords obtained would also occur frequently in the Deep Web databases, returning many matching documents.
- Adaptive: Analyzing the returned documents from the previous queries issued to the Deep Web database, and estimating which keyword would most likely return the most Web pages. The adaptive algorithm learns new keywords from the documents that it downloads and its selection of keywords is driven by a cost model. This cost model can determine the efficiency of each every keyword with regard to the number of links that it returns and select the candidate that will return the most unique documents. This efficiency measurement is identified by maintaining a query statistics table.

The query statistics table is updated with every new query, where more documents are downloaded. Using the above algorithms, an experimental evaluation was conducted. It should be noted that with these algorithms the initial keyword chosen has minimal affect on the performance overall. The experiment focused on three Deep Web sites: The PubMed Medical Library, Amazon, and the Open Directory Project. The first observation made was that the generic frequency and the adaptive policies perform better than the random algorithm. The adaptive policy outperforms the generic-frequency algorithm when the site is topic-specific, i.e., medical, law, historical topics. For example with the PubMed site, 83 queries were issued by the adaptive algorithm to download 80% of the documents stored at PubMed, while the generic-frequency algorithm required 106 queries for the same coverage. The adaptive algorithm also performed much better than the generic-frequency algorithm when visiting the Open Directory Project/art site. Here, the adaptive algorithm returned 99.98% coverage by issuing 471 queries, while the generic-frequency algorithm discovered 72% coverage with the same number of queries. Obviously, as the adaptive algorithm iterates through the Web site material, its keyword selection for queries becomes more accurate. It searches through the downloaded pages and selects keywords that occur most frequently, while the generic-frequency algorithm works from a large generic collection (see Table 9.3).

Table 9.3 Keywords queried to PubMed exclusively by the adaptive policy

Iteration	Keyword	Number of results
23	Departments	2,719,031
34	Patients	1,934,428
53	Clinical	1,198,322
67	Treatment	4,034,565
69	Medical	1,368,200
70	Hospital	503,307
146	Disease	1,520,908
172	Protein	2,620,938

Results from the random algorithm performance were not encouraging. It downloaded a 42% from the PubMed site after 200 queries, while the coverage for Open Directory Project/art site was 22.7% after 471 queries. In conclusion, the adaptive algorithm performed well in all cases. The generic-frequency algorithm proves to be effective also, though less than the adaptive. It is able to retrieve a large proportion of the Deep Web collection, and when the site is not topic-specific it can return the same coverage as that of the adaptive algorithm. The random algorithm should not be considered because it performed poorly overall.

9.3.4 Information Discover, Extraction, and Integration for Hidden Web

Wang and Lochovsky [14] explore methods that discover information sources within the hidden Web. These methods locate the Web site containing the structured data of interest to the Internet user, induce wrappers to extract relevant data objects from discovered Web source pages, label the fields of extracted data, and integrate the various data objects from multiple sources with or without the knowledge of the schema. To integrate data from different Web sites, information extraction systems are utilized. That is, wrappers have been developed which will extract data from a Web page based on their HTML tag structure. Their system called DeLa (Data Extraction and Label Assignment) comprises four components; a form crawler: a wrapper generator, a data aligner, and a label assigner. A form crawler called HiWe is adopted for the first task [13]. This crawler collects labels of the search form and sends queries to the Web site. The wrapper generator is then applied to the retrieved Web pages.

In this results page each book has a book title, zero or more authors, and one or more edition information. The symbols <> represent an unordered list tuple, the symbol { } represent a set and the ? symbol represents an optional attribute (see Figure 9.5). A data aligner then extracts data objects from the Web pages according to the wrapper induced by the wrapper generator. It then filters out the HTML tags and arranges the data instances into a table-like format. After data objects are extracted from the results page, it is rearranged in a table format so that each row of the table represents a data instance, and each column represents a data attribute.

Each Web site's data is represented as a single table to allow easy integration during the later stages (see Figure 9.6). The data aligner must also separate multiple attributes which are represented as one single string, and it must also separate multivalued attributes into two separate rows. A label assigner then labels the column of the table by matching them to the form labels retrieved by the form crawler. The system performance had 90% precision in inducing wrappers and over 80% correctness with assigning meaningful labels to the retrieved data. Anticipating the correct type of queries to enter through search forms has been extensively studied [6]. However, discovering the actual Deep Web would be at a nascent stage [14]. Many systems would assume that the Internet user is aware of the appropriate Web resource and

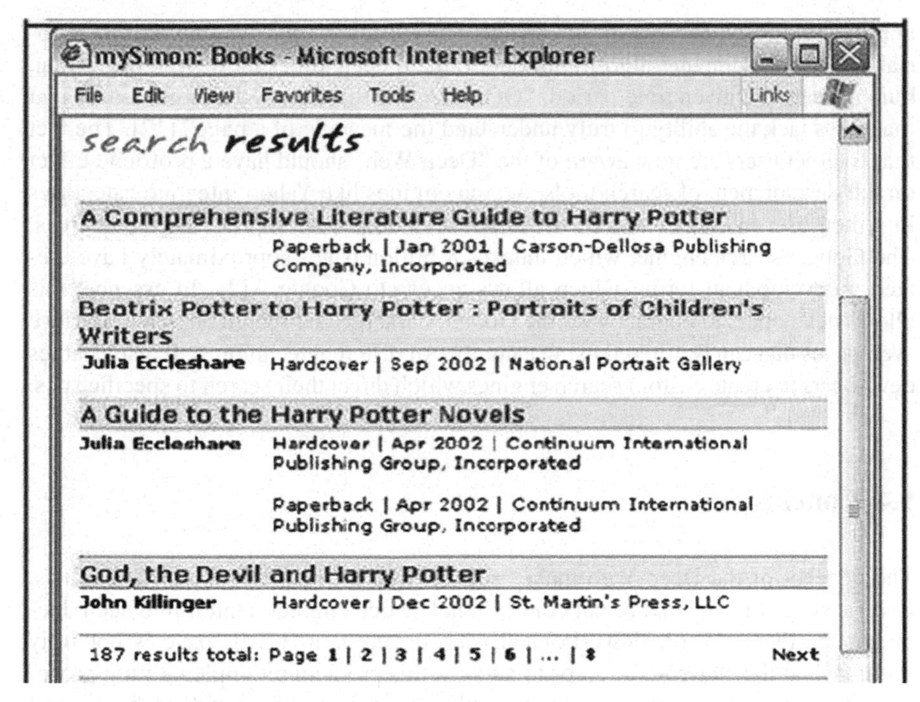

Fig. 9.5 Results page from a book search [14]

..\book_html\example.htm

Title	Author	Format	Date	
A Comprehensive Literat...		Paperback	Jan 2001	Carson-D...
Beatrix Potter to Harry Pot...	Julia Eccleshare	Hardcover	Sep 2002	National ...
A Guide to the Harry Pott...	Julia Eccleshare	Hardcover	Apr 2002	Continuu...
A Guide to the Harry Pott...	Julia Eccleshare	Paperback	Apr 2002	Continuu...
God, the Devil and Harry ...	John Killinger	Hardcover	Dec 2002	St. Martin'...

Fig. 9.6 The table format which was utilized with the given book search

would ignore the need to help users to find the desired Web site that they want to extract information from. Finding the most relevant Web site to a given user query is the main objective of most search engines. Manually administered measures seem to have the highest percentage of success. Google has always favored the algorithmic approach to search because of the obvious scaling advantages over the human editorial approach. Computer algorithms can evaluate many times more Web pages than humans can in a given time period. "Of course the flip side of this argument is that machines lack the ability to truly understand the meaning of a page" [12]. The fact that Internet users are now aware of the "Deep Web" should have a profound effect on the development of search tools. Search engines like Yahoo integrate categories into their user interface, which aids in the selection of appropriate Web resources. The Google search engine, which indexes 8 billion pages approximately have created a development forum which allows access to Google APIs. In essence, this allows developers to interact with the Google database and search engine to retrieve Web pages that can be filtered using a number of different techniques. It also enables developers to create custom search engines which direct their search to specific sites.

9.4 Conclusion

The concept of the Deep Web makes searches quite complex for search engines. Google states that the claim that conventional search engines cannot find such documents as PDFs, Word, PowerPoint, Excel, or any non-HTML page is not fully accurate and they have taken steps to address this problem by implementing procedures to search items such as academic publications, news, blogs, videos, books, and real-time information. However, Google still only provides access to a fraction of the Deep Web. Google has, however, released an exciting feature through its Google custom search engine. By allowing Internet users the capability to filter their own search engine is a positive approach to gaining access to the Deep Web. Typically, Internet users will more freely use this technology to search before attempting to subscribe to individual Deep Web sources. Providing a Google technology that can comparably return results as if from a Deep Web site is invaluable.

References

1. Bergman, M. (2001) The Deep Web: Surfacing Hidden Value, Journal of Elec-tronic Publishing, Publishers: University of Michigan July 2001, http://www.press.umich.edu/jep/07-01/bergman.html
2. Berkeley (2007) What is the Invisible Web, a.k.a. the Deep Web, Berkeley Teaching Guides, http://www.lib.berkeley.edu/TeachingLib/Guides/Internet/InvisibleWeb.html
3. Broder, A., Glassman, S., Manasse, M., and Zweig, G. (1997) Syntatic Clustering of the Web, In 6th International World Wide Web Conference, April 1997. http://proceedings.www6conf.org/HyperNews/get/PAPER205.html
4. Chang, K., Cho, J. (2006) Accessing the web: from search to integration. 25th ACM SIGMOD International Conference on Management of Data / Principles of Database Systems, Chicago, Illinois, USA; June 26-29, 2006. pp: 804-805
5. Cohen, L. (2006) Internet Tutorials, The Deep Web, FNO.com, From Now on, The Educational Technology Journal, Vol. 15, No 3, February 2006

6. Kabra, G., Li, C., Chen-Chuan Chang, K. (2005) Query Routing: Finding Ways in the Maze of the Deep Web, Department of Computer Science, University of Il-linois at Urbana-Champaign
7. Kay R. (2005) QuickStudy: Deep Web, ComputerWorld, June 4th 2007, http://www.computerworld.com/action/article.do?command=viewArticleBasic&articleId=293195&source=rss_dept2
8. Lackie, R. (2006) Those Dark Hiding Places: The Invisible Web Revealed, Rider University, New Jersey, 2006.
9. Lawrence, S. (2001) Online or Invisible, Nature, Volume 411, Number 6837, p. 521, 2001.
10. Lin, K., Chen, H. (2002), Automatic information discovery from the invisible web, Proceedings of the International Conference on Information Technology: Coding and Computing, Washington, DC, p. 332
11. Ntoulas, A., Zerfos, P., Cho, J. (2004) Downloading textual hidden web content through keyword queries Tools & techniques: searching and IR, JCDL'05: Proceedings of the 5th ACM/IEEE-CS Joint Conference on Digital Libraries 2005 p.100–109
12. Seth, S. (2006), Google muscle to power custom search engines, CRIEnglish.com. http://english.cri.cn/2906/2006/10/24/272@154396.htm
13. Raghavan, S., Garcia-Molina, H. (2001) Crawling the Hidden Web, Proceedings of the 27th Intl Conf. on Very large Database, 129–138, 2001
14. Wang, J., Lochovsky, F. (2003) Data Extraction and Label Assignment for Web Databases, Wang & Lochovsky, International World Wide Web Conference, Proceedings of the 12th international conference

Chapter 10
Supporting Web Search with Visualization

Orland Hoeber and Xue Dong Yang

Abstract One of the fundamental goals of Web-based support systems is to promote and support human activities on the Web. The focus of this Chapter is on the specific activities associated with Web search, with special emphasis given to the use of visualization to enhance the cognitive abilities of Web searchers. An overview of information retrieval basics, along with a focus on Web search and the behaviour of Web searchers is provided. Information visualization is introduced as a means for supporting users as they perform their primary Web search tasks. Given the challenge of visualizing the primarily textual information present in Web search, a taxonomy of the information that is available to support these tasks is given. The specific challenges of representing search information are discussed, and a survey of the current state-of-the-art in visual Web search is introduced. This Chapter concludes with our vision for the future of Web search.

10.1 Web Search and Web Support Systems

Web Information Retrieval Support Systems (WIRSS) represent a specific facet within the field of Web Support Systems (WSS). In particular WIRSS focuses on the functionalities that support user-centric tasks performed while conducting Web searches [18, 74]. Systems that consider the human aspect of Web search have the benefit of allowing searchers to take an active role throughout the entire search process, enhancing their abilities to craft and refine queries; and browse, investigate, and explore search results.

O. Hoeber
Department of Computer Science, Memorial University of Newfoundland, St. John's, NL, Canada
e-mail: hoeber@mun.ca

X.D. Yang
Department of Computer Science, University of Regina, 3737 Wascana Parkway, Regina, SK, Canada
e-mail: yang@cs.uregina.ca

J.T. Yao (ed.), *Web-Based Support Systems*, Advanced Information and Knowledge Processing, DOI 10.1007/978-1-84882-628-1_10,

One of the fundamental questions that needs to be considered when conducting research and development in the domain of WIRSS is how to deal with the problems of information complexity and overload. Traditional Web search engines have not adequately addressed the complexity of information, instead providing only simple text boxes for entering queries and simple list-based representations of search results. An outcome of these simple interfaces is that people use them in a simple manner: queries commonly contain only one or two terms [31, 57], and people seldom venture beyond the third page of the search results [55, 57]. While Web search engines may be able to perform well under these conditions when the searcher is seeking to fulfill simple targeted search operations (such as finding the capital city of Tanzania), more complex or exploratory searches are not well served by these interfaces. The focus of WIRSS research is to move beyond the simple functionality provided in these interfaces, supporting the searchers at a deeper, task-oriented level.

The goal of this chapter is to explore how visualization can be used to support the human aspects of Web search. The remainder of the chapter is organized as follows. Section 10.2 provides an overview of Web information retrieval. Section 10.3 introduces information visualization as a means for augmenting the cognitive abilities of users. Section 10.4 presents a taxonomy of the information that is available for supporting Web search processes. Section 10.5 explores the challenges in representing information that is relevant to searching. A survey of both seminal and state-of-the-art work on visual Web search interfaces is provided in Section 10.6. The chapter concludes with our vision for the future of Web search in Section 10.6.

10.2 Web Information Retrieval

Before discussing the features and activities supported by visual Web search systems, a survey of the fundamental features of Web information retrieval are provided. This section delves into the basic foundations of traditional information retrieval, the features of Web information retrieval, and discussions on the interfaces to Web Search and the behaviour of people as they conduct searches on the Web.

10.2.1 Traditional Information Retrieval

Information retrieval deals with the design and study of automated searching within digital collections. In general, the discipline is concerned with finding a set of relevant documents from a document collection that match a user's specified information need. The goal is to automatically find all the relevant documents, while at the same time, selecting as few of the non-relevant documents as possible [53]. Traditionally, the main emphasis of information retrieval has been on document and text retrieval [68], although many of the principles also apply to the retrieval of other types of information, such as images, video, and audio [4].

In general, an information retrieval system has two inputs: the document collection to be searched and the user-supplied query; and one output: the set of documents that match the query. One of the fundamental challenges is to specify data structures for storing documents and queries that are space efficient, yet support effective and accurate matching of queries to documents. Many methods exist for creating such document indexes [53]. Clearly, there is a great benefit to indexing the document collection once, and storing the internal representation for future querying.

Given an indexed document collection and a query that represents the user's information need, the matching process seeks to find documents from the collection that are relevant to the query. This matching process is highly dependent on the methods by which the document collection is indexed, and the methods by which users specify their queries. Common examples are boolean matching, fuzzy matching, vector-based matching, and probabilistic matching [4, 53].

The output of this matching process is a set of references to documents from the collection. It is desirable that the set of matched documents be sorted in order of relevance to the query, such that those documents that at the top of the list are better matches. The underlying assumption is that these top-ranked documents will more likely be relevant to the searcher's information need.

There are two commonly used measures for the success of the information retrieval process: precision and recall. The *precision metric* measures the ratio of the number of relevant documents to the total number of documents retrieved. The precision metric will be high when a large portion of the set of retrieved documents are relevant to the user's information need. The *recall metric* measures the ratio of the number of relevant documents to the total number of relevant documents in the entire collection. The recall metric will be high when the set of retrieved documents contains a large portion of the relevant documents from the collection.

10.2.2 Information Retrieval on the Web

When people today think of information retrieval, text searching, and document searching, they commonly think of Web search engines. The use of Web search engines have become common place among Web users, and is increasingly being used in all aspects of society [39]. Nielsen reported that 88% of Web users start with a Web search engine when provided with a task to complete using the Web [46]. This is in support of earlier studies that reported nearly 85% of Web users find new Web pages using search engines [12]. Jansen et al. [30] classified a large transaction log, finding approximately 80% of the queries to be informational in nature; 10% to be navigational; and 10% to be transactional. Nielsen noted that instead of looking for sites to explore and use in depth, Web searchers are commonly seeking specific answers [46]. In essence, Web search engines have become answer engines.

Given the importance Web users place on search engines, information retrieval on the Web has become a very active research area, both academically and commercially. Although there are unique challenges posed by the Web for information retrieval systems, there are also unique opportunities for improvements over traditional information retrieval systems.

While the Web can be considered a single distributed document collection, this collection has features that make many of the traditional approaches to information retrieval impossible or impractical to implement [73]. These features include the size of the Web (on the order of billions documents, and growing), the diversity of the collection (documents available on virtually any topic), and the potential diversity of individual documents (single documents may discuss multiple distinct topics). Yang noted that information retrieval on the Web "must deal with mostly short and unfocused queries posed against a massive collection of heterogeneous, hyper-linked documents that change dynamically" [73]. Clearly, this is a much more challenging problem than those traditionally addressed by information retrieval research.

However, there are also features of the Web that can be used advantageously to increase information retrieval performance. These features include the link structure of the Web, the implied relevance of the linking text, the structure of the Web documents, and the vast amounts of Web search engine usage statistics. Many Web search engines go beyond traditional term-based information retrieval, and take advantage of these additional features of the Web. A prime example of this is the PageRank algorithm used by the Google search engine [8]. A thorough evaluation of the techniques for conducting information retrieval on the Web is beyond the scope of this Chapter; interested readers are directed to a number of good survey papers and books on this topic [4,36,37,51,73].

10.2.3 Web Search User Interfaces

User interfaces have been an active research area of traditional information retrieval for many years [4, 14]. However, few of these results have been applied to Web information retrieval. Within much of the research literature on Web information retrieval, there is little consideration of the user interface provided by the search engine. The user interface is the gateway to the functionality provided by the system. However, there seems to be a lack of acknowledgment of the importance of the methods by which users are able to specify their queries, or the methods by which search results are provided to the users. It seems that the simple query box and the list-based representations of Web search engines have become so common place that there is little if any discussion on whether these simple interfaces are providing adequate support to the users. We will attempt to address this problem through our discussions on using visualization to support Web search activities and the promotion of visual WIRSS.

As noted by Marchionini, "Much research and development is required to produce interfaces that allow end users to be productive amid such complexity and that protect them from information overload" [42]. Although this comment was made in regards to traditional information retrieval research, it is valid for Web information retrieval as well. This provides a clear motivation for the ongoing research on visualization and interactivity as a means of supporting users in their Web search tasks.

10.2.4 Web Search User Behaviour

A number of studies have been conducted in the past few years evaluating the Web search behaviour patterns of users, as deduced from Web search transaction logs. These studies include those by Silverstein et al. [55], Jansen and Pooch [31], Spink et al. [57], and Jansen and Spink [32].

A number of common themes regarding users' abilities to craft queries have emerged from these studies. In particular, it has been shown that few searchers provide queries containing more than three terms; more common are the short one and two term queries [31, 57]. In addition, few searchers make use of advanced query features; those who do have a tendency to use them incorrectly [57]. There also appears to be a tendency for users to avoid making subsequent modifications or refinements of their queries [55, 57].

A further user trait that has been identified is the lack of willingness to evaluate many Web search results. Even when users were able to effectively craft a query, few Web searchers consider more than three pages of search results [55, 57]. As noted by Spink et al., "the public has a low tolerance for going in depth through what is retrieved" [57]. Users have a strong desire to see relevant documents within the top few search results, and to avoid scrolling or navigating to the next page [36].

In a comparison of the transaction logs from nine search engines over a 5-year period (1997–2002), it was found that the interactions with Web search engines are not becoming more complex over time; in some cases, the interactions are becoming less complex [32]. While this may be an outcome of the improvement of Web search engines for targeted queries, the lack of involvement by the searchers in the information retrieval process remains. Yang noted that: "Web searchers in general do not want to engage in an involved retrieval process. Instead, they expect immediate answers while expending minimum effort" [73]. However, when search engines are not able to satisfy the users' information needs, a deeper involvement in the search process is needed.

An important aspect of any information system is the ability to provide information seeking functions which assist users in defining and articulating their problems, and finding solutions to these problems [42]. It is clear that the current Web search engines provide little of this assistance to users. It is up to the searchers to choose queries that accurately represent their information needs, with little support provided by the Web search engine beyond a box in which to type their queries. When provided with the list of search results, it is up to the users to evaluate these document surrogates one-by-one, with little support for manipulation or exploration in order to find solutions to their information need.

This lack of support for the users' tasks highlights the need for further study on Web search interfaces and the support they provide for the searchers' fundamental tasks of crafting queries that accurately represents their information needs, and evaluating the search results to find relevant documents. The goal is to reduce the cognitive burden of searching the Web [73]. Information visualization, as we will see in the following section, is an effective method for supporting and amplifying the cognition of users [9]. Kobayashi and Takeda predicted that "future systems will have better user interfaces to enable users to visually manipulate retrieved information" [36]. One of the goals of this Chapter is to explore such interfaces for Web search.

10.3 Issues in Information Visualization

Web search is an inherently information-centric process: searchers begin with mental models of the information they are seeking; these models are converted into descriptions of their information needs through the crafting of queries. Once the search results are returned, there are vast amounts of textual information that must be evaluated. Information visualization is a method for taking advantage of the innate visual information processing capabilities of the searcher, resulting in an enhancement of their cognitive ability to process the information that is relevant to their current tasks [9]. It has long since been noted that there is a need to continue to conduct research that focuses on allowing users to be productive amid the complexity and quantity of information being provided by information retrieval systems [42]. With the advent of Web search, things have not improved. To further complicate matters, Web search has allowed information retrieval to move from the domain of trained information analysts to the general public.

In general, information visualization is a technique for creating graphical representations of abstract data or concepts [69]. Moreover, information visualization promotes a cognitive activity in which users are able to gain understanding or insight into the data being graphically displayed by taking advantage of human visual information processing capabilities [56]. In essence, information visualization provides a link between the human vision system and the data being processed within the computer system [76], with the end result being an amplification of cognition [9].

At the most basic level, information visualization techniques are used when one draws graphs to visually represent data sets. However, when these data sets are large, high-dimensional, or complex, generating useful visual representations can become a challenging problem. In general, information visualization techniques allow the display of abstract data in a coherent manner, allowing the viewer to compare and explore the data visually [67].

When graphically representing abstract objects in a visual representation, there are a number of visual features that are available for representing the various dimensions or attributes of the data. These include spatial location, colour, shape, and orientation, among others. Care must be taken to select and use visual features that can be easily decoded and understood by the viewer. For example, while the spatial location and colour of an object can be perceptually separated to represent multiple dimensions in the information, this cannot be done as easily with the shape and orientation of a glyph [69]. Readers interested in the human perception system and its relevance to information visualization are directed to Ware's excellent book on the subject [69].

When designing any information visualization system, careful consideration and restraint must be employed so as to not overload the viewers with unnecessary visual complexity that is difficult to perceive and interpret. Tufte recommends a design strategy of the *smallest effective difference*: "make all visual distinctions as subtle as possible, but still clear and effective" [66]. However, even with this advice, choosing visual features for the representation of data attributes can be challenging. Guidance on this topic is provided in the work of Bertin [7] and Mackinlay [40], along with the excellent textbooks by Ware [69] and Spence [56].

The use of colour often poses an especially challenging dilemma. Although colour is an extremely valuable means for encoding information within a visual display, the overuse of colour has the effect of becoming noise within the information display. As noted by Tufte, the end result of the overuse of strong colours is a "visual war with the heavily encoded information" [65]. Design principles such as those suggested by Tufte [65–67] and Ware [69] are of value, as is an understanding of colour theory [16,62]. We will explore examples of the effective use of colour for conveying Web search information in Section 10.6 of this Chapter.

In order to further aid in the cognition supported by a good visual representation, the interactive manipulation of the visual representations is an integral part of any information visualization system [9, 56]. Clearly, one of the great benefits of computer-generated visualizations are that they allow users to interact with and manipulate the information being graphically represented. However, as with any interactive system, there is a challenge in ensuring that the users remain in control, and that the system provides effective feedback that reflects the results of the users' actions [5]. As such, when discussing any visualization technique, the methods by which users can interact with and manipulate the graphical representation are important in understanding how the information visualization system can support users in conducting their tasks.

While the techniques for information visualization are as numerous as the types of data sets they graphically describe, this Chapter is constrained to information visualization techniques that have been applied to information retrieval systems, and in particular to those that support the visualization of queries and search results, both of which primarily contain textual data. The visualization of such unstructured or semi-structured data is difficult due to the lack of well-defined features upon which to base the visual representation [76]. Often, there is a need to identify features within the data and to construct data structures which support graphical representations. This is the focus of the following Section.

10.4 A Taxonomy of Information to Support Web Search Processes

To facilitate the support of Web search tasks through information visualization, features must be extracted from the information available about the users' information needs. Finding appropriate sources of such information can be challenging, given the highly textual nature of Web search, and the broad range of topics upon which people search. Our research has explored the use of features extracted from the initial set of search results [17, 19, 20, 22–26], as well as the use of external knowledge bases developed for specific domains [21,28,29]. The models generated from this information form a basis for generating visual representations, and provide a critical link between the tasks of query refinement and search results exploration. In this section, we provide an analysis of the types of information that can be used to support Web search processes.

Fundamentally, the type of information available can be divided into two classes: information provided directly by the searcher, and information that is derived either through the Web search process or by other means. In traditional Web search interfaces, the only information provided by the Web searcher is the query itself. However, interactive Web search interfaces can also allow the searcher to provide additional information about their information need through interactive query refinement and interactive search results exploration processes. In addition, there are a number of sources of derived information that can be used to support Web search, including individual document surrogates from the search results set, the actual documents that the document surrogates describe, features of the search results set, and information derived from external knowledge bases.

10.4.1 Attributes of the Query

Fundamentally, the attributes of the query are the specific terms (or phrases) that comprise the query. Every Web search engine allows users to provide simple lists of terms or phrases for querying. Further, they often provide a specific language for constructing advanced queries, including using logical operators and quotes to group query terms together and formulate an expression (similar to the logical expressions defined in common programming languages). Furthermore, they may also provide support to allow users to assign positive or negative attributes to the search terms that will include or exclude documents in the search results respectively. Unfortunately, few searchers make use of the advanced query operators provided by the search engines, instead preferring simple list-of-term queries [32].

Although interactive Web search systems can allow users to exert more control over the crafting of their queries, the result remains a query (with the same types of attributes as noted above). Even systems that allow the search to indicate other terms of interest (such as WordBars [19], which will be discussed in more detail later) provide fundamentally the same information: terms or phrases with either positive or negative relevance to the searcher's information need.

10.4.2 Attributes of the Document Surrogate

For each document surrogate within the search results set, a number of attributes are provided by most search engines. These include the title of the document, a snippet from the document (usually showing the query terms used in context from within the original document), and the URL of the document.

A number of attributes can be further derived from the information within each document surrogate. Based on the URL, the type of the file (e.g., html, pdf, doc, etc.) can be deduced. Using the textual data from the title and snippet, a query term vector can be calculated that represents the frequency of the query terms found within this data. Higher dimensional vectors can also be created based on the occurrence of each unique term within this data.

10.4.3 Attributes of the Document

Since each document surrogate in the search results set provides a URL to the source document, these original documents can easily be retrieved over the Internet. The specific attributes of these documents are dependent on the specific type of document. For example, some documents contain embedded meta-information about the creator, editing history, etc. Some structured documents (such as HTML documents) can lead to specific attributes being derived; others (such as Excel spreadsheets) may provide little additional information that can be automatically processed.

Many of the attributes derived from the document surrogate can also be derived from the actual document, perhaps with higher accuracy. However, the problem with using the document instead of the document surrogate is the delay associated with retrieving each of the documents for analysis purposes.

10.4.4 Attributes of the Search Results Set

With each set of search results provided by the underlying search engine, there is one core attribute that is automatically provided by the search engine: the number of documents in the search results set. Although this data can be useful in determining the generality or specificity of the query, it provides little additional information to the users.

As with the individual document surrogates, other attributes of the search results set can be calculated by analyzing the contents of the top document surrogates (e.g., the top 100 document surrogates). By analyzing the URL, the file type can be determined, resulting in a count for each unique file type. Alternately, the textual contents of the title and snippet for all the document surrogates can be analyzed in order to determine the number of documents that use each of the query terms, as well as all possible combinations of the query terms. This data can also be used to calculate a frequency vector based on the query terms, or even a frequency vector based on all unique terms in the entire set of the top search results.

10.4.5 External Knowledge Bases

External knowledge bases are organized collections of textual information that can be used to provide supplemental information to support the Web search process. Such knowledge bases are generally compact, easily searchable, and provide contextual information about the specific terms, phrases, or concepts which they contain. In general, such external knowledge bases can been used to support the query refinement process; their use for search results exploration remains an open area of research.

A prime example of an external knowledge base is an electronic thesaurus such as WordNet [43]. Although the use of WordNet to support Web search processes is promising, such systems have not proven to be very successful in supporting query refinement. In general, this is attributed to the general nature of the thesaurus and the inability of WordNet to make connections between terms that form different parts of speech [41].

Other well-organized collections of textual information exist on the Web. These include the Open Directory Project [44], and Wikipedia [70], among others. The Open Directory Project contains a hierarchy of topics, along with Web pages that have been classified by humans to belong to those topics. Wikipedia is a collaboratively authored online encyclopedia. Such systems could be dynamically queried during the Web search process, the results of which can be parsed and manipulated in order to assist users in crafting more accurate queries, or guiding their exploration of the search results.

A third class of external knowledge bases are those that have been created specifically to capture knowledge that can be used within information retrieval systems. The concept knowledge base [27] represents knowledge as a weighted graph structure with two classes of nodes: concepts and terms. The edge weight between a concept-term pair represents the degree to which the term can be used to describe the concept. In our work, this concept knowledge base was derived from information in the ACM Computing Classification System [1], and has been used to support interactive query refinement [29].

10.5 Challenges in Search Representations

Although the perceptual issues with visualizing atomic information features is well understood [69], certain types of data are not well suited to visual representation. One such example is textual information, causing special challenges for generating visual representations of the inherently textual nature of Web search. Techniques include either converting the textual information to numerical values, or deducing relationships that can then be visually represented. A further challenge in information visualization is the process of combining the visual elements in such a way that is useful with respect to the ultimate tasks of the users. The evaluations of such systems follow the guidance provided by the field of human–computer interaction.

The most common method for representing Web search results is through a list-based representation. All of the four primary Web search engines (Google,[1] Yahoo,[2] MSN,[3] and AOL[4]) use variations on this representation method. Although the supplemental information provided may differ slightly, these search engines provide a ranked list of ten document surrogates per page. Each document surrogate primarily consists of the title of the document, a snippet showing the context of the query terms, and the URL of the document.

Since the documents themselves are primarily textual, this textual representation of Web search results is a logical method for displaying this information. However, since these lists are static, they provide little ability to manipulate the search results; they only lend themselves to a sequential evaluation of ten document surrogates at a time. This textual interface style requires users to read the information (which can be time consuming), or scan the information (which can be error prone). Further, since text can consume a large amount of space in an interface, the list based representations commonly limit the number of document surrogates displayed per page.

Although graphical representations of search results have been explored within traditional information retrieval research [14,42], few of these techniques have been explored within the Web search setting. One of the challenges in providing a visual representation of Web search results is gaining access to enough information to generate a useful visual representation of the search results. Commonly, in traditional information retrieval systems, the visualization system has access to the entire textual contents of each document. However, for Web search, and especially for Web search interfaces that are meta-search systems, the contents of the entire documents in the search results set are not provided; the data provided is normally limited to the document surrogate (title, snippet, and URL).

If a visualization technique requires access to the entire contents of the documents in the search results set, these documents must be retrieved from the Web. This has the result of introducing significant delays in generating the visual representation, which will not be well received by Web searchers [12]. As such, many of the visualization approaches discussed in the information retrieval literature are not directly applicable to Web information retrieval.

Clustering techniques have also been investigated as a means for supporting users in their search results evaluation tasks. Commonly, these systems add additional interface elements to support users in navigating the clusters. However, they tend

[1] http://www.google.com/

[2] http://www.yahoo.com/

[3] http://www.msn.com/

[4] http://www.aol.com/

to continue to employ the simple list-based representations that are common among the primary Web search engines, limiting the support provided to the searcher to filtering the search results lists.

Another interesting avenue for supporting users in evaluating the results of their Web searchers is through re-sorting or re-ranking of the top Web search results provided by some underlying Web search engine. The re-sorting of search results based on Web search personalization is a rather active research field [50, 63, 64]. These systems generally provide an automated re-sorting and filtering of the search results based on the personalized profiles of the users. Generally, these personalized profiles are created through Web mining techniques that take advantage of the vast amounts of Web search user data [37].

Other work makes use of the Web page categories provided by the Open Directory Project [44] to re-rank Web search results [10]. The end result is similar to the personalization techniques: the search results are provided in a re-ranked order based on other knowledge present in the Web information retrieval system. However, these re-sorting systems continue to provide the search results in static list-based representation. There appears to be little research on interactive tools to allow users to control the re-sorting methods, in personalized systems or otherwise.

Clearly, there are many challenges in generating effective and useful representations of the information present in Web search tasks. Our focus in this Chapter is on visual methods, many of which are explored in the following section.

10.6 Seminal and State-of-the-Art Research in Visual Web Search

In the remainder of this Chapter, we will focus on a survey of both the seminal works and the current state-of-the-art in visual Web search, and provide our vision for the future of visual Web search. We will use the two fundamental tasks of the Web searcher as a means for organizing our discussion, resulting in first discussing query visualization, and then search results visualization. In addition to highlighting important works from the literature, we will also draw attention to our research efforts within this discussion.

10.6.1 Query Visualization

The application of information visualization to queries has primarily been focused on the visualization of Boolean queries [3, 35, 58, 75]. This research direction addresses the great difficulties many users have with crafting correctly specified queries of this type [14]. In the context of visualizing free text queries, very little work has been done. For example, in Zhu and Chen's survey of information visualization techniques for unstructured textual documents [76], the only men-

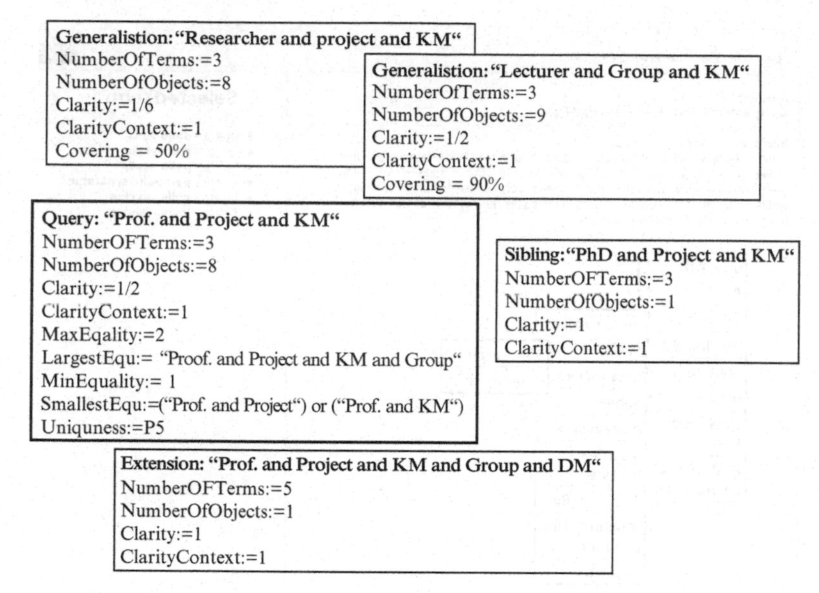

Fig. 10.1 A screenshot from the work of Stojanovic [61]

tion of visual techniques for supporting query specification was through the use of subject hierarchies as a means for suggesting appropriate query terms. However, no specific examples were cited.

Three systems that do provide visual representations of query information are the work by Stojanovic, Joho, and the VisGets system. Stojanovic [60, 61] visually represents the current query within the context of its neighbour queries using a simple box layout (see Figure 10.1). Navigating within this query space can allow users to express their information needs more precisely. Although the visual representation of the query space is secondary to the techniques for determining neighbour queries, a visual representation of the query space provides a meaningful organization of candidate queries. However, a requirement of this work is that the corpus being searched be represented as an annotated information repository. Since this additional information is not available in the common Web search indexes, the applicability of this approach to Web search is limited.

Joho [33, 34] uses a menu structure to visually represent a hierarchy of terms to add to a query in an interactive query expansion system (see Figure 10.2). In this work, hierarchies are automatically generated based on the co-occurrence of terms within a set of retrieved documents [54]. Although this work addresses the problems of the simple list-based representation of potential query terms, it requires that a number of documents returned from the initial search be retrieved in their entirety. For Web search, this may result in the system introducing a significant delay for the users, which is contrary to the preferences of Web users for fast Web search engines [12]. Further, although the visual presentation of candidate query terms is superior to a simple list-base representation, such candidate query terms remain hidden from view until users navigate to the appropriate location.

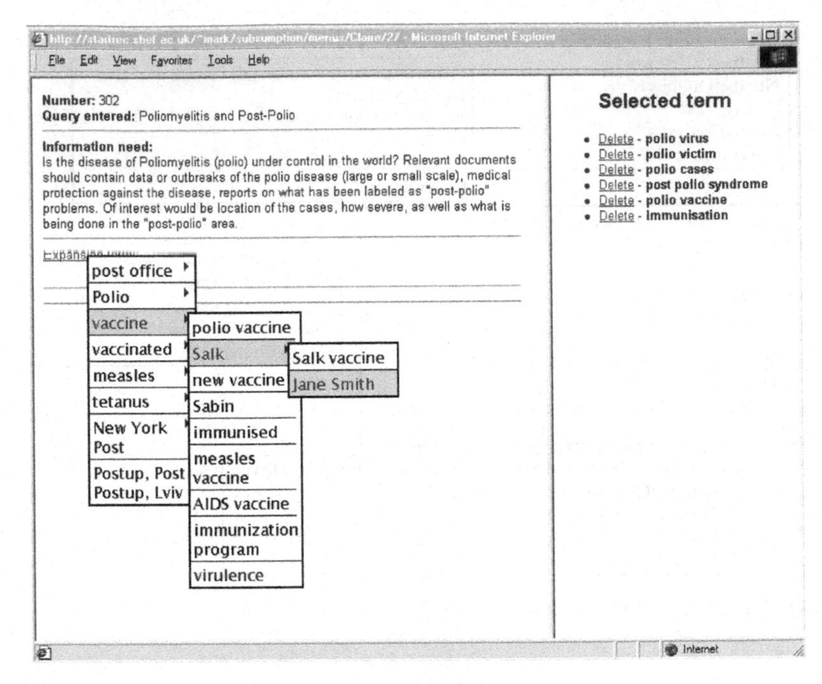

Fig. 10.2 A screenshot from the work of Joho et al. [34]

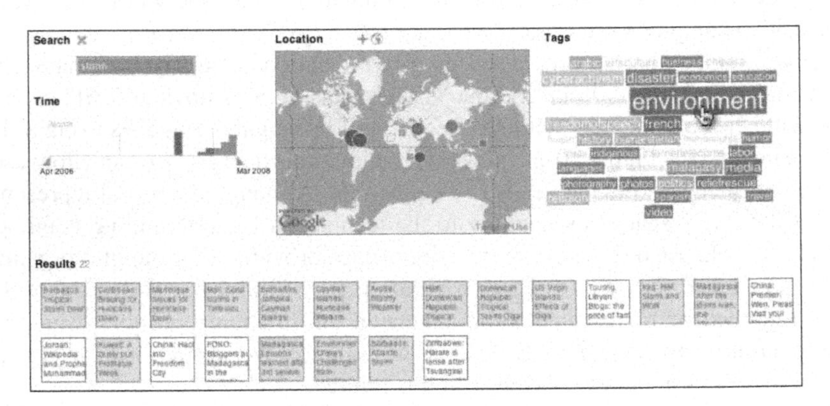

Fig. 10.3 A screenshot of the VisGets query visualization system [11]

VisGets [11] provides a set of coordinated visual representations designed to support the searcher in dynamically refining their query (see Figure 10.3). Three types of information are visually represented and available for manipulation: publication date, geographic location, and author-supplied tags. Searchers are able to visually manipulate these aspects of a query, allowing the identification of relationships between these attributes, as well as the filtering of search results. The system is designed for searching RSS feeds; the extension to Web search is complicated by the general lack of geographical information and author-supplied tagging of content. Preliminary studies showed that users found the ability to manipulate the time dimension of value; the visual manipulation of the spatial and tag components where less useful. However, the combined visual representation of these attributes can be of great value when the information need is exploratory, and includes both temporal and geographic components.

Generating a visualization of a Web search query is challenging since there is often little or no information available upon which to generate a visual representation. For example, in the absence of additional information, how can one generate a visual representation of a set of search keywords entered by a user? Our research efforts have produced two systems that attempt to address this problem: VisiQ and WordBars.

VisiQ [28, 29] makes use of a concept knowledge base as a means of deriving additional information about a query. The user-supplied query is matched to the concept knowledge base, and the concepts that are most similar to the query terms are selected. Additional terms that have also been used to describe these concepts are also selected. Together, these concepts and additional terms form a *query space*. VisiQ generates a visual representation of the query space (see Figure 10.4), allowing users to see how their query terms are related to high-level concepts and other potentially relevant terms. Query refinement is supported by double-clicking on terms to be added or removed from the query. Doing so dynamically updates the query space, and produces a preview of the search results. Once a query has been adequately refined, it can be sent to the Google search engine with a simple click of the "Google This" button.

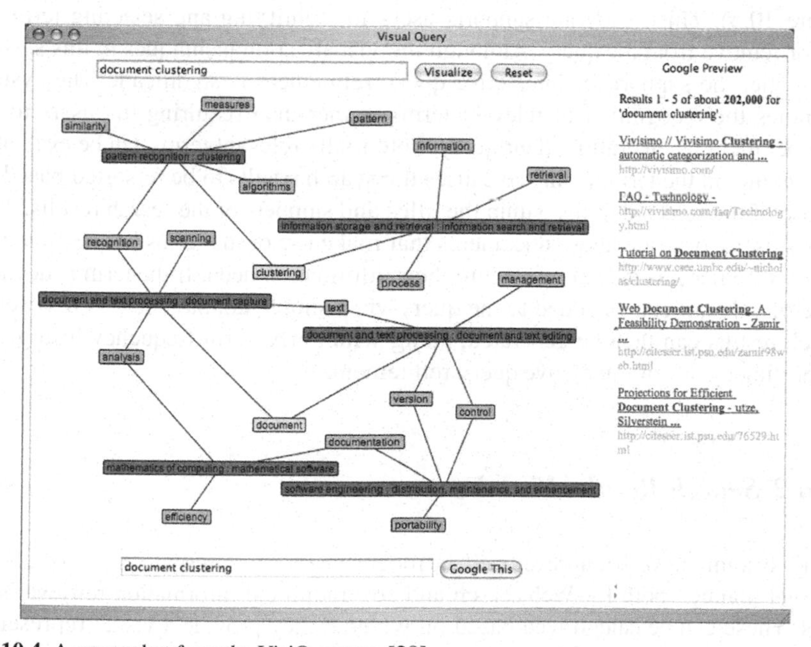

Fig. 10.4 A screenshot from the VisiQ system [29]

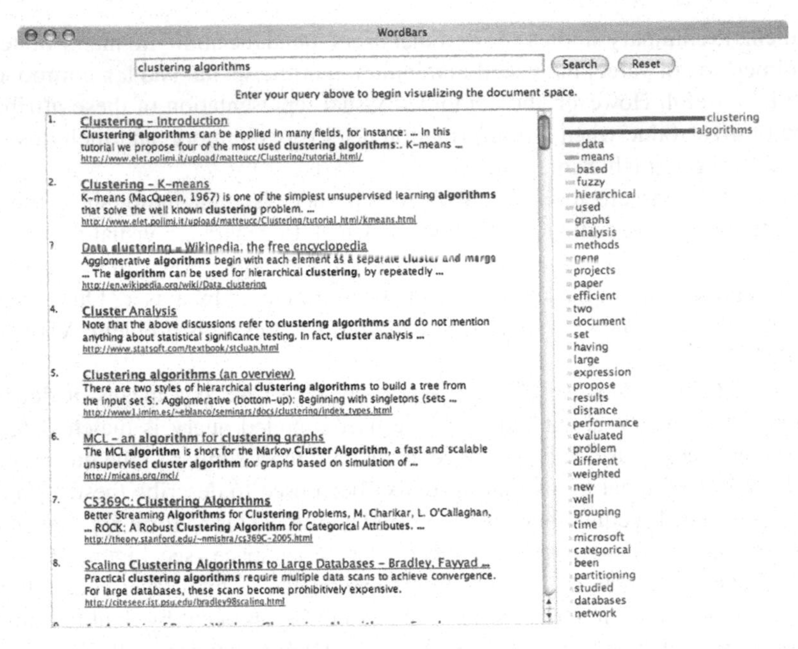

Fig. 10.5 A screenshot from the WordBars system [25]

WordBars [19, 24, 25] visually depicts term frequency histograms generated for all the unique terms found in the top search results from an initial query (see Figure 10.5). This histogram supports users in identifying and selecting terms to add or remove from the query. Although the visualization techniques in this system are subtle, the support for interactive query refinement is significant. The system promotes the recognition of relevant terms, rather than requiring the users to remember them when crafting their query. Potentially relevant terms can be evaluated by clicking on the term. Doing so causes the search results to be re-sorted based on the use of the selected terms within the titles and snippets of the search results. This allows users to review those documents that make use of the terms in question, and decide whether it is indeed relevant to their information need. If the term is deemed to be of value, it can be added to the query via a simple double-click. A new set of search results can then be generated, along with a new term frequency histogram, supporting cycles of interactive query refinement.

10.6.2 Search Results Visualization

Many systems have been developed in recent years to represent search results in a visual manner, both for Web search and for traditional information retrieval systems. These can be categorized based on whether they provide a visual representation based on the entire document contents or a visual representation based on an abstraction of the document (i.e., the document surrogate). While it is not feasible to provide an exhaustive survey and analysis of all the methods for representing search results, a brief review of relevant and interesting techniques is provided, with emphasis given to those techniques which are appropriate for Web information retrieval. Our research efforts in this domain are highlighted at the end.

10.6.2.1 Document Visualization

Document visualization can be defined as the process of converting textual information from a document into graphical representations that can be processed visually rather than read. Since preattentive processing of certain types of graphical information is significantly faster than the non-preattentive processing required for reading [69], there is a great opportunity for taking advantage of the human visual processing capabilities when presenting textual information.

However, the representation of textual information in a visual manner is by no means a simple task. At the most fundamental level, one can think of a document as a collection of terms, represented by a high dimensional vector. Dimensional reduction techniques can be used to map a set of such document vectors into two or three dimensional space, resulting in each document occupying some point in space. The spatial proximity of two documents implies similarity, resulting in a visual clustering of documents.

These vector-based techniques have been used for the visualization of collections of documents in systems such as Galaxy of News [52] (see Figure 10.6), and ThemeScape [71, 72] (see Figure 10.7). However, accessing and viewing the

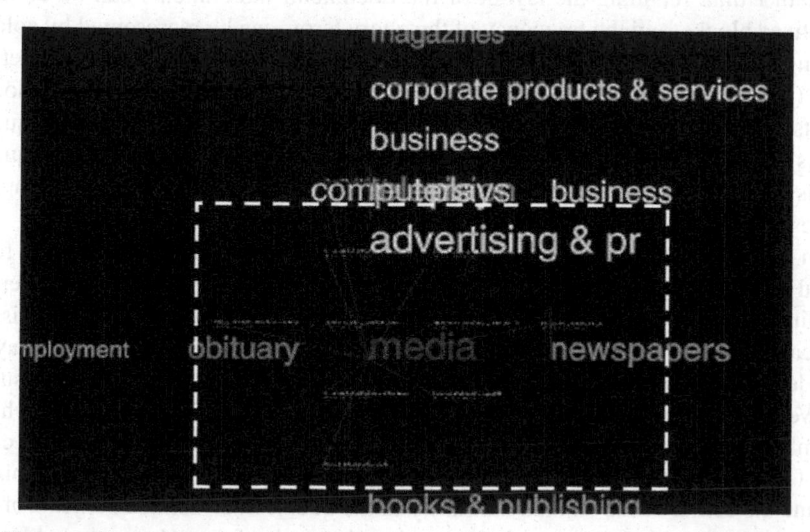

Fig. 10.6 A screenshot from Galaxy of News [52]

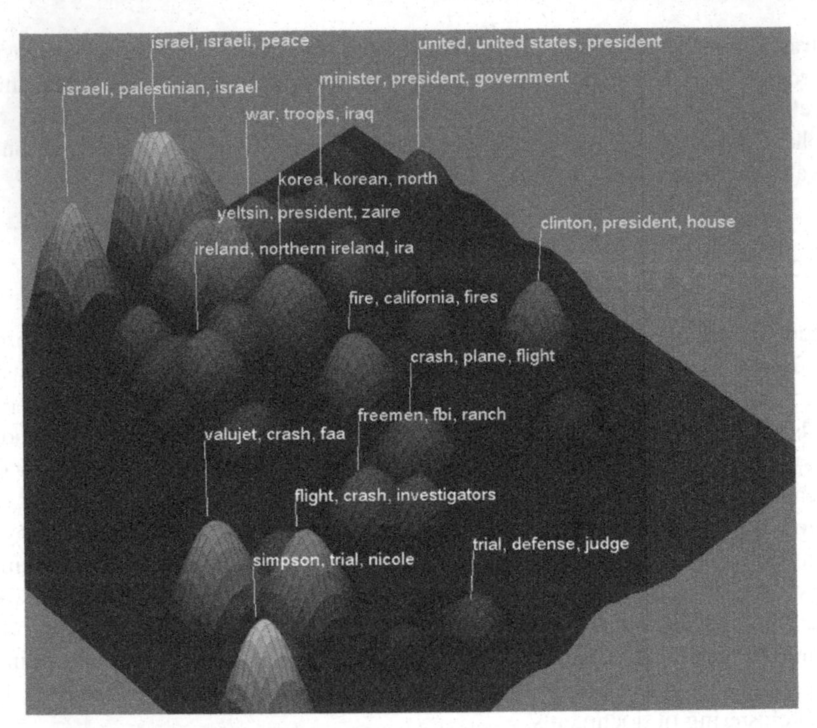

Fig. 10.7 A screenshot from Themescape [72]

information on specific documents is not well supported. Hearst noted that "although intuitively appealing, graphical overviews of large document spaces have yet to be shown to be useful and understandable for users" [14].

Rather than retaining the layout of the document, the contents can be divided into fixed blocks, and the frequency of the query terms can be represented by colour coding in each block, as in TileBars [13] (see Figure 10.8). The result is a set of bars (one for each document) whose widths are relative to the length of the documents, and whose heights are relative to the number of query terms (or sets of query terms). Colour coding within these bars represents the frequency of use of the query terms relative to the location within the document. This results in a more compact representation than the previous example.

In the work by Heimonen and Jhaveri [15], each document is divided into four equal sized blocks (see Figure 10.9). The occurrences of all of the query terms within a 20-word window for each block are counted and depicted using a visual indicator similar to the technique used in TileBars [13]. This indicator is displayed beside each document surrogate in the list-based representation of the search results.

WebSearchViz [45] employs a visual metaphor based on a solar system, which is embedded in a purpose-built Web search browser (see Figure 10.10). At the center of the solar system is the user's query; documents are objects that are organized around the center based on their similarity to the query. User-defined subjects of interest can be specified as external objects placed outside of the solar system. Motion

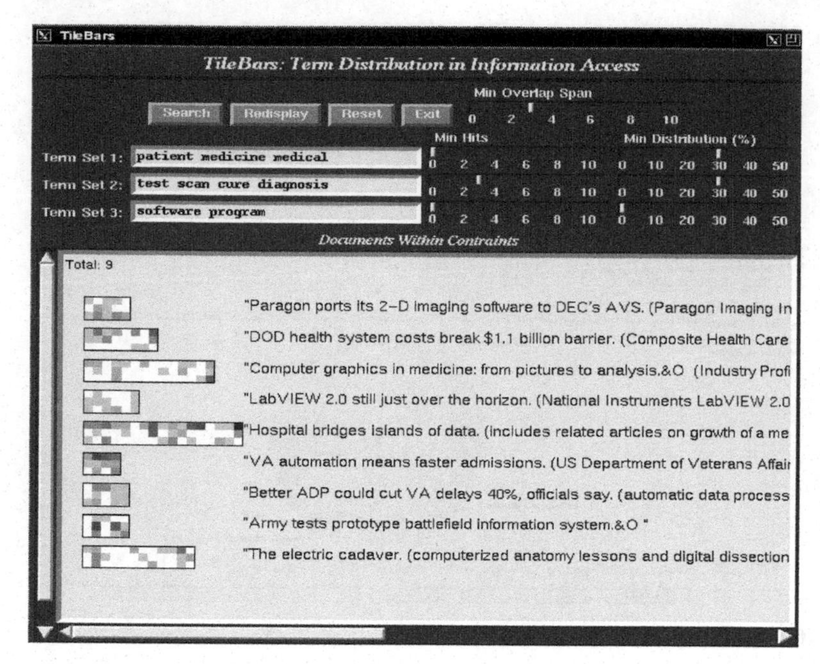

Fig. 10.8 A screenshot from TileBars [13]

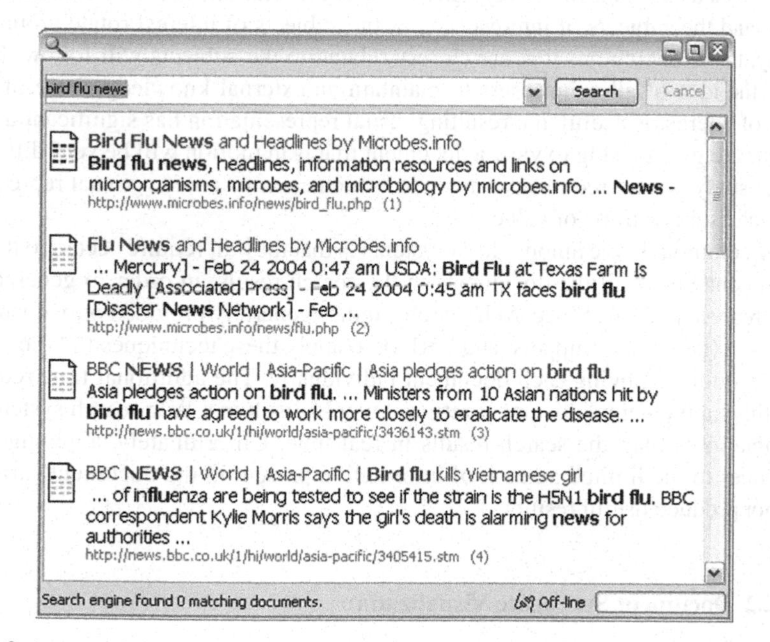

Fig. 10.9 A screenshot from the work of Heimonen and Jhaveri [15]

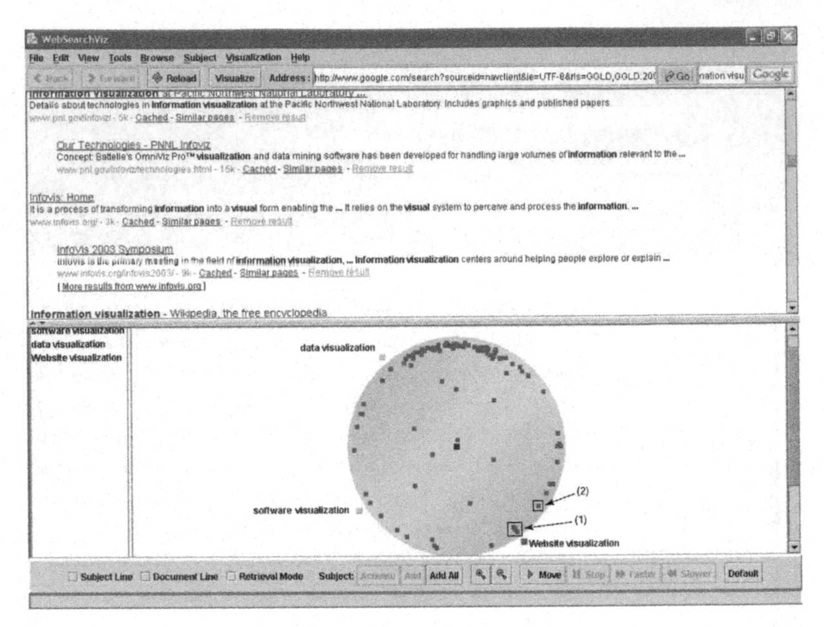

Fig. 10.10 A screenshot of WebSearchViz [45]

is employed as a means of conveying to the user the relationships between the search results and the subjects of interest (i.e., as the subjects of interest rotate around the display, those documents that are closely related to the subjects will follow them). While the idea of allowing users to maintain an external knowledge base of their topics of interest is useful, the resulting visual representation has significant usability issues (e.g., choosing to view a document that is in motion will be very difficult). Further study is required to validate under which circumstances a visual representation such as this will be of value.

One common theme among these systems is that they all require access to the full textual contents of all the documents in the search results in order to generate the visual representations. Since Web search engines generally provided only document surrogates (i.e., titles, snippets, and URLs), to apply these techniques to Web search would require retrieving each document individually. The additional time required to do this supplemental document retrieval would result in a Web search system that is unable to display the search results in real-time. Unfortunately, applying these techniques to the limited amount of information present in the document surrogate may not produce useful results.

10.6.2.2 Document Surrogate Visualization

A meta-search system is a system which makes use of the search results provided by other search engines. Since access to the textual contents of each document is not feasible for a meta-search system, the visual representation of document surrogates

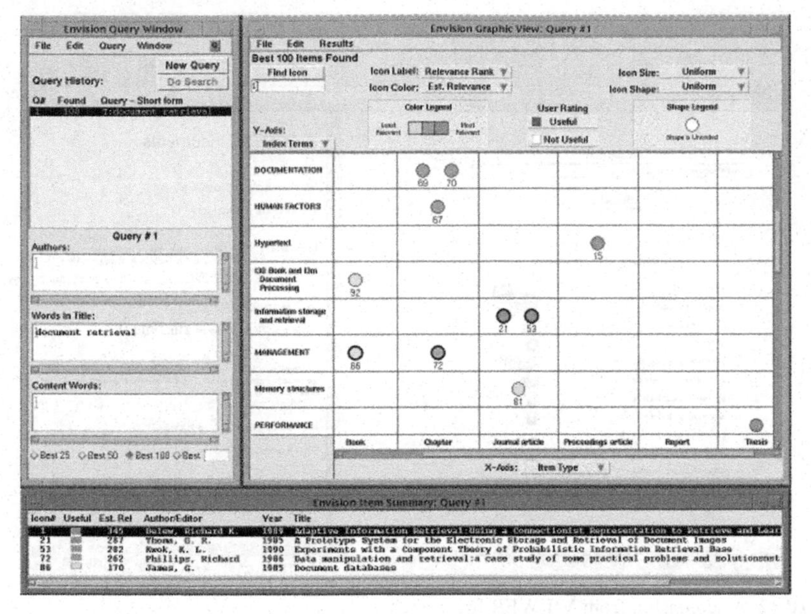

Fig. 10.11 A screenshot from Envision [47]

is a viable alternative. A document surrogate consists of summary information, attributes, and other meta-data that represent the document in the search results. Document surrogates are the primary data objects in the list-based representation used by many Web search engines, where they commonly consist of the title of the document, a snippet showing the query terms in context, the URL of the document, as well as other information.

Envision [47] uses a highly customizable scatterplot and iconic visualization to represent the many different attributes available as part of their custom information retrieval system (see Figure 10.11). Although this visual representation was shown to be very powerful, it makes use of information that is not commonly available in the document surrogates returned by Web search engines. Further, there is an added level of complexity in this interface that may make it too difficult for the general public to use effectively.

In VIEWER [6], all possible combinations of the query terms are generated and searched for in the document surrogates returned by the AltaVista search engine (see Figure 10.12). A histogram of these query term combinations is provided to the users, which can be used to select subsets of the search results for further investigation. Although this system provides valuable information to users in terms of how the query terms are used in the search results set, the interaction with the search results set is limited to filtering.

xFind [2] provides three different interfaces to a custom Web document indexing system: a simple list-based representation; a scatterplot representation similar to that in Envision [47]; and a vector-based spatial clustering representation similar to that in ThemeScape [72] (see Figure 10.13). While these representations of the search

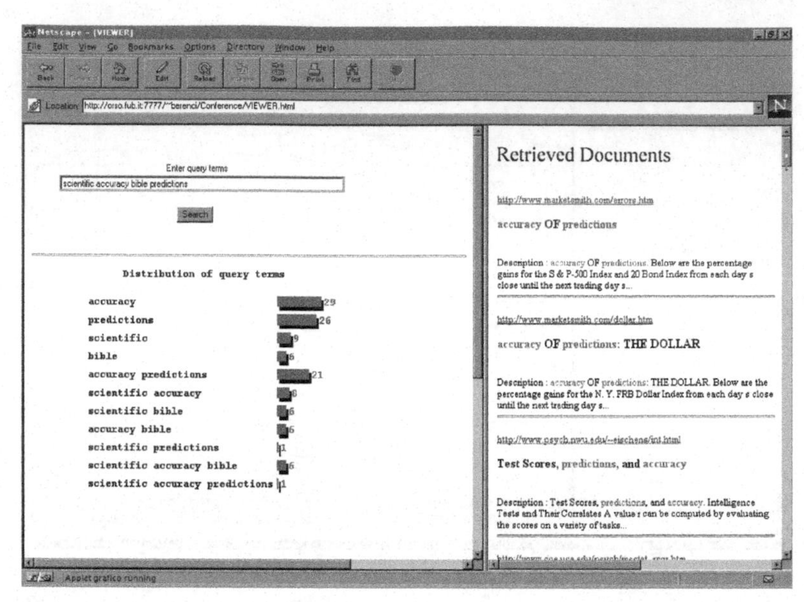

Fig. 10.12 A screenshot from VIEWER [6]

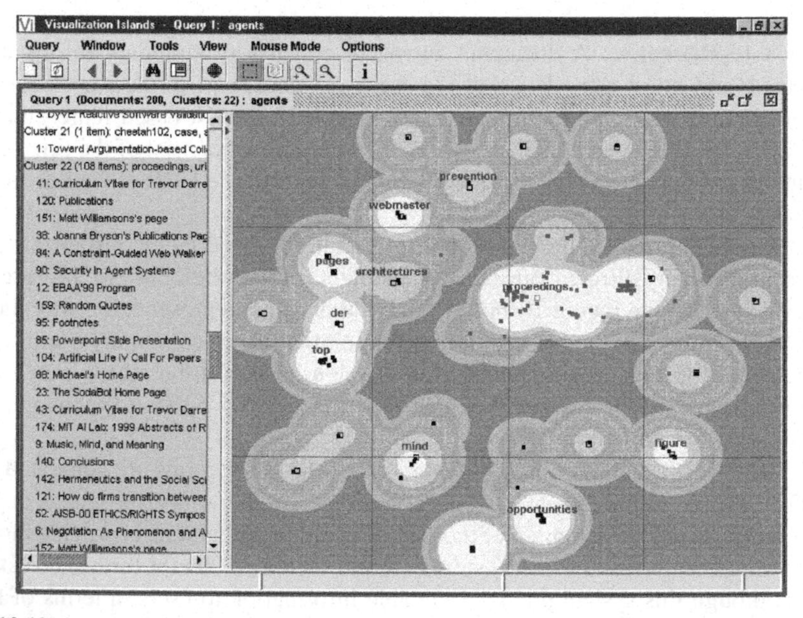

Fig. 10.13 A screenshot from xFind [2]

results take advantage of the extra information that is available through their indexing system, this information is not available with other search engines. Further, the spatial layout of the two visual representations maps the document surrogates

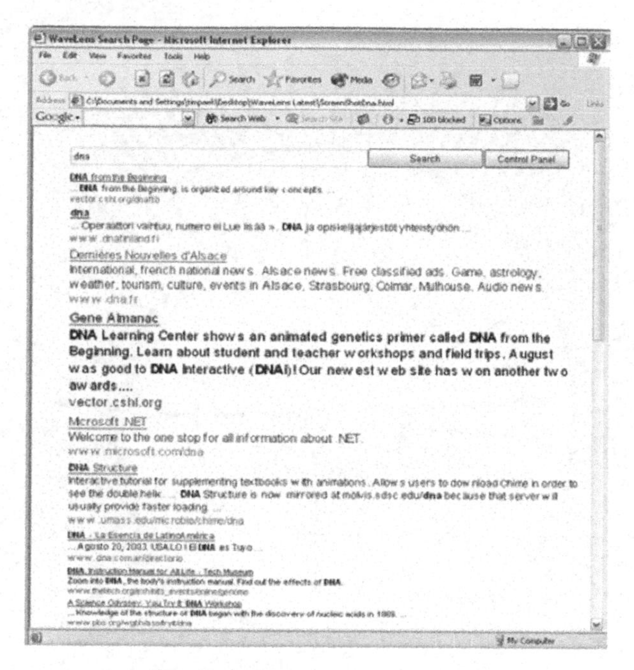

Fig. 10.14 A screenshot from WaveLens [48]

to points in the two-dimensional display, making it difficult to view additional information present in the document surrogate, or to make comparisons between document surrogates.

WaveLens [48] provides a focus+context representation of the search results allowing users to dynamically zoom into document surrogates of interest (see Figure 10.14). The results are provided in the traditional list-based representation. As users move their mouse over a document surrogate, its font size increases, and the font size of the other document surrogates not in focus decrease. This results in a fisheye lens effect. Additional text from the document is dynamically added or removed from a document surrogate by clicking the mouse. While this technique may make it easier for users to read the contents of the list of search results, the primary interaction method remains a sequential evaluation of the document surrogates, with little support for manipulating and exploring the search results.

RankSpiral [59] uses a spiral metaphor to organize the search results provided by multiple search engines (see Figure 10.15). The distance of a search result from the center is a visual indication of the relevance of the document, based on the average rank provided by all the search engines used. Users are able to readily identify documents that were consistently ranked highly, as well as groups of documents that have the same average ranking. Complex icons are used to visually depict which search engines a particular document came from. Although this system does provide a clear indication of groups of documents with similar rank scores, the ability to read the labels in the spiral as well as visually decode the icons is difficult.

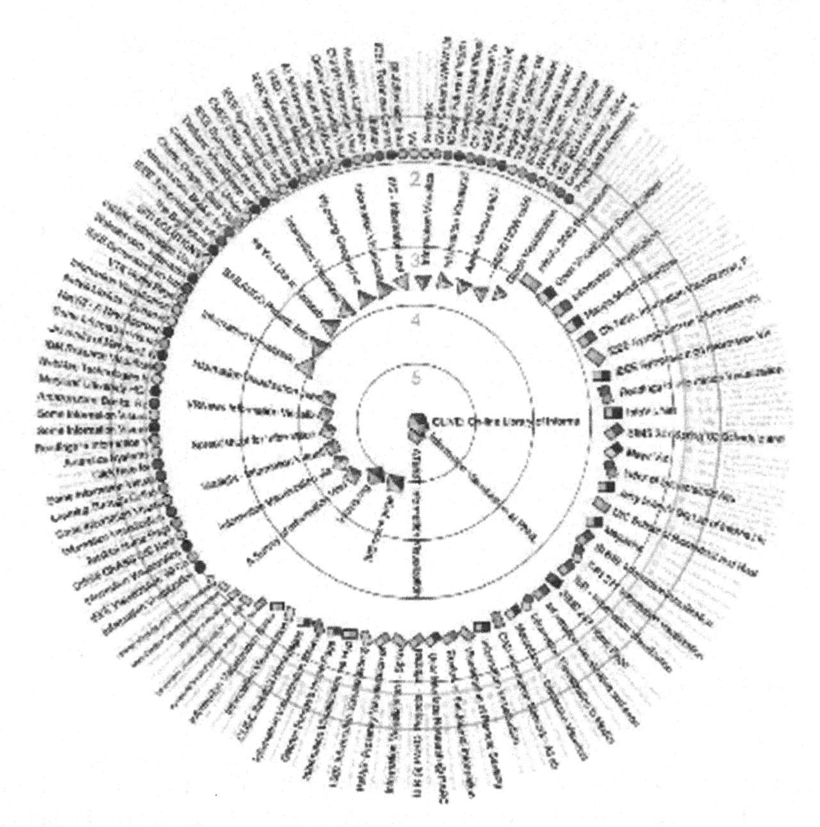

Fig. 10.15 A screenshot from RankSpiral [59]

PubCloud [38] uses tag clouds to visually summarize the abstracts of the search results from the PubMed database (see Figure 10.16). Within the tag cloud, the size of the term represents its frequency within the abstracts of the documents, and the colour of the term represents the average recency of the document (i.e., whether the term is used in recently published documents). Selecting a tag allows the searcher to navigate to those documents that make use of the term. While useful for summarizing the search results set, the utility of tag clouds for manipulating and exploring search results remains questionable.

Many other visualization systems exist for representing the document surrogates returned by Web search engines, including a number of publicly accessible clustering search engines such as Kartoo,[5] Mooter,[6] and Grokker.[7] An evaluation of the merits and problems with these systems is beyond the scope of this dissertation, although one review indicated that some of these systems do not add any support for users in assimilating or processing the information [49].

[5] http://www.kartoo.com/

[6] http://www.mooter.com/

[7] http://www.grokker.com/

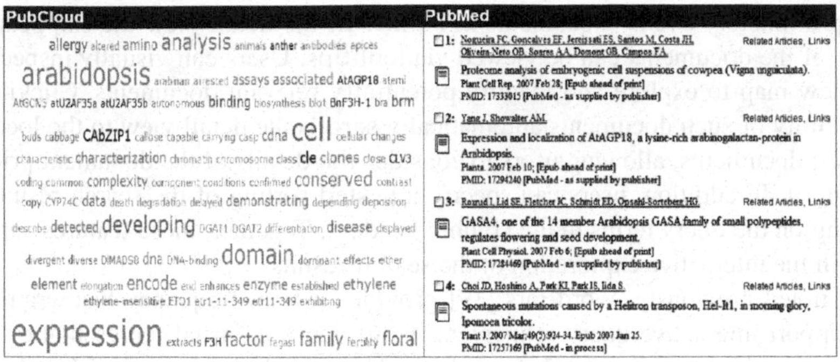

Fig. 10.16 A screenshot from PubCloud [38]

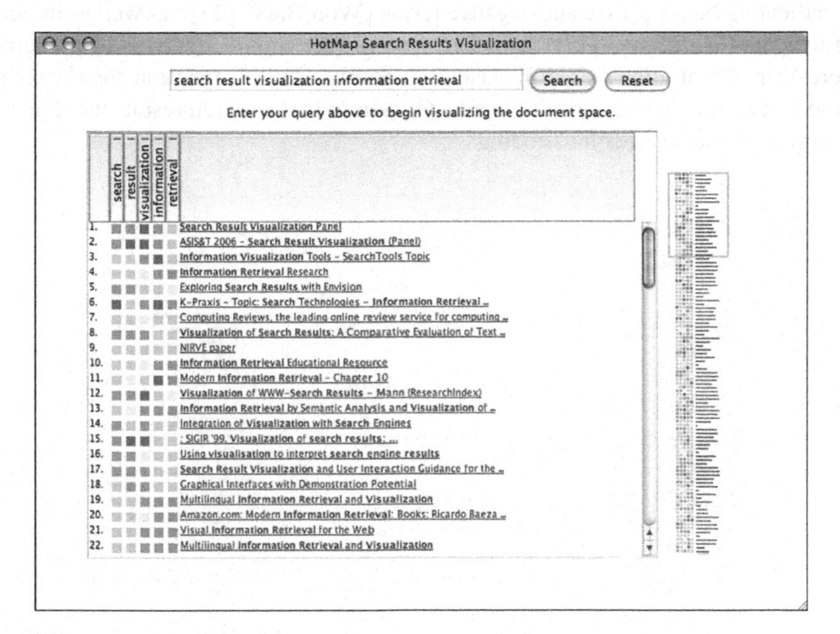

Fig. 10.17 A screenshot from HotMap [20]

In the course of our research efforts, we have developed two prototype systems that make use of visualization techniques based on the document surrogates: HotMap and WordBars3. Note that the original WordBars system was previously discussed in terms of supporting query expansion; the focus here is on a refinement to this system providing advanced search results exploration features.

In HotMap [20, 26], query term frequencies are visually represented at two levels of detail using colour coding on a heat scale (see Figure 10.17). In both the overview map and the detailed view, each column represents a term in the users' queries; each row represents a search result. The overview map shows a zoomed-out, abstract view of 100 documents at a time; the detailed view shows the titles of approximately twenty documents at a time. In the detail view the snippets and URLs of the documents can be viewed via tool tips. Users can visually inspect the overview map to explore and identify potentially relevant documents. Clicking in the vicinity of such documents automatically scrolls the detail view to the location of those documents, allowing users to consider the details of the document prior to viewing it. In addition, users can specify a nested sorting of the search results by clicking on the query terms in the column headers. Together, these features support users in the interactive exploration of the search results.

As noted previously, WordBars [19] provides a term frequency histogram that can support interactive query refinement. However, this visual representation also provides a visual overview of the search results set. Selecting terms within the histogram causes the search results to be re-sorted, supporting interactive search results exploration processes. Subsequent modifications of the system have added support for indicating both positive and negative terms (WordBars2 [23]), as well as indicating weights (WordBars3 [17]). This last system (see Figure 10.18) is of particular interest, in that it allows users to visually specify a utility function for re-sorting the search results, which visually represents where in the search results the terms of interest are located via colour coding.

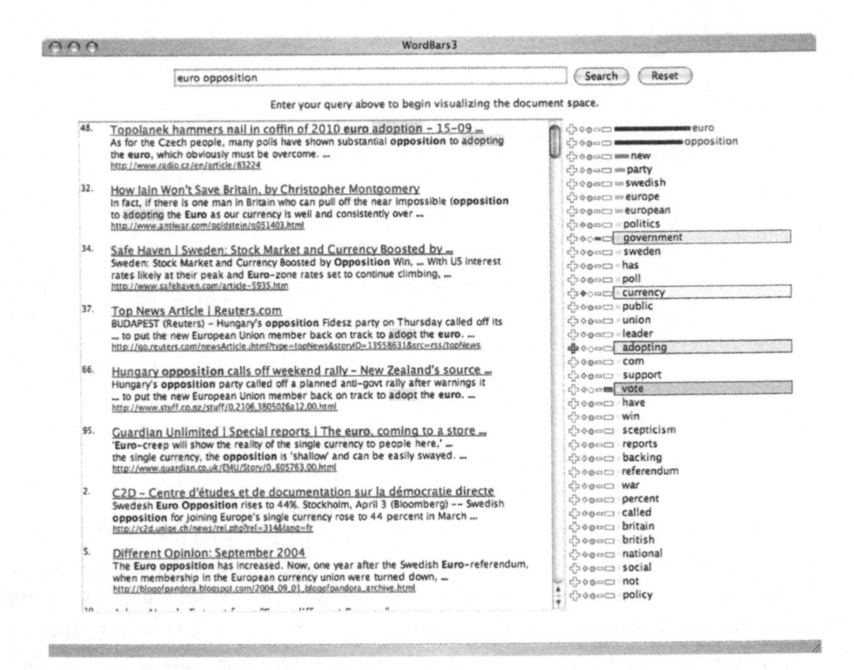

Fig. 10.18 A screenshot from WordBars3 [17]

10.6.3 Revisiting the Taxonomy of Information

Previously, a taxonomy of the information that can be used to support Web search processes was presented. In this section, we re-visit this taxonomy and discuss how the systems discussed in this Chapter can be classified based on this taxonomy.

Figure 10.19 shows the classification of these works. Each row in this table represents one type of information identified in the taxonomy. The first column represents the query visualization techniques; the second and third represent the two different types of search results visualization (i.e., document visualization and document surrogate visualization).

The result is not a strict classification, in that specific systems can take advantage of multiple types of information. For example, WordBars [25] represents information based on both the attributes of the search results set (i.e., generating the term frequency histogram), and the attributes of the query (i.e., showing which terms in the

	Query Visualization	Document Visualization	Document Surrogate Visualization
Attributes of the Query	Stojanovic [61] WordBars [25]	WebSearchViz [45]	WordBars3 [17]
Attributes of the Document Surrogate	VisGets [11]		Envision [47] VIEWER [6] xFind [2] WaveLens [48] PubCloud [38] HotMap [20] WordBars3 [17]
Attributes of the Document	Joho et al. [34]	Galaxy of News [52] Themescape [72] Tilebars [13] Heimonen & Jhaverl [15] WebSearchViz [45]	
Attributes of the Search Results Set	WordBars [25]		Envision [47] xFind [2] WordBars3 [17] RankSpiral [59]
External Knowledge Bases	VisiQ [29] VisGets [11]	WebSearchViz [45]	

Fig. 10.19 A classification of the systems discussed in this chapter with respect to the taxonomy of information that can support Web search processes

histogram were in the query). WordBars3 [17] extends this to include representing attributes of the document surrogates (i.e., showing where in the document surrogates the selected terms are present). Both Envision and xFind make use of additional information provided by their custom information retrieval systems, resulting in them both being classified according to two types of information.

Of note in this classification are the row that represents the use of attributes of the document as the source of the information, and the column that represents document visualization. Clearly, these two aspects are closely related to one another. In order to visualize document information, the attributes of the document are required. However, accessing the attributes of each individual document in a search results set is only feasible when the inner workings of a Web search engine are available. The alternative is to retrieve the textual contents of each document one-by-one, resulting in significant delays. Of the systems described, only the work by Heimonen and Jhaveri [15] and WebSearchViz [45] were applied to the problem of Web search visualization; the rest provide visual interfaces to traditional information retrieval systems.

An important gap in this classification exists at the intersection of external knowledge bases and document surrogate visualization. This represents an open area of research in which there appears to be little work being done. It may be possible to take external knowledge bases such as the concept knowledge base used in VisiQ [29], or the information in Wikipedia [70], and use this to support the visual representation of document surrogates. For example, one might match each document surrogate in the search results to one or more topics in Wikiepedia, and then use this information to visually represent the relationships between the individual document surrogates.

10.7 Conclusions

As the amount of information on the Web continues to grow, search engines will continue to be the primary method by which people find information. The advances that Web search companies make in their algorithms and infrastructure has and will continue to allow them index the Web as it grows, yet still return the results of a search in fractions of a second. Other advances in Web search will include indexing the "deep Web", the personalization of Web search, and improvements in deducing the potential relevance of documents.

One aspect that will have a significant impact on the utility of Web search engines of the future will be the interface provided to the users. As people become more accustomed to searching and using the Web in their daily activities, they will begin to demand more powerful tools to support their search activities. The visual interfaces for Web search discussed in this chapter are on the forefront of a movement towards tools that allow users to take an interactive role in the Web search process.

The development and study Web information retrieval support systems that provide visual and interactive interfaces to the Web search process will form an important step towards the next generation of Web search. Such systems will provide visual representations of queries and support the interactive refinement of such queries. As users exert more control in the crafting of their queries through the support of the visual query interfaces, the quality of their queries will increase. However, even with well crafted queries that accurately capture their information needs, users will still need to explore the search results due to the high volume of information returned from such queries. The visual representations of features of the search results will assist users in this regard, allowing them to interactively explore the search results as they seek relevant documents.

References

1. ACM: ACM computing classification system (2004). URL http://www.acm.org/class/
2. Andrews, K., Gutl, C., Moser, J., Sabol, V.: Search result visualization with xFind. In: Proceedings of the Second International Workshop on User Interfaces to Data Intensive Systems (2001)
3. Anick, P., Brennan, J., Flynn, R., Hanssen, D., Alvey, B., Robbins, J.: A direct manipulation interface for boolean information retrieval via natural language query. In: Proceedings of the ACM SIGIR Conference on Research and Development in Information Retrieval (1990)
4. Baeza-Yates, R., Ribeiro-Neto, B.: Modern Information Retrieval. Addison-Wesley (1999)
5. Benderson, B.B., Shneiderman, B.: The Craft of Information Visualization. Morgan Kaufmann (2003)
6. Berenci, E., Carpineto, C., Giannini, V., Mizzaro, S.: Effectiveness of keyword-based display and selection of retrieval results for interactive searches. International Journal on Digital Libraries **3**(3), 249–260 (2000)
7. Bertin, J.: Semiology of Graphics. University of Wisconsin Press (1983)
8. Brin, S., Page, L.: The anatomy of a large-scale hypertextual Web search engine. In: Proceedings of the Seventh International World Wide Web Conference (1998)
9. Card, S.K., Mackinlay, J.D., Shneiderman, B.: Readings in Information Visualization. Morgan Kaufmann (1999)
10. Chirita, P.A., Nejdl, W., Paiu, R., Kohlschutter, C.: Using ODP metadata to personalize search. In: Proceedings of the ACM SIGIR Conference on Research and Development in Information Retrieval (2005)
11. Dörk, M., Carpendale, S., Collins, C., Williamson, C.: VisGets: Coordinated visualizations of Web-based information exploration and discovery. IEEE Transactions on Visualization and Computer Graphics **14**(6), 1205–1212 (2008)
12. Graphics, Visualization, & Usability Center: GVU's 10th WWW user survey (1998). URL http://www.gvu.gatech.edu/user_surveys/survey-1998-10/
13. Hearst, M.: TileBars: Visualization of term distribution information in full text information access. In: Proceedings of the ACM Conference on Human Factors in Computing Systems (1995)
14. Hearst, M.: User interfaces and visualization. In: R. Baeza-Yates, B. Ribeiro-Neto (eds.) Modern Information Retrieval. Addison-Wesley (1999)
15. Heimonen, T., Jhaveri, N.: Visualizing query occurrence in search result lists. In: Proceedings of the International Conference on Information Visualization (2005)
16. Hering, E.: Outlines of a Theory of Light Sense (Grundzge der Lehr von Lichtsinn, 1920). Harvard University Press (1964)
17. Hoeber, O.: Exploring Web search results by visually specifying utility functions. In: Proceedings of the IEEE/WIC/ACM International Conference on Web Intelligence (2007)

18. Hoeber, O.: Web information retrieval support systems: The future of Web search. In: Proceedings of the IEEE/WIC/ACM International Conference on Web Intelligence – Workshops (International Workshop on Web Information Retrieval Support Systems) (2008)
19. Hoeber, O., Yang, X.D.: Interactive Web information retrieval using WordBars. In: Proceedings of the IEEE/WIC/ACM International Conference on Web Intelligence (2006)
20. Hoeber, O., Yang, X.D.: The visual exploration of Web search results using HotMap. In: Proceedings of the International Conference on Information Visualization (2006)
21. Hoeber, O., Yang, X.D.: Visually exploring concept-based fuzzy clusters in Web search results. In: Proceedings of the Atlantic Web Intelligence Conference (2006)
22. Hoeber, O., Yang, X.D.: A unified interface for visual and interactive Web search. In: Proceedings of the IASTED International Conference on Communications, Internet, and Information Technology (2007)
23. Hoeber, O., Yang, X.D.: Visual support for exploration within Web search results lists. In: Poster Compendium of the IEEE Information Visualization Conference (2007)
24. Hoeber, O., Yang, X.D.: Evaluating the effectiveness of term frequency histograms for supporting interactive Web search tasks. In: Proceedings of the ACM Conference on Desigining Interactive Systems (2008)
25. Hoeber, O., Yang, X.D.: Evaluating WordBars in exploratory Web search scenarios. Information Processing and Management **44**(2), 485–510 (2008)
26. Hoeber, O., Yang, X.D.: HotMap: Supporting visual explorations of Web search results. Journal of the American Society for Information Science and Technology **60**(1), 90–110 (2009)
27. Hoeber, O., Yang, X.D., Yao, Y.: Conceptual query expansion. In: Proceedings of the Atlantic Web Intelligence Conference (2005)
28. Hoeber, O., Yang, X.D., Yao, Y.: Visualization support for interactive query refinement. In: Proceedings of the IEEE/WIC/ACM International Conference on Web Intelligence (2005)
29. Hoeber, O., Yang, X.D., Yao, Y.: VisiQ: Supporting visual and interactive query refinement. Web Intelligence and Agent Systems: An International Journal **5**(3), 311–329 (2007)
30. Jansen, B.J., Booth, D.L., Spink, A.: Determining the user intent of Web search engine queries. In: Proceedings of the International World Wide Web Conference (2007)
31. Jansen, B.J., Pooch, U.: A review of Web searching studies and a framework for future research. Journal of the American Society for Information Science and Technology **52**(3), 235–246 (2001)
32. Jansen, B.J., Spink, A.: How are we searching the World Wide Web? a comparison of nine search engine transaction logs. Information Processing and Management **42**(1), 248–263 (2006)
33. Joho, H., Coverson, C., Sanderson, M., Beaulieu, M.: Hierarchical presentation of expansion terms. In: Proceedings of the ACM Symposium on Applied Computing (2002)
34. Joho, H., Sanderson, M., Beaulieu, M.: A study of user interaction with a concept-based interactive query expansion support tool. In: Proceedings of the European Conference on IR Research (2004)
35. Jones, S.: Graphical query specification and dynamic results previews for a digital library. In: Proceedings of the ACM Symposium on User Interface Software and Technology (1998)
36. Kobayashi, M., Takeda, K.: Information retrieval on the Web. ACM Computing Surveys **32**(2), 114–173 (2000)
37. Kosala, R., Blockeel, H.: Web mining research: A survey. SIGKDD Explorations **2**(1), 1–15 (2000)
38. Kuo, B.Y.L., Hentrich, T., Good, B.M., Wilkinson, M.D.: Tag clouds for summarizing Web search results. In: Proceedings of the International World Wide Web Conference (2007)
39. Lawrence, S., Giles, C.L.: Accessibility of information on the Web. Nature **400**, 107–109 (1999)
40. Mackinlay, J.: Automating the design of graphical presentations of relational information. ACM Transactions on Graphics **5**(2), 110–141 (1986)

41. Mandala, R., Tokunaga, T., Tanaka, H.: The use of WordNet in information retrieval. In: Proceedings of the COLING/ACL Workshop on Usage of WordNet in Natural Language Processing Systems (2000)
42. Marchionini, G.: Interfaces for end-user information seeking. Journal of the American Society for Information Science **43**(2), 156–163 (1992)
43. Miller, G.A.: Wordnet: An online lexical database. International Journal of Lexicography **3**(4), 235–244 (1990)
44. Netscape: The open directory project (2004). URL http://www.dmoz.org/
45. Nguyen, T.N., Zhang, J.: A novel visualization model for Web search results. IEEE Transactions on Visualization and Computer Graphics **12**(5), 981–988 (2006)
46. Nielsen, J.: When search engines become answer engines. Alertbox (2004). URL http://www.useit.com/alertbox/20040816.html
47. Nowell, L., France, R., Hix, D., Heath, L., Fox, E.: Visualizing search results: Some alternatives to query-document similarity. In: Proceedings of the ACM SIGIR Conference on Research and Development in Information Retrieval (1996)
48. Paek, T., Dumais, S., Logan, R.: Wavelens: A new view onto internet search results. In: Proceedings of the ACM Conference on Human Factors in Computing Systems (2004)
49. Powers, D.M., Pfitzner, D.: The magic science of visualization. In: Proceedings of the Joint International Conference on Cognitive Science (2003)
50. Radlinski, F., Dumais, S.: Improving personalized Web search using result diversification. In: Proceedings of the ACM SIGIR Conference on Research and Development in Information Retrieval (2006)
51. Rasmussen, E.M.: Indexing and retrieval for the Web. Annual Review of Information Science and Technology **37**(1), 91–124 (2003)
52. Rennison, E.: Galaxy of news: An approach to visualizing and understanding expansive news landscapes. In: Proceedings of the 7th Annual ACM Symposium on User Interface Software and Technology (1994)
53. van Rijsbergen, C.J.: Information Retrieval. Butterworths (1979)
54. Sanderson, M., Croft, B.: Deriving concept hierarchies from text. In: Proceedings of the ACM SIGIR Conference on Research and Development in Information Retrieval (1999)
55. Silverstein, C., Henzinger, M., Marais, H., Moricz, M.: Analysis of a very large Web search engine query log. SIGIR Forum **33**(1), 6–12 (1999)
56. Spence, R.: Information Visualization: Design for Interaction, 2nd edn. Pearson Education (2007)
57. Spink, A., Wolfram, D., Jansen, B.J., Saracevic, T.: Searching the Web: The public and their queries. Journal of the American Society for Information Science and Technology **52**(3), 226–234 (2001)
58. Spoerri, A.: InfoCrystal: A visualization tool for information retrieval. In: Proceedings of IEEE Visualization (1993)
59. Spoerri, A.: RankSpiral: Toward enhancing search results visualizations. In: Proceedings of the IEEE Symposium on Information Visualization (2004)
60. Stojanovic, N.: Information-need driven query refinement. In: Proceedings of the IEEE/WIC International Conference on Web Intelligence (2003)
61. Stojanovic, N.: Information-need driven query refinement. Web Intelligence and Agent Systems: An International Journal **3**(3), 155–169 (2005)
62. Stone, M.C.: A Field Guide to Digital Color. A. K. Peters (2003)
63. Sugiyama, K., Hatano, K., Yoshikawa, M.: Adaptive Web search based on user profile construction without any effort from users. In: Proceedings of the 2004 World Wide Web Conference (2004)
64. Teevan, J., Dumais, S., Horvitz, E.: Personalizing search via automated analysis of interests and activities. In: Proceedings of the ACM SIGIR Conference on Research and Development in Information Retrieval (2005)
65. Tufte, E.: Envisioning Information. Graphics Press (1990)
66. Tufte, E.: Visual Explanations. Graphics Press (1997)

67. Tufte, E.: The Visual Display of Quantitative Information. Graphics Press (2001)
68. Voorhees, E.M.: Overview of TREC 2004. In: Proceedings of the Thirteenth Text Retrieval Conference (2004)
69. Ware, C.: Information Visualization: Perception for Design. Morgan Kaufmann (2004)
70. Wikimedia Foundation: Wikipedia – the free encyclopedia (2006). URL http://en.wikipedia.org/
71. Wise, J.A.: The ecological approach to text visualization. Journal of the American Society for Information Science **50**(13), 1223–1233 (1999)
72. Wise, J.A., Thomas, J.J., Pennock, K., Lantrip, D., Pottier, M., Schur, A., Crow, V.: Visualizing the non-visual: Spatial analysis and interaction with information from text documents. In: Proceedings of IEEE Information Visualization (1995)
73. Yang, K.: Information retrieval on the Web. Annual Review of Information Science and Technology **39**(1), 33 – 80 (2005)
74. Yao, Y.: Information retrieval support systems. In: Proceedings of the 2002 IEEE World Congress on Computational Intelligence (2002)
75. Young, D., Shneiderman, B.: A graphical filter/flow model for boolean queries: An implementation and experiment. Journal of the American Society for Information Science **44**(6), 327–339 (1993)
76. Zhu, B., Chen, H.: Information visualization. Annual Review of Information Science and Technology **39**(1), 139 – 177 (2005)

Chapter 11
XML Based Markup Languages for Specific Domains

Aparna Varde, Elke Rundensteiner, and Sally Fahrenholz

Abstract A challenging area in web based support systems is the study of human activities in connection with the web, especially with reference to certain domains. This includes capturing human reasoning in information retrieval, facilitating the exchange of domain-specific knowledge through a common platform and developing tools for the analysis of data on the web from a domain expert's angle. Among the techniques and standards related to such work, we have XML, the eXtensible Markup Language. This serves as a medium of communication for storing and publishing textual, numeric and other forms of data seamlessly. XML tag sets are such that they preserve semantics and simplify the understanding of stored information by users. Often domain-specific markup languages are designed using XML, with a user-centric perspective. Standardization bodies and research communities may extend these to include additional semantics of areas within and related to the domain. This chapter outlines the issues to be considered in developing domain-specific markup languages: the motivation for development, the semantic considerations, the syntactic constraints and other relevant aspects, especially taking into account human factors. Illustrating examples are provided from domains such as Medicine, Finance and Materials Science. Particular emphasis in these examples is on the Materials Markup Language MatML and the semantics of one of its areas, namely, the Heat Treating of Materials. The focus of this chapter, however, is not the design of one particular language but rather the generic issues concerning the development of domain-specific markup languages.

A. Varde
Assistant Professor, Department of Computer Science, Montclair State University, Montclair, New Jersey, NJ 07043, USA
e-mail: vardea@montclair.edu

E. Rundensteiner
Full Professor, Department of Computer Science, Worcester Polytechnic Institute, Worcester, Massachusetts, MA 01609, USA
e-mail: rundenst@cs.wpi.edu

S. Fahrenholz
Publishing and Content Management Professional, Northeast Ohio, OH 44073, USA
e-mail: spfahren@gmail.com

J.T. Yao (ed.), *Web-Based Support Systems*, Advanced Information and Knowledge Processing, DOI 10.1007/978-1-84882-628-1_11,

11.1 Background

Web based support systems encompass a wide range of areas such as decision support, information retrieval, knowledge management, web services, user interface design and visualization. A critical element in such systems is the incorporation of human factors particularly from a domain based angle. This is where XML and domain-specific markup languages play an important role. XML has self-explanatory tag sets that are very easily understandable by users even if they have a limited knowledge of web technology and standards.

Markup languages designed using XML tag sets to capture the semantics of specific domains make it even easier for users of the respective domains to exchange information across a common platform. Moreover, this facilitates storage and publishing of information worldwide without having to use a medium of conversion (as against information stored in formats such as relational databases). It also enhances the retrieval and analysis of data from a user perspective. For instance, information stored in XML based markup languages can be easily accessed using XQuery, XSLT and XPath, thus facilitating query processing and further analysis.

In this chapter, we will provide illustrating examples to demonstrate the power of markup languages in terms of being able to capture semantic features, use XML constraints and provide easy access to information. All these factors help in the development of advanced web based support systems by enhancing storage, retrieval and analysis of information from a user standpoint. Given this background, we now provide an overview of XML, followed by details on the development of domain-specific markup languages.

11.1.1 XML: The eXtensible Markup Language

XML, the eXtensible Markup Language is becoming a widespread standard in web publishing. We briefly introduce the need for XML and the terminology used in XML development [11].

11.1.1.1 Need for XML

Traditionally, the markup language typically used by Internet browsers to display a web page has been HTML or the Hyper Text Markup Language. Each HTML tag usually contains an instruction, commanding the browser how to display images and words [3]. HTML is standardized by the World Wide Web Consortium (W3C) and is followed by most of the leading browsers.

Standard web browsers, such as Internet Explorer or Mozilla Firefox, feature a built in presentation layer that interprets each HTML tag and presents it in a web page. An example of HTML tagging appears in Figure 11.1.

```
<HTML>
<HEAD>
<TITLE>WPI CS Graduate Course Information</TITLE>
<LINK REL=stylesheet TYPE=text/css HREF="/cs-style.css">
</HEAD>
<BODY background="/images/parchment.gif">
<FONT FACE="Helvetica,Arial,sans-serif">
<p align=center>
<IMG SRC="/images/banner.gif"
 ALT="WPI Computer Science Department" BORDER=0 WIDTH=418 HEIGHT=40>
<H1 align=center>Graduate Course Information</H1>
<a href="http://www.wpi.edu/Pubs/Catalogs/Grad/sect15.html#courses">
  Graduate Level Courses</a>
</BODY>
</HTML>
```

Fig. 11.1 Example of HTML tagging

However HTML has its limitations [3]. The tags in HTML do not capture semantics. They stress on displaying the form or structure of the information without addressing the details of its content. This poses issues in interpretation and interoperability. Moreover, HTML has a fixed set of tags. They are not extensible to any application-specific domain.

In order to overcome these limitations, the World Wide Web Consortium has developed the eXtensible Markup Language, popularly known as XML [11]. XML is designed to improve the functionality of the Web by providing more flexible and adaptable information interpretation. XML is extensible in the sense that it does not adhere to a rigid format as opposed to HTML which is a single predefined language with a non-descriptive fixed tag set. Instead, XML is a meta-language that can be used for the design of customized markup languages. It serves a twofold purpose of providing a data model for storing information and a medium of communication for exchanging information over the worldwide web. XML has this capacity because it is developed from SGML (Standard Generalized Markup Language) [8], which is an international standard meta-language for text markup systems (ISO 8879). Figure 11.2 shows an XML example. This example describes sales data in an industry.

11.1.1.2 XML Terminology

The common terms used within XML development are elements, attributes, relationships and namespaces [4, 11]. These are explained below.

- Element: An XML element is everything from (including) the a start tag to (including) an end tag that typically represents the entity associated with it. An element can contain other elements, simple text or a mixture of both [11]. Elements can also have attributes.

```
<salescube>
        <summary>...</summary>
        <summary year="...">...</summary>
        ...
        <summary region="...">...</summary>
        ...
        <summary product="...">...</summary>
        ...
        <summary year="..." region="...">...</summary>
        ...
        <summary year="..." product="...">...</summary>
        ...
        <summary region="..." product="...">...</summary>
        ...
        <summary year="..." region="..." product="...">...</summary>
        ...
</salescube>
```

Fig. 11.2 An XML sample

- Attribute: XML attributes are normally used to describe XML elements, or to provide additional information about elements. Attributes are always contained within the start tag of an element. Often attribute data is more important to the XML parser than to the reader [11].
- Relationship: This describes the nesting of the elements within each other. Relationships could be of various types such as parent-child, sibling and so forth which are captured by the hierarchy of corresponding the tag set [11].
- Namespace: An XML namespace provides a simple method for qualifying element and attribute names used in XML documents by associating them with names that are identified by URL references. Document constructs should have names that avoid clashes between different markup vocabularies [4]. XML namespaces serve to provide a mechanism which accomplishes this with reference to context.

These terms are commonly referred to as a set of XML tags. This tag set forms the XML schema which provides the structure or the syntax for the language analogous to the grammar in a natural language such as English. The XML schema thus serves as the formal definition of the language which is usually standardized.

11.1.2 Domain-Specific Markup Languages

Communities of specific users have identified certain needs for tagging and structuring information within a given domain while dealing with web-based information. Some examples of targeted users are industries, universities, standards bodies, publishers and research groups. It is important to facilitate storage, retrieval and exchange of information among these users. This serves as the motivation for the development of XML based markup languages geared towards specific domains [22].

Consequently, it is found that many languages are defined within the context of XML. These follow the XML syntax and encompass the semantics of a given domain. They are called domain-specific markup languages. Such languages typically include XML based tags for storage that capture the domain semantics. The inclusion of semantic tags in a document makes a document self-describing. Hence, this helps to enhance document storage and exchange.

11.1.2.1 Examples of Domain-Specific Markup Languages

A notable example of a domain-specific markup language is the Medical Markup Language (MML) developed in Japan [12]. MML is designed to create a set of standards by which medical data, within Japan and hopefully worldwide, can be stored, accessed and exchanged. The following MML module contents are defined at the present time: patient information, health insurance information, diagnosis information, lifestyle information, basic clinic information, particular information at the time of first visit, progress course information, surgery record information and clinical summary information [12]. These form the XML elements. They are of use to primary care physicians, general surgeons, their patients and related entities.

However, specific information, for example opthalmological details such as eye-diseases, spectacle prescriptions and blindness, cannot be stored using these tags. Thus, more specific markup languages can be developed within the context of MML to include the semantics of opthalmology within medicine.

Other examples of XML based domain-specific markup languages include [5, 10, 13]:

- AniML: Analytical Information Markup Language
- CML: Chemical Markup Language
- femML: Finite Element Modeling Markup Language
- MathML: Mathematics Markup Language
- ThermoML: Thermodynamic Markup Language
- WML: Wireless Markup Language
- UnitsML: Units Markup Language

These markup languages enable the storage, retrieval and display of information using a non-proprietary format. They describe the given domain in an understandable language that is common to its members and interested parties, thus enabling effective information exchange.

Let us now take a closer look at a particular language called the Materials Markup Language abbreviated as MatML. We will use several examples from MatML to explain the concepts in this chapter.

11.1.2.2 MatML: The Materials Markup Language

MatML is the Materials Markup Language serving as the XML for Materials property data [1]. More specifically, MatML was created to equip the materials data marketplace with the following:

(a) Common materials data exchange format
(b) Non-proprietary *Esperanto*
(c) Direct program to program interoperability
(d) Flexible, extensible markup language

The MatML effort was pioneered at NIST, the National Institute of Standards and Technology. Its development began in October 1999. It later expanded to the MatML Coordination Committee, spearheaded by ASM International, the Materials Information Society, and comprising a consortium of experts from academia and industry. Standardization of MatML has thereafter been governed by OASIS, the Organization for the Advancement of Structured Information Standards [19].

The MatML elements correspond to the entities that define materials property data. Currently, the MatML structure is as follows [1].

Structure of MatML, the Materials Markup Language

- < *Materials* >: This element stores data about the material and has the following sub-elements.
  - < *BulkDetails* >: This stores generic information about the material as a whole.
  - < *ComponentDetails* >: This is used to provide information on the chemical components that make up the material.
  - < *Graphs* >: This provides for the storage of two-dimensional graphical data plotting parameters related to materials processes.
  - < *Glossary* >: As several domain-specific terms may apply to the data, the glossary serves to clarify their meaning.
- *Metadata*: As the name implies, this element stores data about the data, i.e, background information related to the material.
  - < *AuthorityDetails* >: This stores data related to ownership and authority.
  - < *DataSourceDetails* >: This has information acknowledging the sources from where the data was collected.
  - < *MeasurementTechniqueDetails* >: This states which specific technique was used in measuring the concerned property.
  - < *ParameterDetails* >: Information on the process parameters pertaining to materials is stored here.
  - < *PropertyDetails* >: Additional background data related to material properties is stored here.

- < *SourceDetails* >: This gives information on the source of the material itself.
- < *SpecimenDetails* >: Data about the specific instance of the given material specimen is stored here.
- < *TestConditionDetails* >: This indicates the conditions under which a test is conducted on the concerned material.

All these MatML elements have their respective attributes. This nested structure of elements and attributes forms the MatML schema. This provides a means to store the information related to the properties of various materials such as metals, ceramics and plastics. For example, the chemical composition of a particular alloy would be stored under the element < *Materials* > within its (sub) element < *ComponentDetails* >, the actual details of the composition forming the individual attributes. Figure 11.3 shows an example of data storage using MatML [19].

Given MatML as a base language for materials property data, specific markups can be defined within its context in order to capture the semantics of the individual fields within Materials Science. For example, one such field is the Heat Treating of Materials [14]. Heat Treating involves the controlled heating and cooling of materials to achieve desired mechanical and thermal properties. An important step in the Heat Treating operations is a process called *Quenching* [18] which is the rapid cooling of the material in a liquid or gas medium. There are entities in the Quenching process such as the Quenchant, i.e., the cooling medium. These have properties, e.g., viscosity refers to the capacity of the Quenchant to flow.

```
<Properties>
    <PropertyDetails>
      <Name>Critical Current Density</Name>
      <Units>kA/cm<sup>2</sup></Units>
      <DataSource>Journal</DataSource>
      <DataType>Evaluated</DataType>
    </PropertyDetails>
    <Value>3040</Value>
    <Parameters>
      <Name>Magnetic Field</Name>
      <Value type="integer">0</Value>
      <Units>T</Units>

      <Name>Temperature</Name>
      <Value type="integer">3</Value>
      <Units>K</Units>
    </Parameters>
</Properties>
```

Fig. 11.3 MatML data storage

In order to store the details on the Heat Treating of Materials, in particular, Quenching, a markup language called QuenchML [20, 21] has been developed by the Center for Heat Treating Excellence, CHTE. Likewise, other languages can also be developed using XML specifications. We now discuss the general steps involved in developing such markup languages, using suitable examples.

11.2 Development of Markup Languages

Markup language development has several steps which can be considered analogous to the design of software systems [15]. These are listed below and discussed in the following subsections. It should be noted that steps 1 through 3 serve as precursors for development while steps 4 through 7 are the actual developmental phases, often iterative. Steps 4 through 7 typically corresponding to the pilot version, alpha version, beta version and release version of a software respectively [15].

Steps in Markup Language Development

(1) Acquisition of Domain Knowledge
(2) Data Modeling
(3) Requirements Specification
(4) Ontology Creation
(5) Revision of the Ontology
(6) Schema Definition
(7) Reiteration of the Schema

11.2.1 Acquisition of Domain Knowledge

It is important to study the domain thoroughly and know the terminology. This helps to determine the tags that are essential to store the data in the domain. Also, it is necessary to be well-acquainted with related existing markup languages in the domain [5, 12, 13] to find out if the newly developed markup should be an extension to an existing language or an independent language. Let us now consider this with reference to Materials Science domain. We present the motivation for the development of a new markup language [19].

Motivating Example for Markup Language Development

MatML is generic to the entire Materials Science field. The complete tag set specified in MatML literature is not sufficient to capture the details of the Heat

Treating processes, in particular its rapid cooling step called *Quenching*. For example, information related to the viscosity of a quenching medium used in the process, or the geometry of the part being quenched cannot be represented using MatML tags. This motivates the development of a Markup Language for Quenching, namely QuenchML.

In order to maintain consistency with the above example, in the remaining subsections, we will present examples from the development of QuenchML: The Quenching Markup Language [20]. These examples are presented at various stages of markup language development.

11.2.2 Data Modeling

A data model is a generalized, user-defined view of the data related to applications that describes how the data is to be represented for manipulation by humans or computer programs [16]. In the development of Markup Languages, it is useful to model the data since this also serves as a medium of communication to capture domain knowledge. Thus, data modeling is not imperative but certainly desirable.

In general, tools such as Entity Relationship (E-R) diagrams, petri-nets, Unified Modeling Language (UML) and formal specifications are used in modeling the data [15, 16]. For the purpose of Markup Languages, it is advisable to use E-R diagrams. This is because the entities and relationships they represent can be mapped to the corresponding entity names and the respective relationships between tags in the Markup Language. E-R diagrams are explained below.

11.2.2.1 Entity Relationship Diagram

An E-R diagram is a formal method for arranging data to mimic the behavior of the real world entities represented [16]. This helps to create a picture of the entities in the domain, view their attributes and understand their relationships with each other. Figure 11.4 shows a subset of an E-R diagram capturing the relationships between the entities in Quenching in the Heat Treating of Materials [18].

Entity Relationship diagrams used as data models provide useful insights into Markup Language development in terms of designing the tag set with respect to the nomenclature as well as the nesting. This will be evident in the steps to follow.

11.2.3 Requirements Specification

The needs of the potential users of the markup language must be identified. Hence it is necessary to gather their requirements through a clear set of specifications. A good

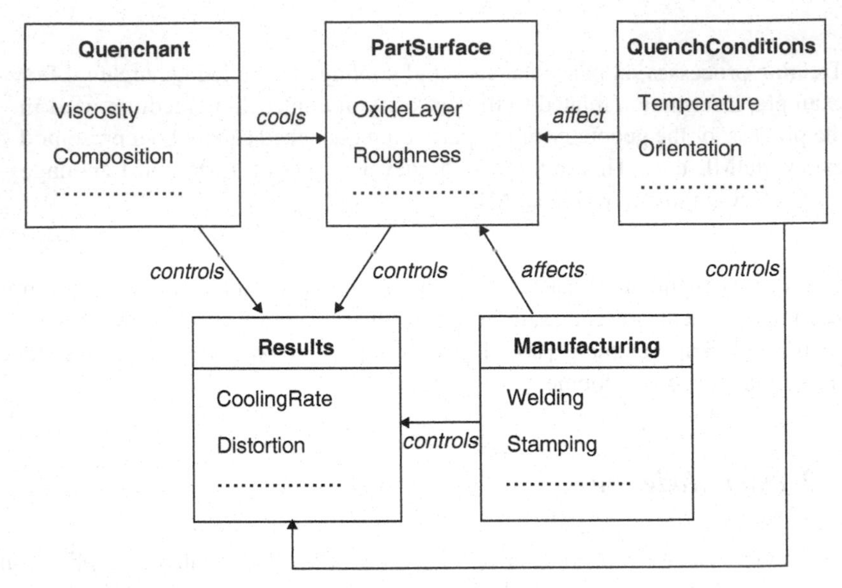

Fig. 11.4 E-R diagram for the quenching process

means to address this is by conducting detailed interviews with domain experts who can adequately represent the needs of the targeted users. Potential users (as stated earlier in this chapter) include industries, universities, standards bodies, publishers, research groups, domain experts and others. Typically, the needs of the potential users are well-identified by the domain experts.

Other means such as formal user surveys may also be used although they are often more time-consuming. It is to be noted that users are needed at a later stage for testing the developed Markup Language where they will need significant time commitment. Hence, if domain experts can serve to identify user needs at earlier stages of development, that is definitely advisable.

Requirements specification helps to select the entities and their attributes to be included in the language. It also helps to capture the required relationships between the entities needed for designing the Markup Language. Hence this step is an important precursor to the developmental phases of the Markup Language which start with the creation of the ontology as explained next.

11.2.4 Ontology Creation

Ontology is a system of nomenclature used within a given domain, colloquially termed as the established *lingo* for the members of the domain [4, 11]. The ontology thus serves as the study of what there is, i.e., an inventory of what exists. An ontological commitment is a commitment to an existence claim. Hence, after understanding the domain and conducting interviews with experts, defining the ontology is essential in order to proceed with the design.

Two important considerations in ontology are synonyms and homographs as explained below.

- Synonyms: These are two or more words having the same meaning. For example in the financial domain [17], the terms *salary* and *income* mean the same and are synonyms.
- Homographs: These refer to one word having multiple meanings. For example, the term *share* can have two connotations in this domain. It could mean: assets belonging to or due to or contributed by an individual person or group, or: any of the equal portions into which the capital stock of a corporation is divided and ownership of which is evidenced by a stock certificate [17]. Thus, the term *share* is a homograph in the financial domain.

Issues such as these need to be addressed through the ontology. With reference to Quenching in Heat Treating, the terms *part*, *probe* and *workpiece* are synonyms, i.e., they refer to the same entity [18]. Such terms need to be clarified with reference to the context. This is done through the ontology as explained in the following example.

We present a specific example of ontology creation with reference to the development of QuenchML the Heat Treating domain [20]. This is a high level ontology that relates to the entities identified by the E-R diagram above.

High Level Ontology for the Quenching Markup Language

- *Quenchant*: This refers to the medium used for cooling in the heat treatment process of rapid cooling or Quenching.
  - Alternative Term(s): *CoolingMedium*
- *PartSurface*: The characteristics pertaining to the surface of the part undergoing heat treatment are recorded here.
  - Alternative Term(s): *ProbeSurface, WorkpieceSurface*
- *Manufacturing*: The details of the processes used in the production of the concerned part such as welding and stamping are stored here.
  - Alternative Term(s): *Production*
- *QuenchConditions*: This records the input parameters under which the Quenching process occurs, e.g., the temperature of the cooling medium, or the extent to which the medium is agitated.
  - Alternative Term(s): *InputConditions, InputParameters, QuenchParameters*

- *Results*: This stores the output of the Quenching process with respect to properties such as cooling rate (change in part temperature with respect to time) and heat transfer coefficient (measurement of heat extraction capacity of the whole process of rapid cooling).
  - Alternative Term(s): *Output, Outcome*

The ontology creation phase in the development of Markup Languages corresponds to the pilot version of a tool in the design of software systems [15]. It serves to provide a working draft to help proceed with further development just as a pilot tool serves to give an initial system layout setting the stage for enhancement. The ontology once created could go through substantial revision as discussed in the next step.

11.2.5 Revision of the Ontology

Once the ontology has been established, it is useful to have additional discussions with domain experts to make the required changes if any. For example, it may seem necessary to create new entities for clarification or remove existing entities to avoid ambiguity.

The created ontology is revised accordingly based on critical reviews. This is usually done internally by additional discussions with domain experts and internal users. Often the designers themselves run test cases with sample data to find out if the ontology is satisfactory to continue with the formal definition of a schema. Thus revision of the ontology can go through several rounds of discussion and testing.

The ontology revision can be considered analogous to the alpha version of a tool in software development [15]. Just as the alpha version is a working tool that is internally tested, so is the revised ontology prior to the schema definition phase which is described next.

11.2.6 Schema Definition

The schema provides the structure of the Markup Language. In other words it defines the grammar for the language. Once the ontology is formally approved by a team of experts, the first draft of the schema is outlined.

The E-R model, requirements specification and ontology serve as the basis for schema design. Each entity in an E-R model considered significant in the requirements specification typically corresponds to one schema element. The ontology helps to include the tags in the schema as needed based on the established system of nomenclature.

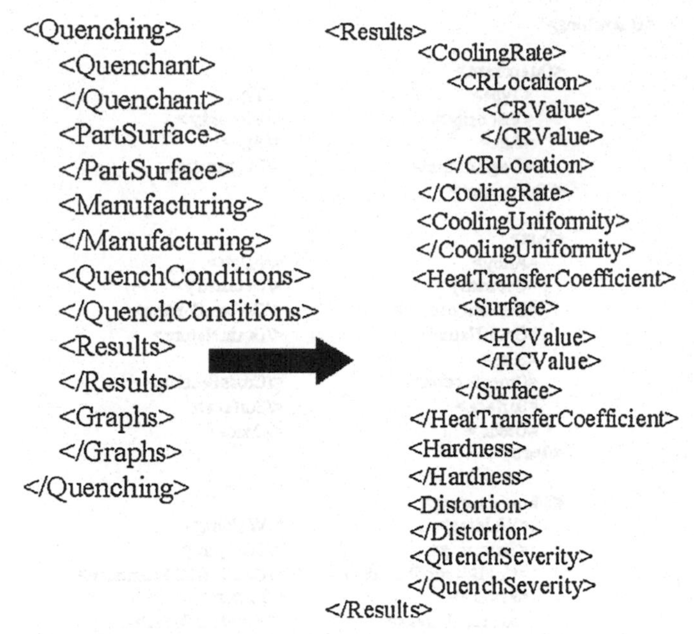

Fig. 11.5 An extract from the first draft of QuenchML

Figure 11.5 shows a partial snapshot of a first schema draft for the Heat Treating of Materials [20]. This is based on the ontology created and revised in the previous two steps.

Note that the schema definition phase is analogous to the beta version of a software tool [15]. In this phase, the developers interact with the external users as well to get their opinion on the draft. Since the schema is a formal draft, it is highly recommended that the users get involved at this stage in carrying out tests for storing data using the structure provided by the schema. This is similar to the user involvement in the beta testing of software systems.

11.2.7 Reiteration of the Schema

The initial schema serves as the medium of communication between the designers and the potential users of the markup language. This is subject to revision until the users are satisfied that this adequately represents their needs. Schema revision may involve several iterations, some of which are the outcome of discussions with standards bodies such as NIST (National Institute of Standards and Technology).

Thus, this stage goes beyond communication with domain experts and targeted users. In order for the proposed extension to be accepted as a standard for

```
<Quenching>

    <Quenchant>
        <Type>                          </Type>
        <Viscosity>                     </Viscosity>
        <Age>                           </Age>
        <Degradation>                   </Degradation>
    </Quenchant>

    <Part>
        <Alloy>                         </Alloy>
        <Geometry>                      </Geometry>
        <CarbonContent>                 </CarbonContent>
        <GrainNature>                   </GrainNature>
        <Size>                          </Size>
        <CrossSection>                  </CrossSection>
        <Surface>                       </Surface>
        <Oxide>                         </Oxide>
    </Part>

    <Manufacturing>
        <Welding>                       </Welding>
        <Stamping>                      </Stamping>
        <ColdPlasticFormation>          </ColdPlasticFormation>
        <Fixture>                       </Fixture>
        <ResidualStress>                </ResidualStress>
    </Manufacturing>

    <QuenchConditions>
        <Temperature>                   </Temperature>
          <Agitation>                   </Agitation>
          <SpeedImprovers>              </SpeedImprovers>
          <Impellers>                   </Impellers>
          <GMQuenchometer>              </GMQuenchometer>
    </QuenchConditions>

    <Results>
        <CoolingRate>                   </CoolingRate>
        <CoolingUniformity>             </CoolingUniformity>
        <SlopeGraph>                    </SlopeGraph>
        <HeatTransferCoefficient>       </HeatTransferCoefficient>
        <Hardness>                      </Hardness>
        <Distortion>                    </Distortion>
        <Cracking>                      </Cracking>
        <Bowing>                        </Bowing>
        <Warpage>                       </Warpage>
        <QuenchSeverity>                </QuenchSeverity>
    </Results>

</Quenching>
```

Fig. 11.6 Partial snapshot of final draft of QuenchML

communication worldwide and be incorporated into the existing Markup Language, it is important to have it thoroughly reviewed by well-established standards bodies.

Figure 11.6 shows a partial snapshot of the final QuenchML schema resulting from discussions with the standards bodies and concerned user groups [21].

The final schema is an outcome of the last phase of Markup Language development and corresponds to the release version of a software tool [15]. The software once released is out in the market for potential consumers and buyers. Likewise, the final schema draft is ready for use in terms of serving as a standard for storage and exchange of information in the respective domain.

Thus, it is clear that the development of software systems does have analogy with the development of Markup Languages. This is based on the phases involved in development that have been described in detail here.

11.3 Desired Properties of Markup Languages

In order to be powerful enough to capture semantics and yet cater to the needs of simplicity, the markup language needs to incorporate the following features.

(a) Avoidance of Redundancy
(b) Non-Ambiguous Presentation of Information
(c) Easy Interpretability of Information
(d) Incorporation of Domain-Specific Requirements
(e) Potential for Extensibility

We discuss these in the subsections to follow.

11.3.1 Avoidance of Redundancy

The structure of the Markup Language should be such that duplicate information should not be stored, i.e., if information about a particular entity or attribute is already stored using an existing Markup Language, then it should not be repeated in the new Markup Language.

For example, in Heat Treating, consider a material property such as thermal conductivity [14]. If the structure of MatML, the Materials Markup Language [1] already provides for the storage of this property, then the data on that property should not be stored again using a newly developed Markup Language such as QuenchML [21]. Thus, the structure of the new Markup Language should avoid such redundant storage.

11.3.2 Non-ambiguous Presentation of Information

The information stored using the markup language should be clear and should avoid ambiguity. Consider the concepts of synonyms and homographs described earlier.

Recall that synonyms are two or more words having the same meanings, while the term homograph on the other hand refers to one word having multiple meanings.

The Markup Language needs to be precise in order to avoid confusion while reading the stored information. The structure of the Markup Language should be such that the reader does not get confused about the meanings of the tags that may refer to terms which could be synonyms or homographs.

Consider the example given earlier of the term *share* in the financial domain. Since this term has multiple connotations, its exact meaning in a given document should be clear with reference to context. The Markup Language needs to cater to such issues related to ambiguity. Usually, the ontology is helpful here.

11.3.3 Easy Interpretability of Information

The markup language should be such that readers are able to understand and interpret stored information without much reference to related documentation. For example, in science and engineering domains [12,22], the details of the input conditions of a performed experiment should be stored close to its results and preferably before its results in order to enhance readability.

In addition, using terms that are easily understandable in the domain is very helpful. It is advisable to use a tag set that is designed from a user perspective rather than a developer perspective. The tags should be such that they are easy for a non Computer Scientist to understand and are geared more towards the users of the domain. Thus, the markup language tags need to be meaningful enough to capture the semantics of the domain.

11.3.4 Incorporation of Domain-Specific Requirements

There are typically certain requirements imposed by the domain that need to be captured in the schema using XML constraints [9]. A simple example is the primary key constraint. A primary key serves to uniquely identify an entity [16]. In addition, XML provides a choice constraint that allows the declaration of mutually exclusive elements [9].

For example, in the financial domain a person could be either an *insolvent* (bankrupt) or an *asset-holder*, but not both [17]. Thus, these two terms are mutually exclusive. Other XML constraints are sequence constraints to declare a list of elements in order, and occurrence constraints that define the minimum and maximum occurrences of an element [9]. The Markup Language needs to make use of these as needed in order to adequately represent the domain semantics.

11.3.5 Potential for Extensibility

The issue of extensibility is often debatable. Some language developers argue that it is desirable to construct the tag set in such a manner that it is easy for users of related domains to augment it with additional semantics, thereby proposing new elements, attributes and relationships as an extension to the existing Markup Language. On the other hand, there are counter-arguments that too much extensibility is an obstacle in the path of standardization. Standards bodies are typically not in favor of an existing standard being updated too frequently.

Hence, in theory the potential for extensibility is good for capturing additional semantics. However, in practice extensibility may not always be feasible. On the basis of the discussions with the developers of the languages described in this chapter, we conclude that extensibility is a desired property of XML based Markup Languages.

Given these desired properties of Markup Languages, we now proceed to discuss the XML features (mainly constraints) that are useful in development and help to provide some of these properties.

11.4 Application of XML Features in Language Development

Among the XML features available for Markup Language development, by far the most important ones are certain XML constraints [9,22]. Thus, we focus our discussion on XML constraints in this chapter. Constraints are mechanisms in XML that enable the storage of information adhering to specific rules such as enforcing order and declaring mutually exclusive elements [9]. Some of these constraints relevant to developing domain-specific Markup Languages are:

(a) Sequence Constraint
(b) Choice Constraint
(c) Key Constraint
(d) Occurrence Constraint

These are described below. We use suitable examples from the Materials Science Markup Language, MatML [1] and the Quenching Markup Language QuenchML [21] in explaining these constraints.

11.4.1 Sequence Constraint

This constraint is used to declare that a list of elements occur in a particular order. Enclosing the concerned elements within $< xsd : sequence >$ tags provides this constraint [9].

For example, the element *QuenchConditions* must appear before the element *Results*. This is required by the domain to enhance readability. This is because

```
<xsd:element name="Quenching">
    <xsd:complexType>
        <xsd:sequence>
            ............................
            <xsd:element name="QuenchConditions">
            .....
            </xsd:element>
            <xsd:element name="Results"/>
            .....
            </xsd:element>
            ....................
        </xsd:sequence>
    </xsd:complex Type>
</xsd:element>
```

Fig. 11.7 Example of sequence constraint

the element *QuenchConditions* denotes the input conditions of the process, while *Results* denotes the observations. The input conditions of the Quenching process affect the observations. It is thus necessary for a user to read the input conditions first in order to understand how they have an impact on the corresponding observations. Figure 11.7 shows an example of a sequence constraint with reference to this discussion.

11.4.2 Choice Constraint

This is used to declare mutually exclusive elements, i.e., the elements that are such that only one of them can exist. It is declared using $< xsd : choice >$ in the schema [9, 20].

Consider the following example. In Materials Science, a part can be manufactured by either a casting process or a powder metallurgy process but not both [14]. Thus the corresponding sub-elements *Casting* and *PowderMetallurgy* are mutually exclusive and are enclosed within $< xsd : choice >$ tags as shown. Figure 11.8 illustrates an example of the choice constraint.

11.4.3 Key Constraint

A key constraint is analogous to a primary key in relational databases [16]. It is used to declare an attribute to be a primary key. This implies that the attribute must have

```
<xsd:element name="Manufacturing">
    <xsd:complex Type>
        <xsd:choice>
            <xsd:element ref="Casting"/>
            <xsd:element ref="PowerMetallurgy"/>
        </xsd:choice/>
        .........................
    </xsd:complexType>
</xsd:element>
```

Fig. 11.8 Example of choice constraint

```
<xsd:element name="Quenchant">
  <xsd:complexType>
    <xsd:attrubute name = "id" type ="xsd:ID" use="required"/>

    .......................................
  </xsd:complexType>
</xsd:element>
```

Fig. 11.9 Example of key constraint

unique values and cannot have empty or null values. This is indicated in the schema by declaring the corresponding attribute as type $< xsd : ID >$ and declaring its use as "*required*" [9].

As an example, consider the element *Quenchant* in QuenchML, which refers to the cooling medium used in a Quenching process in the heat treatment of materials. Its "*id*" is crucial since it serves to uniquely identify the medium [18]. In other words, in storing the details of the Quenching process, it is required that the id or name of the cooling medium be stored. This is because the purpose of conducting these experiments is to characterize the Quenchants. Figure 11.9 presents an example of the key constraint discussed here.

11.4.4 Occurrence Constraint

This constraint is used to declare the minimum and maximum permissible occurrences of an element. It is written as "$minOccurs = x$" and "$maxOccurs = y$" where "x" and "y" denote the minimum and maximum occurrences respectively [9, 20]. A "*maxOccurs*" value of "*unbounded*" implies that there is no limit on the number

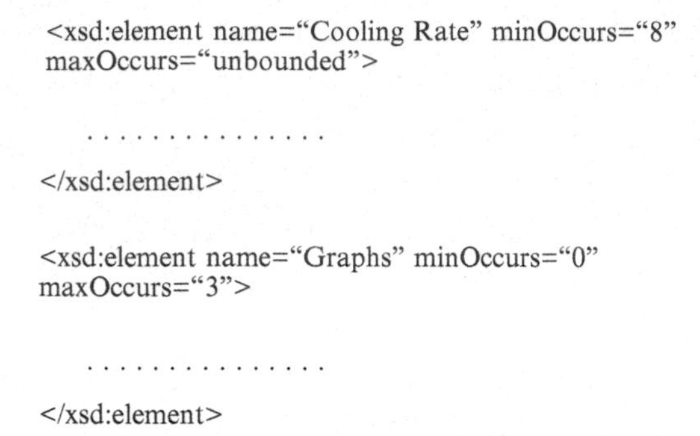

Fig. 11.10 Example of occurrence constraint

of times this element can occur within a schema. A "*minOccurs*" value greater than "0" implies that it is necessary to include this element within the schema at least once. Figure 11.10 depicts an example of the occurrence constraint with reference to QuenchML [21].

With reference to this example, it is clear that the *Cooling Rate* element must occur at least 8 times and that there is no upper bound on the number of times it can occur. This is because in the domain, the value of cooling rate must be stored at least at 8 points in order to adequately capture the details of the Quenching process. However, cooling rate values may be recorded at hundreds or even thousands of points and thus there is no upper limit on the number of values that can be stored for cooling rate [18, 20].

On the other hand, in the case of the *Graphs* element, it is not necessary that at least one graph be stored. It is essential though to keep the number of graphs stored less than three per instance of the process [18, 20]. This is required as per the domain. Generally the two graphs stored in Quenching are the cooling rate curve and heat transfer coefficient curve. In addition, a cooling curve may be stored. A cooling curve is a plot of temperature (T) versus time (t) during Quenching or rapid cooling. A cooling rate curve is a plot of cooling rate (dT/dt) versus time (t). A heat transfer coefficient curve is a plot of heat transfer coefficient (hc) versus temperature (T), where a heat transfer coefficient characterizes a Quenching process [18].

In this manner the occurrence constraint serves to capture such information on the lower and upper bounds required in data storage in Markup Language.

The use of XML features in storing data with domain-specific Markup Languages provides convenient access for information retrieval purposes. We now provide an overview of this in the next section.

11.5 Convenient Access to Information

Using XML data storage provides customized access to information through the Markup Language. This serves as very convenient high level access for users. There are many forms of declarative access in the XML family, e.g., languages such as XQuery, that are well-suited for retrieving the information stored using the Markup Language. These are listed below.

(a) XQuery: XML Query Language
(b) XSLT: XML Style Sheet Language Transformations
(c) XPath: XML Path Language

We discuss each of these declarative forms of access to XML based information in the subsections to follow.

11.5.1 XQuery: XML Query Language

XQuery is a query language for XML developed by the World Wide Web Consortium (W3C). XQuery can be used to retrieve XML data. Hence it can query information stored using a domain-specific markup language that has been designed with XML tags. It is thus advisable to design the extension to the markup language to facilitate retrieval using XQuery [2]. A few suggestions for achieving this are as follows.

- Storing Data in a Case Sensitive Manner: XQuery is case-sensitive [2]. Hence it is useful to place emphasis on case when storing the data using the domain-specific Markup Language and its potential extensions if any. This helps to obtain correct retrieval of information.
- Using Tags to Enhance Querying Efficiency: In many domains, it is possible to anticipate a typical user query. For example, in Heat Treating, a user is very likely to retrieve the details of a Quenchant in terms of its name, type and manufacturer without requesting information about the Quenchant properties. Hence, it is advisable to add a level of abstraction around the name-related tags and the property-related tags. This is done using additional tags such as $< NameDetails >$ and $< PropertyDetails >$ [20]. The XQuery expression to retrieve information for a name-related query can then be constructed such that it gets the name details in a single traversal of the path. In the absence of this abstraction, the XQuery expression to get these details would have contained a greater number of tags, with additional levels of nesting. Therefore, introducing abstraction by anticipating typical user queries enhances the efficiency of querying.

11.5.2 XSLT: XML Style Sheet Language Transformations

XSLT, which stands for XML Style Sheet Language Transformations, is a language for transforming XML documents into other XML documents [6]. XSLT is designed for use as part of XSL, which is a stylesheet language for XML. In addition to XSLT, XSL includes an XML vocabulary for specifying formatting. XSL specifies the styling of an XML document by using XSLT to describe how the document is transformed into another XML document that uses the formatting vocabulary. XSLT is also designed to be used independently of XSL.

However, XSLT is not intended as a completely general-purpose XML transformation language. Rather it is designed primarily for the kinds of transformations that are needed when XSLT is used as part of XSL. Thus, information stored using an XML based Markup Language is easily accessible through XSLT [6]. This provides a convenient form of information retrieval.

11.5.3 XPath: XML Path Language

XPath, namely, the XML Path Language, is a language for addressing parts of an XML document [7]. The primary purpose of XPath is to refer to specific portions of an XML document. In support of this primary purpose, it also provides basic facilities for manipulation of strings, numbers and booleans. XPath uses a compact, non-XML syntax to facilitate use of XPath within URLs and XML attribute values. XPath operates on the abstract, logical structure of an XML document, rather than its surface syntax. XPath gets its name from its use of a path notation as in URLs for navigating through the hierarchical structure of an XML document.

XPath models an XML document as a tree of nodes. There are different types of nodes, including element nodes, attribute nodes and text nodes. XPath defines a way to compute a string-value for each type of node [7]. Some types of nodes also have names. XPath fully supports XML Namespaces [4]. This further enhances the retrieval of information with reference to context.

11.6 Conclusions

In this chapter we have described the need for domain-specific Markup Languages within the context of XML. We have provided a detailed outline of the steps involved in markup language development and given the desired properties of Markup Languages. We have also explained how XML features can be applied in language development and how the developed markup provides convenient information access. Suitable real-world examples have been included from existing Markup Languages in the literature for an in-depth understanding of the concepts in the chapter.

It is anticipated that this chapter serves useful to potential developers and users of domain-specific XML based Markup Languages.

Acknowledgements The authors thank the National Institute of Standards and Technology in Maryland, USA and the members of the international MatML steering committee for useful references to the Materials Markup Language. The feedback of the Database Systems Research Group in the Department of Computer Science at Worcester Polytechnic Institute, Massachusetts, USA is also acknowledged. In addition, we value the inputs of the researchers at the Center for Heat Treating Excellence, part of the Metal Processing Institute, an international industry-university consortium.

References

1. Begley, E.F.: MatML Version 3.0 Schema. NIST 6939, National Institute of Standards and Technology Report, USA (January 2003).
2. Boag, S., Fernandez, M., Florescu, D., Robie J. and Simeon, J.: XQuery 1.0: An XML Query Language. W3C Working Draft (November 2003).
3. Bouvier, D.J.: Versions and Standards of HTML. In: ACM SIGAPP Applied Computing Review, Vol. 3, No. 2, pp. 9–15 (October 1995).
4. Bray, T., Hollander, D., Layman, A. and Tobin, R.: Namespaces in XML 1.0 (Second Edition). W3C Recommendation (August 2006).
5. Carlisle, D., Ion, P., Miner, R., and Poppelier, N.: Mathematical Markup Language (MathML). World Wide Web Consortium (2001).
6. Clark, J.: XSL Transformations (XSLT) Version 1.0. W3C Recommendation, (November 1999).
7. Clark, J. and DeRose, S. XML Path Language (XPath) Version 1.0. W3C Recommendation (November 1999).
8. Connolly, D.: Overview of SGML Resources. World Wide Web Consortium (1995).
9. Davidson, S., Fan, W., Hara, C. and Qin, J.: Propagating XML Constraints to Relations. In: International Conference on Data Engineering, pp. 543–552 (March 2003).
10. Fahrenholz, S.: Materials Properties Thesaurus Development: A Resource for Materials Markup Language. In: ASM Aeromat Conference (May 2006).
11. Flynn, P.: The XML FAQ. World Wide Web Consortium Special Interest Group (2002).
12. Guo, J., Araki, K., Tanaka, K., Sato, J., Suzuki, M., Takada, A., Suzuki, T., Nakashima, Y. and Yoshihara, H.: The Latest MML (Medical Markup Language) Version 2.3 — XML based Standard for Medical Data Exchange / Storage. In: Journal of Medical Systems, Vol. 27, No. 4, pp. 357–366 (August 2003).
13. Murray-Rust, P. and Rzepa, H.S.: Chemical Markup, XML, and the Worldwide Web Basic Principles. In: Journal of Chemical Informatics and Computer Science, Vol. 39, pp. 928-942 (1999).
14. Pellack, L.J.: Introduction to Materials Science. Science and Technology Librarianship (2002).
15. Pressman, R.: Software Engineering: A Practitioner's Approach. McGraw Hill (2005).
16. Ramakrishnan, R. and Gehrke, J.: Database Management Systems. McGraw Hill Companies (2000).
17. Seward, J. and Logue, D.: Handbook of Modern Finance. WG and L Financial (2004).
18. Totten, G., Bates, C. and Clinton, N.: Handbook of Quench Technology and Quenchants. ASM International (1993).
19. Varde, A., Begley, E. and Fahrenholz, S.: MatML: XML for Information Exchange with Materials Property Data In: ACM International Conference on Knowledge Discovery and Data Mining, Workshop on Data Mining Standards, Services and Protocols (August 2006).

20. Varde, A., Rundensteiner, E., Mani, M., Maniruzzaman, M. and Sisson Jr., R.: Augmenting MatML with Heat Treating Semantics. In: ASM International's Symposium on Web-Based Materials Property Databases (October 2004).
21. Varde, A., Maniruzzaman, M. and Sisson Jr., R.: QuenchML: The Quenching Markup Language. Submitted to Journal of ASTM International (2007).
22. Yokota, K., Kunishima, T. and Liu, B.: Semantic Extensions of XML for Advanced Applications. In: IEEE Australian Computer Science Communications Proceedings of Workshop on Information Technology for Virtual Enterprises, Vol. 23, No. 6, pp. 49–57 (January 2001).

Chapter 12
Evaluation, Analysis and Adaptation of Web Prefetching Techniques in Current Web

Josep Domènech, Ana Pont-Sanjuán, Julio Sahuquillo, and José A. Gil

Abstract The basics of web prefetching are to preprocess user requests before they are actually demanded. Therefore, the time that the user must wait for the requested documents can be reduced by hiding the request latencies. Prefetching is usually transparent to the user, that is, there is no interaction between the prefetching system and the user. For this reason, systems speculate on the following user's requests and thus the prediction can fail. In such a case, web prefetching increases the resources requirements, so it should be applied carefully.

This chapter is aimed at describing a methodology in order to evaluate, analyze and improve the performance of web prefetching algorithms. Moreover, we show how this methodology can be used for improving the existing algorithms by considering current workload characteristics.

To do so, in a first step a solid framework and methodology to evaluate web prefetching techniques from the user's point of view are presented. In a second step, we analyze how prefetching algorithms can be improved from the user's point of view. We take benefit of the characteristics of current web in order to design a new algorithm that outperforms those existing in the open literature. Finally, we also explore the performance limits of web prefetching to know the potential benefits of this technique depending on the architecture in which it is implemented.

12.1 Introduction to Web Prefetching

The goal of web prefetching is to preprocess user requests before the user demands them explicitly, so reducing the user-perceived latency. The main prefetching mechanisms proposed in the literature are transparent to the user and, consequently, they are necessarily speculative. As any speculative technique, prefetch predictions can

---

J. Domènech, A. Pont-Sanjuán, J. Sahuquillo, and J.A. Gil
Universitat Politècnica de València, Camí de Vera, s/n 46022 València, Spain
e-mail: jdomenech@ai2.upv.es; {apont,jsahuqui,jagil}@disca.upv.es

J.T. Yao (ed.), *Web-Based Support Systems*, Advanced Information and Knowledge Processing, DOI 10.1007/978-1-84882-628-1_12,

fail, i.e., downloading an object that will not be demanded later. That means that prefetching can involve many hazards because if the prediction is not accurate, it can pollute the cache, waste bandwidth and overload the original server. This implies that prefetching must be carefully applied, by using, for example, idle times in order to avoid performance degradation [16]. Despite these hazards, it has been shown that many prefetching algorithms [3,8,30,46,50] can be used to considerably reduce the latency perceived by users.

The prefetch predictions can be performed by the server, by the proxy, or by the client itself. Some research studies suggest to perform the predictions at the server [46, 50] because it is visited by a high number of users and it has enough information about how they visit the site, therefore its predictions can be quite accurate. Some other studies suggest that proxy servers can perform more accurate predictions because their users are much more homogeneous than in an original server, and they can also predict cross-server links which can reach about 29% of the requests [29]. Finally, other authors consider that predictions must be performed by the client browser, because it knows users' preferences better [40, 43]. There are also some studies indicating that the different parts of the web architecture (users, proxies and servers) must collaborate when performing predictions [46].

12.1.1 Generic Web Architecture

Prefetching systems are usually based on a generic web architecture. This technique is implemented by means of two extra elements, the prediction and prefetching engines, that can be located in the same or different elements of the system. In this section, we describe the elements that compose a generic web architecture and how it can be extended and used to carry out prefetching.

There are two main elements in a generic web architecture: (i) user agents, or clients, that is, the software employed by users to access the Web, and (ii) web servers, which contain the information that users demand. The generic web architecture is an example of the client-server paradigm, which works as follows. Human users tell the client which page they want to retrieve by writing down a URI or by clicking a hyperlink on a previously loaded page. Then, the client demands each object the page consists of. Finally, the whole page is displayed to the user. Optionally, there may be more elements between clients and servers, as Figure 12.1 shows. A proxy is usually located near a group of clients to cache the most popular objects accessed by that group. By doing so, the user-perceived latency and the network traffic between the proxy and the servers can be reduced. Surrogates, also called reverse proxies, are proxies located at the server side. They cache the most popular server responses and are usually transparent to the clients, which access to the surrogate as if they accessed to the web servers.

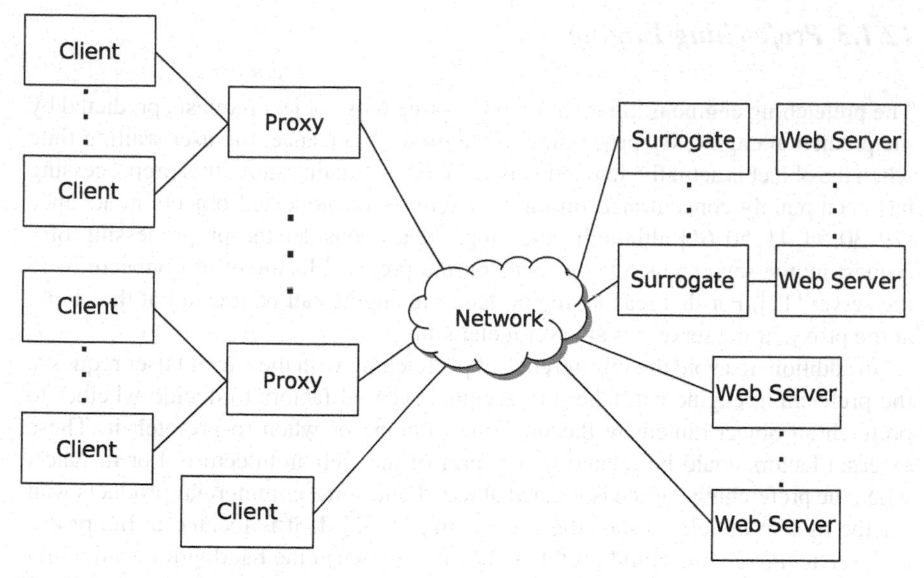

Fig. 12.1 Generic Web architecture

12.1.2 Prediction Engine

The prediction engine is the part of the prefetching system aimed at guessing the following user accesses. This engine can be located at any part of the web architecture: clients [40,43], proxies [6,30], and servers [21,44,51,55], or even working in a collaborative way between several elements [46]. To make the predictions, a high amount of algorithms that learn from the past access patterns to predict future ones have appeared in the literature, e.g., [21,54,57]. These patterns of user accesses differ depending on the element of the architecture in which the prediction engine is implemented because the information that each one can gather is completely different. For instance, a predictor located at the web server can only gather transitions between pages of the same server, but not cross-server transitions.

The output of the prediction engine is the hint list, which is composed of a set of URIs which are likely to be requested by the user in a near future. Nevertheless, this list can also consist of a set of servers if only the connection is going to be pre-established. Due to the fact that in many proposals the hint list is included in the HTTP headers of the server response, the time taken by the predictor to provide this list should be short enough to avoid delaying every user request and, therefore, degrade overall performance.

The predictor must distinguish between those objects that can be prefetched (and therefore predicted) and those that cannot. For this reason, the hint list can only contain prefetchable URIs. According to the definition proposed in [17], a web resource is prefetchable if and only if it is cacheable and its retrieval is safe. Browsers based on Mozilla [32] consider that a resource is not prefetchable if the URI contains a query string.

12.1.3 Prefetching Engine

The prefetching engine is aimed at preprocessing those object requests predicted by the prediction engine. By processing the requests in advance, the user waiting time when the object is actually demanded is reduced. In the literature, this preprocessing has been mainly concentrated on the transference of requested objects in advance [29, 30, 34, 41, 50, 61] although other approaches consider the preprocessing of a request by the server [44, 55], or focus on the pre-establishment of connections to the server [12]. For this reason, the prefetching engine can be located at the client, at the proxy, at the server, or at several elements.

In addition, to avoid the interference of prefetching with the current user requests, the prefetching engine can take into account external factors to decide whether to prefetch an object hinted by the prediction engine or when to prefetch it. These external factors could be related to any part of the web architecture. For instance, when the prefetching engine is located at the client, some commercial products wait for the user to be idle to start the prefetching [1, 32]. If it is located at the proxy, the prefetching engine could prefetch objects only when the bandwidth availability is higher than a given threshold. The case of the predictor located at the server, it could prefetch only when its current load permits an increase of requests.

The capability of reducing the user-perceived latency by means of web prefetching depends on where this engine is located. In this way, the closer to the client the prefetching engine is, the more latency is avoided. Therefore, a prefetching engine located at the client can reduce the whole user-perceived latency. Besides, it is the current trend as it is included in commercial products like Mozilla Firefox and Google Web Accelerator.

12.1.4 Web Prediction Algorithms

A quick and accurate prediction of user accesses is the key of web prefetching. We have classified the large variety of prediction algorithms that can be found in the literature into two main categories according to the data taken into account to make the predictions.

12.1.4.1 Prediction from the Access Pattern

Algorithms in this category lie in the statistical inference about future user accesses from the past ones. According to the amount of research works found, this is the most important category. The prediction of user accesses has usually been made by adapting prediction algorithms from other fields of computing to deal with web accesses. These kind of algorithms has been classified in three subcategories depending on the statistical technique used to compute the predictions: (i) those based on Markov models, (ii) those using data mining techniques and (iii) other techniques.

The algorithms that fit in the first subcategory are the ones that require less CPU time to make predictions. Consequently, the algorithms based on Markov models are the main candidates to be implemented in a web prefetching system to achieve the best reduction of user-perceived latency.

12.1.4.2 Prediction from Web Content

The second subset of web prediction algorithms are those whose predictions are made from the analysis of the content of the web pages (usually HTMLs) that the user has requested instead of using only the sequence of accesses. As a consequence, the complexity of these algorithms is significantly higher than the complexity of just analyzing the references and, usually requires extra information to make the predictions; for instance, the usage profiles that are generated by the web servers. Techniques like neural networks to extract the knowledge from the HTML content has been also proposed.

12.2 Performance Evaluation

There is a large amount of research works focusing on web prefetching in the literature, but comparative studies are rare. Different proposals cannot be easily compared because of four main reasons: (i) the proposed approaches are applied and tested in different baseline systems, (ii) they use workloads that do not represent the current web, (iii) different performance metrics are measured to quantify the benefits, and (iv) studies do not usually focus on the user's perspective or use a cost–benefit analysis. This fact leads to the inability to quantify in a real working environment which proposal is better for the user.

In this section we show how these limitations have been overcome to provide the appropriate framework and methodology to evaluate and compare web prefetching algorithms from the user's point of view.

12.2.1 Experimental Framework

Although the open literature presents very interesting proposals about prefetching techniques, it has not been possible to reproduce them yet (or to propose new ones) using similar environmental conditions in order to compare their main features, advantages and disadvantages. Despite the fact that there are some simulator environments to check cache replacement algorithms [7, 15] as well as network features [18, 58], there is a lack of simulation environments which concentrate on prefetching. This fact, together with the potential performance benefits that prefetching can reach, motivates us to propose a simulation environment to check the performance of

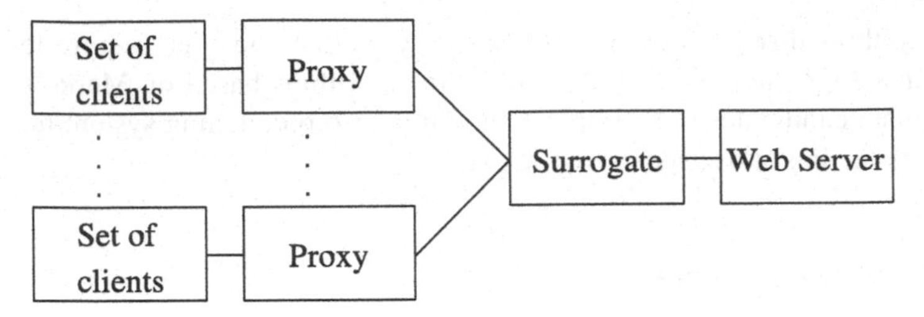

Fig. 12.2 Architecture of the simulation environment

different prefetching algorithms under similar conditions. Our environment permits the implementation of prefetching algorithms on any part of the web architecture (user browser, proxy or server), either individually or working together.

The simulation environment (Figure 12.2) is composed of three main parts: the server, the client, and the proxy. We have developed a hybrid implementation which combines both real and simulated parts in order to provide the environment flexibility and accuracy.

The back end part has both the real web server and the surrogate server. The server is external to our environment and the system can access to it through the surrogate where the prefetching engine is implemented. The surrogate, which does not include cache functions in our framework, is implemented on a real machine offering fully representativeness.

The front end contains the client component, which represents the user behavior in a prefetch enabled browser. It is possible to use real traces or a synthetic workload generator for each set of users accessing concurrently to the same server.

The intermediate part, which is located between the clients and the surrogate system, models one or more proxy servers. Each one can implement the prefetching independently.

This organization provides a useful and operative environment since it permits the implementation of prefetching algorithms working in any part of the system by modifying only few parts of the structure. In addition, the environment provides the appropriate interfaces to communicate the different modules between them in order to implement easily those prefetching algorithms in which the different elements collaborate. Our simulator also includes a module to provide performance results and statistics.

12.2.1.1 Surrogate

A surrogate proxy is an optional element located close to the origin server that generally acts as a cache of the most popular server responses ([33, 59] or accelerator mode in Squid [52]). Usually surrogates are not used to perform prefetching predictions. Although it is equivalent to perform predictions from the server

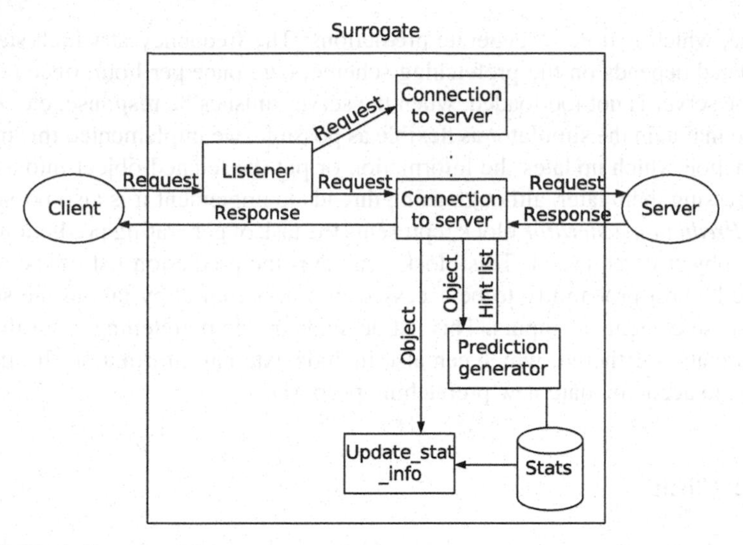

Fig. 12.3 Block diagram of the surrogate

or the surrogate, the latter one allows much easier and more flexible software implementations. Therefore, we perform our implementation on the surrogate.

The surrogate hints clients about what objects should be prefetched by means of an optional header offered by the HTTP/1.1 standard [31] in the same way as Mozilla (from version 1.2 on) admits [32, 47]. This implementation is based on the one described in [50]. The list of hints that the surrogate proxy piggybacks is generated for each response.

Figure 12.3 shows the block diagram of the proposed surrogate proxy. Each block represents a functional unit. Some of them are implemented as functions and others as threads, keeping separately all their functionality.

The listener block, which is the simplest one, is implemented as a thread that waits for clients to connect its socket, acting as a concurrent TCP server that listens at a proxy port (generally at port 80). By default, there are no restrictions on the number of simultaneous incoming connections. The main goal of this block is to create a new *Connection to server* thread.

The *Connection to server* block, which is the main one, handles the connections. The connection involves a request that is transferred to the web server, which processes it and starts the transference. In that moment, the module first parses the response until the full header is received and then, asks for a prediction to the *Prediction generator* block. The prediction consists of a list of hints whose content depends on the implemented algorithm. This list will be piggybacked on the response headers, so the client can prefetch the object if the client application considers it appropriate.

When the whole object is received, the listener calls the *Update stat info* block to notify that event as well as its headers (object size, content type, prefetch or user request, etc.). This information is used to update the contents of the statistical database, which is used to generate predictions. The frequency at which statistics are updated depends on the prefetching scheme; e.g., once per hour, once per day, when the server is not too loaded, when the server finishes its response, etc. As our aim is to maintain the simulator as flexible as possible, we implemented this module as a function which updates the information or put the fetched object into a queue for processing them later, although using threads to implement it is also possible.

The *Prediction generator* block represents the task of generating predictions each time an object is requested. This block generates the prediction list of the objects with the highest probability to be accessed in a near future, by taking the statistical database contents as inputs. This list depends on the prefetching algorithm and the information gathered, which can also include external information. It offers an interface to accommodate new prefetching proposals.

12.2.1.2 Client

The client part of the system reproduces the behavior of a set of users browsing the WWW with a prefetching enabled web browser. The client is fed with real traces collected by a Squid proxy. This module can be replaced either by other log format or even by any synthetic workload generator.

In order to simulate the behavior of the client, it is important to respect the time intervals between two successive requests. Those times are obtained from timestamp differences observed on logs. As timestamps in logs are made at the end of the transfer time, we need to subtract the duration field to the timestamp on logs to obtain the start time of the transfer. For this purpose, the environment includes a simple *chgtime* script.

According to the real behavior of a prefetching enabled browser like Mozilla, and as several research studies suggest [16,41], the prefetching of objects should be performed only when the user is not downloading objects. Furthermore, those proposals suggest stopping the prefetch connection when a user request rises during the prefetch transference. The client simulator permits to stop the prefetch transference, if wanted, by setting the corresponding parameter (*StopPrefetch*) in the configuration file. This option aborts the prefetch, so losing the connection.

The client part has also the ability to introduce latency for each downloaded object simulating the time a client waits to get an object. There are three parameters in the simulator affecting the response time of an object: the connection time, the server processing time and the available bandwidth.

These parameters are not enough to model closely a real network. To this end, an external delay generator can be included in the simulator, like the ns2 proposed in [58]. By using a generator it is possible to simulate scenarios having any particular kind of clients; e.g., systems with wireless clients losing a lot of packets and/or cable users with a high bandwidth. Analogously, it is possible to model different network conditions on the server; e.g., congestion and packet loss.

When simulating cache clients, the simulator only stores the object references in order to avoid the fetching of the object from the server later. Note that, for

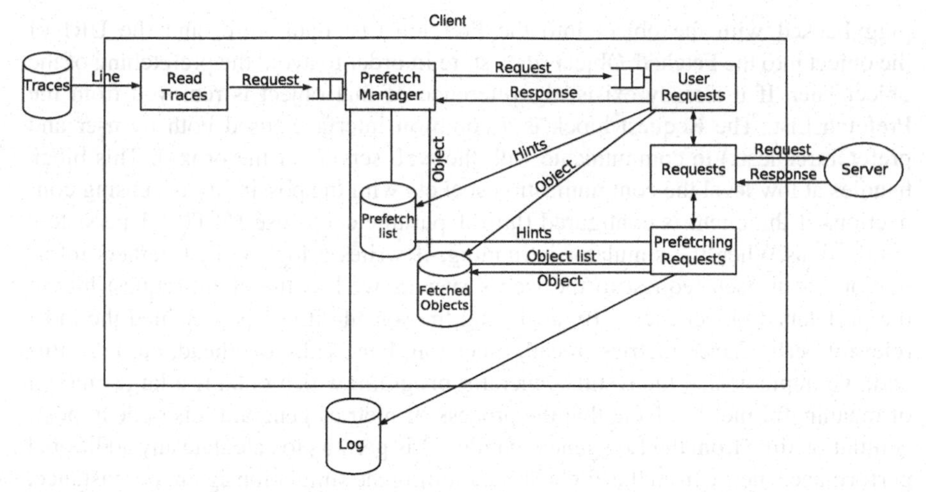

Fig. 12.4 Block diagram of the client

our purposes, it is not necessary to store the content of each object. However, as explained below, this must be considered when working in the client part of the proxy cache.

Once the hints attached to the prefetched objects are received, they must be included into the prefetching list. A parameter states the moment when they are included having two alternatives: (i) when received, and (ii) when accessed the prefetched object.

Figure 12.4 shows the block diagram of the client simulator. As observed, the client part communicates only with the Squid traces, the log file (for analyzing the results achieved by prefetching later), and the server. The latter, depending on the modeled system, could be replaced by one or more proxies.

There are two data stores inside the client. One contains the list of objects that the server hints the client to prefetch. These hints have been piggybacked with the requested object by the user, and accumulated for each burst of user requests. When there is a new user request after an idle period, the remaining hints in the list are removed. The other data store keeps all the fetched objects, and, if they were prefetched, whether the user has accessed them later or not. A block (Read_Traces) reads the users request from a Squid log. The Prefetch_Manager block reads requests from the Read_Traces queue and asks for the object if it has not been prefetched yet (by checking the Fetched_Objects data store). In order to request files to the server, the Prefetch_Manager has many User_Request threads created and waiting for a new request. When the object is received, if its URI is in the Prefetch_List, it is removed. Then, that URI is inserted into Fetched_Objects data store in order to avoid prefetching the object later.

The User Requests block receives requests from the Prefetch Manager and redirects them to the Requests block, reusing the socket when using the HTTP 1.1 version. When an object is received, this block inserts the prefetching hints piggybacked with the object into the Prefetch_List data store, and the URI of the object into the Fetched_Objects data store in order to avoid the prefetching of the object later. If it already exists, the reference of that object is removed from the Prefetch_List. The Request block is a common interface (used both by user and prefetch requests) to communicate with the web server (or the proxy). This block handles at low level the communication sockets with the possibility of reusing connections if the client is configured (http11 parameter) to use HTTP/1.1 persistent connections.When the simulator is running, it writes a log file that gathers information about each request to the web server as well as the client request hits to the prefetched object cache. By analyzing the log file it can be obtained the most relevant performance metrics (recall, precision, bandwidth overhead, etc.). To this end, we implemented the results generator program, which obtains a large amount of meaningful metrics. Note that the process of metrics generation is done at post-simulation time from the Log generated file. This permits to calculate any additional performance metric from this file without running the simulation again. For instance, metrics grouped by client or by file type.

12.2.1.3 Proxy Server

Prefetching in the proxy server has many advantages. First, it allows users of non-prefetching browsers to benefit from server prefetch hints, in the same way we discussed above with respect to the surrogate. Second, it has information from more sources (the clients and the servers), so it is the one that can make more complex and precise predictions about how many and which objects should be prefetched.

When the different parts of the system support prefetch, many different behaviors can be chosen. For example, the proxy server could receive hints from the server and, after processing them, it could start a prefetch connection to cache the selected objects. Then, the proxy could redirect the received hints to the client. These hints could also be removed or even updated (omitting or adding new hints) by applying its own proxy prediction algorithm, since the proxy has a better knowledge of its users than the server has. In addition, the proxy can observe cross-server links.

The structure of the prefetching proxy server can be seen as a combination of a surrogate and a client. On one hand, the proxy waits for connections and performs predictions like the surrogate; on the other hand, the proxy requests files to the server and processes hints like the client. From this point of view, only few changes should be performed, (i) the objects must be stored, and (ii) requests to the server must be generated from client requests (instead of from log traces as done in the client part).

By having all these three parts (surrogate, client, and proxy) we are able to: (i) study, test, and evaluate any proposal already appeared in the literature and (ii) create novel test environments which inter-relate the three parts of the global system. In addition, the proxy side can be split in more parts to check the performance of new organizations; e.g., prefetching between proxies.

Simulations can be run by just setting up one set of clients and one server, which is the simplest system that can be modeled. The system can become more and more complex to represent more realistic scenarios.

To run simulations, the three parts of the modeled system can be placed in a single computer or placed distributed among computers of the same or different LANs. The number of clients that a computer machine can simulate depends on the CPU requirements of the client prefetching algorithm.

12.2.2 Performance Key Metrics

Among the research works found in the literature, a large set of performance key metrics are used to evaluate web prefetching benefits. For the correctness of the performance evaluation studies of prefetching techniques we have identified and classified the most meaningful indexes.

To better understand the performance key metrics selected, first we define some basic concepts related to this technique:

- *Predictions*: amount of objects predicted by the prediction engine.
- *Prefetchs*: amount of objects prefetched by the prefetching engine.
- *GoodPredictions*: amount of predicted objects that are subsequently demanded by the user.
- *BadPredictions*: those predictions that do not result in good predictions.
- *PrefetchHits*: amount of prefetched objects that are subsequently demanded by the user.
- *ObjectsNotUsed*: amount of prefetched objects never demanded by the user.
- *UserRequests*: amount of objects that the user demands.

Figure 12.5 shows graphically the relations between the variables defined above. A represents an object requested by the user but neither predicted nor prefetched. B represents a GoodPrediction that has not been prefetched, while C represents a PrefetchHit, i.e., a GoodPrediction that has been prefetched. D is an ObjectNotUsed,

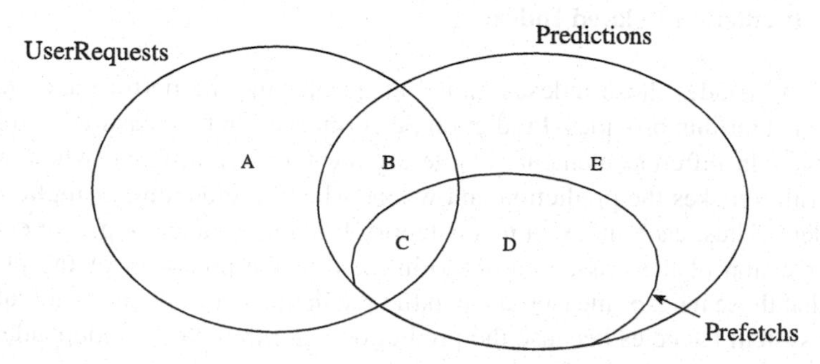

Fig. 12.5 Sets representing the relation between *UserRequests*, *Predictions* and *Prefetchs*

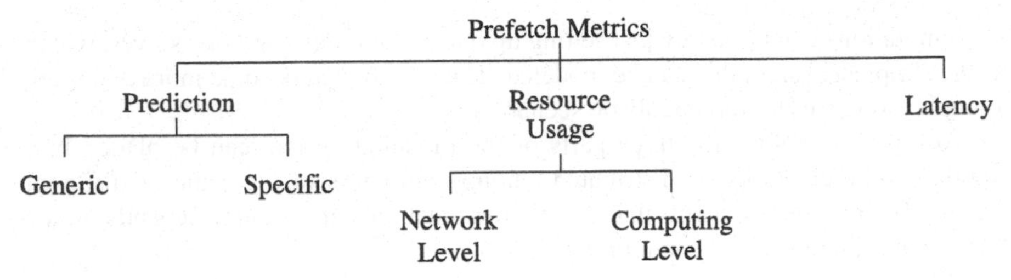

Fig. 12.6 Prefetching metrics taxonomy

which is also a BadPrediction. Finally, E represents a BadPrediction. Analogously, we can define byte related variables (Predictions_B, Prefetchs_B, and so on) by replacing the objects with the corresponding size in bytes in their definition.

To the better understanding of the meaning of the performance indexes, we classify them into three main categories (see Figure 12.6), according to the system feature they evaluate:

- Category 1: prediction related indexes
- Category 2: resource usage indexes
- Category 3: end-to-end perceived latency indexes

The first category is the main one when comparing prediction algorithms performance. It includes those indexes which quantify both the efficiency and the efficacy of the algorithm (e.g., precision). The second category quantifies the additional cost that prefetching incurs (e.g., traffic increase or processor time). Finally, the third category summarizes the performance achieved by the system from the user's point of view. Notice that prefetching techniques must pay attention to the cost increase because they can negatively impact on the overall system performance (traffic increase, user-perceived latencies). Therefore, the three categories are closely related since in order to achieve a good overall performance (Category 3), prefetching systems must trade off the aggressiveness of the algorithm (Category 1) and the cost increase due to prefetching (Category 2).

12.2.2.1 Prediction Related Indexes

This group includes those indexes aimed at quantifying the performance that the prediction algorithm provides. Prediction performance can be measured at different moments or in different elements of the architecture, for instance when (where) the algorithm makes the prediction and when (where) prefetching is applied in the real system. Thus, each index in this category has a *dual* index; e.g., we can refer to the precision of the prediction algorithm and to the precision of the prefetch. Notice that those indexes measured when the prediction list is given do not take into account system latencies because the prediction algorithm works independently of the underlying network and the user restrictions.

Generic Prediction Indexes

Precision (Pc)

Precision measures the ratio of good predictions to the number of predictions [2, 4, 13, 19, 53] (see Equation 12.1). Precision, defined in this way, just evaluates the algorithm without considering physical system restrictions; e.g., cache, network or time restrictions; therefore, it can be seen as a theoretical index.

There are other research works that measure the impact on performance of the precision [28, 45]. In these cases, the number of prefetched objects and prefetch hits are used instead of the number of predictions and good predictions respectively (see Equation 12.2).

$$Pc^{\text{pred}} = \frac{GoodPredictions}{Predictions} \tag{12.1}$$

$$Pc^{\text{pref}} = \frac{PrefetchHits}{Prefetchs} \tag{12.2}$$

Recall (Rc)

Recall measures the percentage of user requested objects that were previously prefetched [2, 13, 53]. *Recall* quantifies the weight of the predicted (see Equation 12.3) or prefetched objects (see Equation 12.4) over the amount of objects requested by the user.

$$Rc^{\text{pred}} = \frac{GoodPredictions}{UserRequests} \tag{12.3}$$

$$Rc^{\text{pref}} = \frac{PrefetchHits}{UserRequests} \tag{12.4}$$

Precision alone or together with *recall* has been the most widely used index to evaluate the goodness of prediction algorithms. The time taken between the client request making and the response receiving consists of four main components; i.e., connection establishment time, request transference time, request processing time, and response transference time. Both *precision* and *recall* are closely related to the three first components. Therefore, we consider that a complete comparison study about prediction algorithms should also include byte related indexes to quantify the last component. In this sense, similarly to some web proxy caching indexes (e.g., the *byte hit ratio* [11]), we propose the use of *byte precision* and *byte recall* as indexes to estimate the impact of prefetching on the time that the user wastes when waiting for the bytes of the requested objects.

Byte Precision (Pc_B)

Byte precision measures the percentage of predicted (or prefetched) bytes that are subsequently requested. It can be calculated by replacing the number of predicted objects with their size in bytes in Equation (12.1).

$$Pc_B^{\text{pred}} = \frac{GoodPredictions_B}{Predictions_B} \tag{12.5}$$

Byte Recall (Rc_B)

Byte recall measures the percentage of demanded bytes that were previously predicted (or prefetched). As mentioned above, this index quantifies how many accurate predictions are made, measured in transferred bytes.

This index becomes more helpful than the previously mentioned *recall*, when the transmission time is an important component of the overall user-perceived latency.

$$Rc_B^{\text{pred}} = \frac{GoodPredictions_B}{UserRequests_B} \tag{12.6}$$

Specific Prediction Indexes

Request savings [30], *Miss rate ratio* [3] or *Probability of making a prediction* are examples of specific indexes suggested for authors to evaluate their proposals.

12.2.2.2 Resource Usage Indexes

The benefits of prefetching are achieved at the expense of using additional resources. This overhead, as mentioned above, must be quantified because they can negatively impact on performance.

Although some prediction algorithms may require huge memory or processor time (e.g., high order Markov models), it is not the current general trend, where the main prefetching bottleneck is the network traffic. Therefore, we divide indexes in this category into two subgroups: network level and computing level.

Network Level

Traffic Increase (ΔTr_B)

Traffic increase quantifies the increase of traffic (in bytes) due to unsuccessfully prefetched documents [46] (see Equation 12.7).

When using the prefetch, the network traffic usually increases due to two side effects of the prefetch: the objects not used and the extra information interchanged. Objects that are not used waste network bandwidth because these objects are never requested by the user. On the other hand, the network traffic increases due to the prefetch related information interchange, called *Network overhead* by [29]).

$$\Delta Tr_B = \frac{ObjectsNotUsed_B + NetworkOverhead_B + UserRequests_B}{UserRequests_B} \quad (12.7)$$

Extra Control Data Per Byte

Extra control data per byte quantifies the extra bytes transferred that are related to the prefetch information, averaged per each byte requested by the user. It is the *Network overhead* referred in [29], but quantified per requested bytes, and will be used below when relating *Traffic increase* to the prediction indexes.

$$ExtraControlData_B = \frac{NetworkOverhead_B}{UserRequests_B} \quad (12.8)$$

Object Traffic Increase (ΔTr_{ob})

Object traffic increase quantifies in which percentage the number of documents that clients get when using prefetching increases.

As Equation (12.9) shows, this index estimates the ratio of the amount of prefetched objects never used with respect to the total user requests. It is analogous to the *traffic increase*, but it measures the overhead in number of objects.

$$\Delta Tr_{ob} = \frac{ObjectsNotUsed + UserRequests}{UserRequests} \quad (12.9)$$

Computing Level

Server Load Ratio

Server load ratio is defined as the ratio between the number of requests for service when speculation is employed to the number of requests for service when speculation is not employed [3].

Space and Time Overhead

In addition to the server load, some research works discuss how the overhead impacts on performance. For instance, [29] discusses the memory and processor time that the prefetch could need.

Prediction Time

This index quantifies the time that the predictor takes to make a prediction. It is used in [28] to compare different predicting algorithms.

12.2.2.3 Latency Related Indexes

Indexes belonging to this category include those aimed at quantifying the end-to-end latencies; e.g., user or proxy related latencies. The main drawback of these indexes is that they include several time components, some of them difficult to quantify. Researchers do not usually detail which components they measure, although they use a typical index name; i.e., latency. Several names have been used instead; for instance, *access time* [50], *service time* [3] and *responsiveness* [39, 56]. This situation is not the best for the research community, due to the fact that the different proposals cannot be fairly compared among them.

Through the different research works, latencies are measured both per page and per object. The *latency per page* (L_p) is calculated by comparing the time between browser's initiation of an HTML page GET and browser's reception of the last byte of the last embedded image or object for that page [29]. Analogously, the *latency per object* (L_{ob}) can be defined as the elapsed time since the browser requests an object until it receives the last byte of that object. In order to illustrate the benefits of prefetching, researchers calculate the ratio of the latency that prefetching achieves (either per page [29, 30, 45] or per object [42]) to the latency with no prefetching.

Unfortunately, some proposals that use *latency* when measuring performance do not specify to which latency they are referring; e.g., [3, 6, 39]. This fact can be misleading because both indexes do not perform in the same way as demonstrated in [22].

12.2.3 Comparison Methodology

Despite the large amount of research works focusing on web prefetching, comparative and evaluation studies from the user's point of view are rare. Some papers comparing the performance of prefetching algorithms have been published [5,6,9,28,49] but they mainly concentrate on predictive performance [5, 9, 28, 49].

In addition, performance comparisons are rarely made using a useful cost–benefit analysis, i.e., latency reduction as a function of the traffic increase. As examples of some timid attempts, Dongshan and Junyi [28] compare the accuracy, the model-building time, and the prediction time in three versions of a predictor based on Markov chains. Another current work by Chen and Zhang [9] implements three variants of the PPM predictor by measuring the hit ratio and traffic under different assumptions.

Nanopoulos et al. [49] show a cost–benefit analysis of the performance of four prediction algorithms by comparing the precision and the recall to the traffic increase. Nevertheless, they ignore how the prediction performance affects the final user. Bouras et al. in [5] show the performance achieved by two configurations of the PPM algorithm and three of the *n*-most popular algorithms. They quantify the usefulness (recall), the hit ratio (precision) and the traffic increase but they present a low number of experiments, which make it difficult to obtain conclusions. In a more recent work [6] they also show an estimated upper bound of the latency reduction for the same experiments.

Consequently, we propose a cost–benefit methodology to perform fair comparisons of web prefetching algorithms from the user's point of view.

As it was mentioned in the previous section one of the most important steps in a performance evaluation study is the correct choice of the performance indexes. As the main goal of prefetching is to reduce the user-perceived latency when navigating the web, the evaluation and comparison of prediction algorithm and/or prefetching techniques should be made from the user's point of view and using a cost–benefit analysis. Therefore, we propose the use of the main metrics related to the user-perceived performance and the prefetching costs: Latency per page ratio, Traffic increase and Object traffic increase.

When predictions fail, the prefetched objects waste user and server resources, which can lead to a performance degradation either to the user himself or to the rest of users. Since in most proposals the client downloads the predicted objects in advance, the main cost of the latency reduction in prefetching systems is the increment of the network load due to the transference of objects that will not be used. This increment has two sides: the first is the increase in the amount of bytes transferred (measured through the *Traffic Increase* metric), and the second is the increase of the server requests (measured through the *Object Traffic Increase* metric). As a consequence, the performance analysis should consider the benefit of reducing the user-perceived latency at the expense of increasing the network traffic and the amount of requests to the server.

Each simulation experiment on a prefetching system takes as input the user behavior, its available bandwidth and the prefetching parameters. The main results obtained are traffic increase, object traffic increase and latency per page ratio values.

Comparisons of two different algorithms only can be fairly done if either the benefit or the cost have the same or close value. For instance, when two algorithms present the same or very close values of traffic increase, the best proposal is the one that presents less user perceived latency, and vice versa. For this reason, in our research works performance comparisons are made through curves that include different pairs of traffic increase and latency per page ratio for each algorithm. To obtain each point in the curve, the aggressiveness of the algorithm is varied, i.e., how much an algorithm will predict. This aggressiveness is usually controlled by a threshold parameter in those algorithms that support it or by the number of returned predictions in those based on the top-*n*.

A plot can gather the curves obtained for each algorithm in order to be compared. By drawing a line over the desired latency reduction in this plot, one can obtain the traffic increase of each algorithm. The best algorithm for achieving that latency per page is the one having less traffic increase. In a similar way, we can proceed with the object traffic increase metric.

The methodology developed in this section was presented in [23–25].

12.2.4 Workload

The benefits of a web prefetching system depend mostly on the workload chosen to evaluate its performance. For this reason, the choice of this workload is of paramount importance in order to estimate its benefits appropriately. By analyzing the open literature, we found that the workload used to evaluate prefetching is quite heterogeneous. A high amount of research works takes real traces (usually from a web server or from a proxy) to conduct the experiments [10, 19, 36, 37, 60, 62], whereas some others use synthetically generated workloads [35, 38, 48, 57]. Synthetic workloads do not usually take into account the structure of the web, i.e., container and embedded objects [48], while real traces are often stale when they are evaluated. For instance, [19] runs a 9-year old trace and [10] a 6-year old trace.

In other research works [20, 27] we ran experiments to compare the performance of web prefetching in old and current web from two perspectives: user-perceived latency and prediction engine. Results show that there is no reason to think that prefetching studies should be conducted using old traces, since the performance of prefetching algorithms is quite different from both the prediction and the user's point of view. Hence, conclusions obtained for old workload cannot be directly moved to current web.

The behavior pattern of users for the experiments presented in this chapter was taken from a news web server, although in our research work we used a wide spectrum of web traces. To perform the experiments, a mirror of the origin server was located close to the surrogate to avoid external interferences. The main characteristics of the trace is shown in Table 12.1. The training length has been adjusted to optimize the perceived latency reduction of the prefetching.

Table 12.1 Trace characteristics

Characteristics	
Year	2003
Users	300
Page accesses	2,263
Object accesses	65,569
Training accesses	35,000
Average objects per page	28.97
Bytes transferred (MB)	218.09
Average object size (KB)	3.41
Average page size (KB)	98.68
Average HTML size (KB)	32.77
Average image size (KB)	2.36
Average page download time at 1 Mbps (s)	3.01
Average page download time at 8 Mbps (s)	2.95

12.3 Evaluation of Prefetching Algorithms

Some papers comparing the performance of prefetching algorithms have been published [5, 6, 9, 28, 49, 61]. These studies mainly concentrate on the performance of the predictor engine but, unfortunately, only [6] attempts to evaluate the user-perceived latency. In spite of being an essential part of the prefetching techniques, performance measured at the prediction level does not completely explain the performance perceived by the user, since the perceived latency is affected by a high amount of factors. In addition, performance comparisons are rarely made by means of a useful cost–benefit analysis, e.g., latency reduction as a function of the traffic increase.

Using the methodology previously described we evaluate and compare the performance of the widest used prediction algorithms in the literature against a novel algorithm proposed by the authors that has been designed to take benefit of the current web structure. In our study each algorithm is represented through a curve that relates the user-perceived latency reduction (by means of the *latency per page ratio* index) achieved by the algorithm to the cost of achieving such benefit, i.e., *traffic increase* or *object traffic increase*. A deep analysis through the predicted objects has also been carried out in order to identify the main reasons why performance among prediction algorithms differs.

12.3.1 Prefetching Algorithms Description

The Dependency Graph Algorithm

Dependency Graph (DG) algorithm proposed in [50], constructs a dependency graph that depicts the pattern of accesses to the objects. The graph has a node for every object that has ever been accessed. There is an arc from node A to B if and only if at some point in time a client accessed to B within w accesses after A, where w is the *lookahead window* size. The weight of the arc is the ratio of the number of accesses to B within a window after A to the number of accesses to A itself. The prefetching aggressiveness can be controlled by applying a cutoff threshold parameter to the weight of the arcs.

For illustrative purposes, we use a working example. Let us suppose that the algorithms are trained by two user sessions. The first one contains the following accesses: HTML1, IMG1, HTML2, IMG2. The second session includes the accesses: HTML1, IMG1, HTML3, IMG2. Note that IMG2 is embedded both in HTML2 and in HTML3. We found this characteristic common through the analyzed workload, where different pieces of news (i.e., HTML files) contain the same embedded images, since they are included in the site structure. Figure 12.7(a) shows the state of the graph of the DG algorithm for a lookahead window size of 2 corresponding to the mentioned training. Each node in the graph represents an object whereas the weight of each arc is the confidence level of the transition.

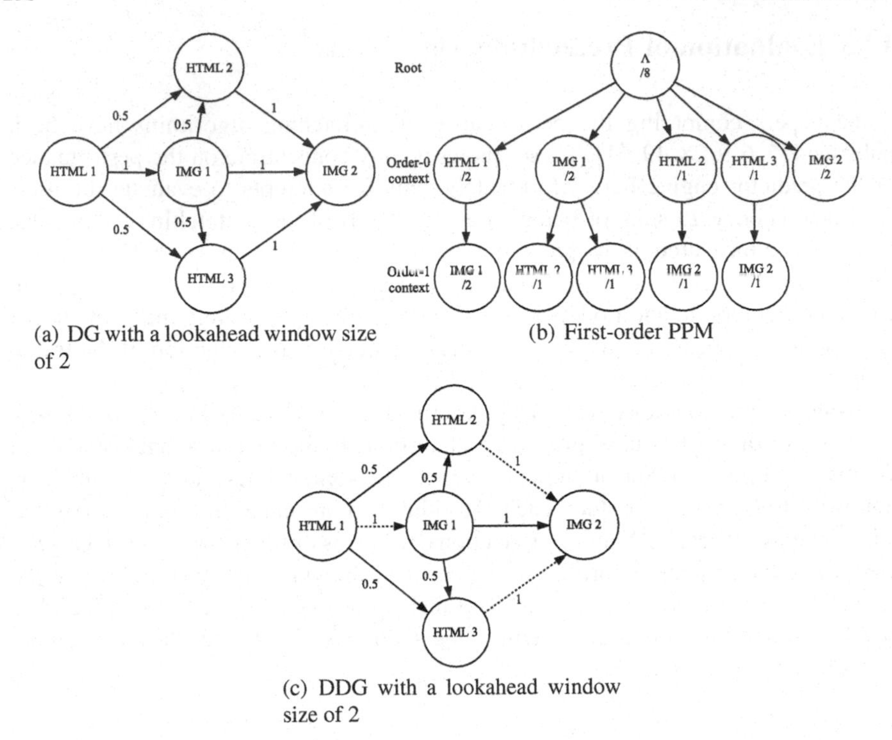

(a) DG with a lookahead window size of 2

(b) First-order PPM

(c) DDG with a lookahead window size of 2

Fig. 12.7 State of the graphs of the algorithms after the accesses HTML1, IMG1, HTML2, IMG2 by one user; and HTML1, IMG1, HTML3, IMG2 by other user

The Prediction by Partial Matching Algorithm

Prediction by Partial Match (PPM) proposed by [51], uses Markov models of m orders to store previous contexts. Predictions are obtained by comparing the current context to each Markov model, where each context consists of the n last accesses of the user, where n is the order of the context plus 1. To select which objects are predicted, the algorithm uses a confidence threshold.

Figure 12.7(b) shows the graph obtained when the PPM algorithm is applied to the training used in the previous example. Each node represents a context, where the root node is in the first row, the order-0 context is in the second, and the order-1 context is in the third one. The label of each node also includes the number of times that a context appears, so one can obtain the confidence of a transition by dividing the counter of a node by the counter of its parent. The arcs indicate the possible transitions. For instance, the label of the IMG2 in order-0 context is 2 because IMG2 appeared twice in the training; once after HTML2 and another after HTML3. IMG2 has two nodes in the order-1 context, i.e., one per each HTML on which it depends.

The Double Dependency Graph Algorithm

The Double Dependency Graph (DDG) algorithm [21] is based on a graph that keeps track of the dependences among the objects accessed by the user. It depicts the same graph (nodes and arcs) as DG. However, to keep track of the web site structure, it distinguishes two classes of arcs to differentiate the dependences between objects of the same page from the dependences between objects of different pages. There is an arc from node A to B if and only if at some point in time a client accessed to B within w accesses to A after B, where w is the *lookahead window* size. The arc is a primary arc if A and B are objects of subsequent pages, that is, either B is an HTML object or the user accessed one HTML object between A and B. When there are more than one HTML between A and B, no arc is included in the prediction model. On the contrary, if there are no HTML accesses between A and B, the arc is secondary.

The confidence of each primary or secondary transition (i.e., the confidence of each arc) is calculated by dividing the counter of the arc by the amount of appearances of the node, both for primary and for secondary arcs. This algorithm has the same order of complexity as the DG, since it makes the same graph but distinguishing two classes of arcs.

Figure 12.7(c) shows the state of the DDG algorithm corresponding to the working example. Arrows with continuous lines represent primary arcs while dashed lines represent secondary arcs.

In order to obtain the predictions, we firstly apply a cutoff *threshold* to the weight of the primary arcs that leave the node of the last user access. Then, to predict the embedded objects of the following page, we apply a *secondary threshold* to the secondary arcs that leave the nodes of the objects predicted in the first step.

The DDG algorithm includes the option of disabling the prediction of HTML objects. It might be useful not to predict HTMLs when working with dynamically generated HTMLs or in those cases in which the cost of downloading an HTML object (i.e., its size) is very high when compared to the contained objects (e.g., images). This fact occurs in the workload used in this section, since HTMLs are, on average, 13 times larger than the embedded objects. The algorithm determines if a given object is an HTML file by looking at its MIME type in the response header given by the web server.

The configuration parameters of each algorithm is highly dependent on the web site characteristics. All three algorithms were tuned to achieve their best performance. The best configuration of DG and PPM is the one with less complexity, that is, DG with a lookahead window size of 2, and a first-order PPM. In a previous work we showed that extra complexity in these algorithms does not involve better performance [20] in current web due to the increase in the amount of embedded objects in each page. The best configuration for DDG under the considered workload is a lookahead window size of 12 and a secondary threshold value of 0.3.

12.3.2 Experimental Results

12.3.2.1 Latency Per Page Ratio

Figure 12.8(a) shows that for any value of traffic increase in bytes, the algorithm that provides the lowest user-perceived latency is always the DDG. Similar results can be extracted when considering as main cost the traffic increase measured in amount of objects (see Figure 12.8(b)): DDG curve always falls below the other algorithms curves. Figure 12.9(a) shows the relative improvement of the DDG algorithm compared to the best of the other two algorithms (i.e., the DG). Figure 12.9 shows that given a value of latency per page ratio, DDG requires between 25% and 60% less of bytes transference and between 15% and 65% less requests to the server than the DG algorithm. On the other hand, given a value of traffic increase in bytes, DDG

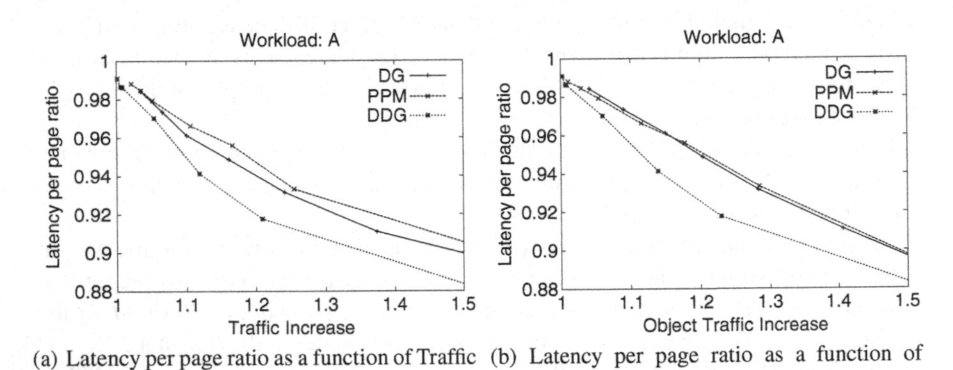

(a) Latency per page ratio as a function of Traffic increase (b) Latency per page ratio as a function of Object Traffic increase

Fig. 12.8 Algorithms comparison

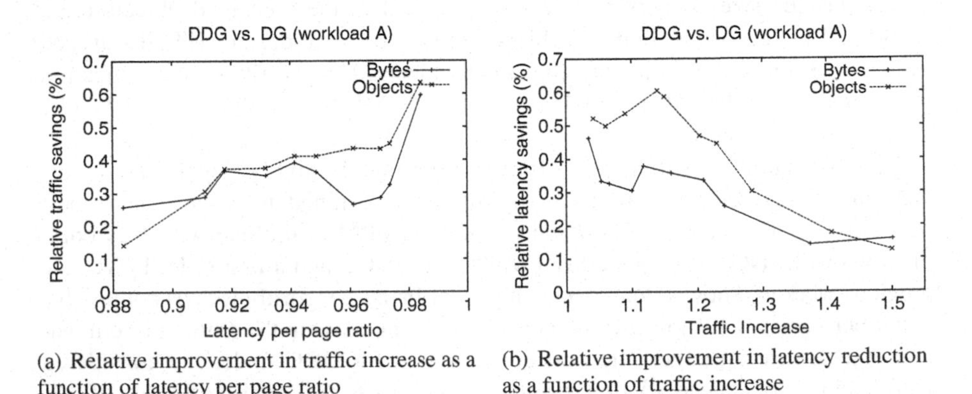

(a) Relative improvement in traffic increase as a function of latency per page ratio (b) Relative improvement in latency reduction as a function of traffic increase

Fig. 12.9 Relative performance comparison of DDG vs DG algorithms. The higher values the better is the DDG against DG

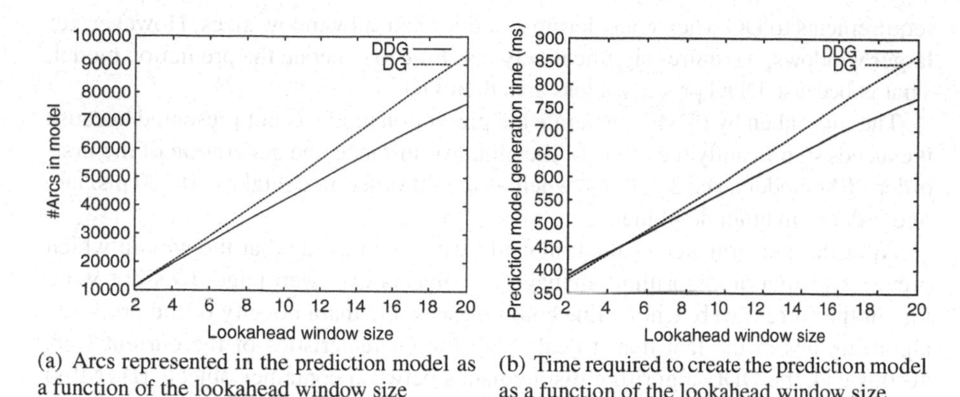

(a) Arcs represented in the prediction model as a function of the lookahead window size

(b) Time required to create the prediction model as a function of the lookahead window size

Fig. 12.10 Comparison of resources used by DG and DDG algorithms

achieves a latency reduction between 15% and 45% higher than DG (see the continuous curve in Figure 12.9(b)). The benefits could rise up to 60% of extra perceived latency reduction when taking into account the object traffic increase, as the dashed curve in Figure 12.9(b) shows.

12.3.2.2 Space

The space overhead required by PPM is usually measured by means of the number of nodes required in the prediction model. However, this metric is not valid for analyzing DG and DDG because they require, for any lookahead window size, as many nodes as different files are in the server. In these algorithms, the amount of arcs in the model is the main metric of space overhead. Figure 12.10(a) shows that DDG requires fewer arcs than DG. Differences are higher as larger the lookahead window size is, reaching about 20% for a size of 20. That is due to the limitation introduced by DDG in which there is no arc between two objects in the same lookahead window when the arc goes through more than one HTML.

On the other hand, PPM has much more memory requirements than DG and DDG. For instance, a first-order PPM model requires, under the considered workload, 12,678 nodes. Moreover, this value increases up to 743,065 nodes when considering a 20-order model, that is, about one order of magnitude higher than the DG and DDG algorithms.

12.3.2.3 Processor Time

Looking at the description of the DDG algorithm, one can realize that this algorithm has the same order of complexity as DG. Figure 12.10(b) shows the processor time taken to generate the prediction model. DDG has similar processor time requirements to DG when considering small lookahead window sizes. However, for larger windows, it requires significatively less time to generate the prediction model. That is because DDG processes less arcs than DG.

The time taken by PPM to generate the prediction model is not presented because it exceeds significantly the range of the plot. For instance, the generation of the first-order PPM model takes 2,300 ms, whereas the 20-order model takes 20,200 ms, i.e., one order of magnitude higher.

After this performance evaluation study we can conclude that the way in which previous prediction algorithms manage the objects of a web page does not work fine in the current web. Unlike this common way, the main novelty of the proposed algorithm lies in the fact that it deals with the characteristics of the current web. To this end, the DDG algorithm distinguishes between container objects (HTMLs) and embedded objects (e.g., images) to create the prediction model and to make the predictions.

12.3.3 Summary

Experimental results have shown that given an amount of requests to the server, the proposed algorithm outperforms the existing ones by reducing the perceived latency between 15% and 60% more with the same extra requests. Furthermore, these results were achieved not only without increasing the order of complexity with respect to existing algorithms, but also by reducing the processor time and memory requirements.

12.4 Theoretical Limits on Performance

Finally this section analyzes the impact of the web prefetching architecture on the theoretical limits of reducing the user-perceived latency. To do so, we examine the factors that limit the predictive power of each prefetching architecture and quantify these theoretical limits both in amount of requests that can be predicted and in user-perceived latency that can be reduced. This study addresses two main issues: (i) to identify the best architecture to reduce user-perceived latency when performing prefetch, and (ii) to provide insights about the goodness of a given proposal by comparing it to the performance limits.

12.4.1 Metrics

There exist two situations in which a user access cannot be predicted. To quantify and analyze these situations, which may strongly limit the performance, new performance indexes are defined.

The first scenario in which a user access cannot be predicted is when accessing for the first time to the element of the web architecture in which the prediction engine is located. For instance, a predictor located at the web server cannot predict the first access of a user in a session. We quantify this limitation by means of the *session first-access ratio* (SFAR), described below.

The second situation in which the prediction is limited is when the prediction engine has never seen the object before. This is particularly important when the prediction engine is located at the client. We use the *first seen ratio* (FSR) to evaluate the impact of this limiting factor on the predictive performance. This index has two variants, FSR_{object} and $\text{FSR}_{latency}$, depending on whether the limits on performance are measured in number of objects that can be potentially predicted or in potential latency savings. The FSR_{object} is defined as the ratio of the amount of different objects to the total amount of requests (see Equation 12.10). When aggregating data from different clients, the FSR_{object} is calculated as the ratio of different pairs {URI, client IP} to the total amount of requests. $\text{FSR}_{latency}$ is defined as the ratio of the sum of the latency of the first appearance of each object to the latency of the total amount of accesses.

$$FSR_{object} = \frac{\#DifferentObjects}{\#ObjectRequests} \tag{12.10}$$

Like the FSR, the SFAR has different variants depending on how the limiting factor is quantified: object, latency and page. The index uses the *session* concept, defined as a sequence of accesses made by a single user with idle times between them no longer than t seconds. Since each session has always a first page, which cannot be predicted, we define the SFAR_{page} as the ratio of the number of sessions to the total amount of page accesses (see Equation 12.11). SFAR_{object} is used to quantify the amount of objects that this factor makes it impossible to predict. It is defined as the ratio of session first-accesses to the total amount of object requests, where a session first-access is a client request that is included in the first page of a session. Finally, to measure the impact of the SFAR factor on the perceived latency, the $\text{SFAR}_{latency}$ is calculated as the ratio of the latency of the session first-accesses to the latency of the total amount of accesses.

$$SFAR_{page} = \frac{\#Sessions}{\#PageAccesses} \tag{12.11}$$

Since the location of the prediction engine makes the pattern of accesses seen by the predictor vary, the described indexes are evaluated in each prefetching architecture to quantify the two situations in which the predictive performance is limited.

The workload used to quantify the limits of the reduction of web latencies when using prefetching was obtained from the log of the Squid proxy of the Polytechnic University of Valencia. The log includes 43,418,555 web accesses of 9,359 different users to 160,296 different servers. It is 7 day long, from May 31st to Jun 6th 2006. In the analysis, the first 6 days were taken as a training set while the trace of the last day was used to calculate performance metrics.

12.4.2 Predicting at the Server

The main advantage of locating the prediction engine at the server is that it is the element that has the most knowledge about the objects that it contains. In addition, it is possible to avoid interferences with the user requests that prefetching might produce by varying the prediction aggressiveness as a function of the server load. However, from the server point of view, it is not possible to gather dependencies between accesses to other servers, increasing the SFAR measured at the servers and, therefore, decreasing the potential reduction of the user-perceived latency.

Figure 12.11 shows the three variants of SFAR metric as a function of the maximum idle time allowed in a session (t parameter). As observed, for a t value of 1,800 s (as used in [14]), the first pages of a session seen from the server point of view ($SFAR_{page}$) represent about 35% of the total pages visited by users. However, these pages contain about 45% of the accessed objects, as shown in the $SFAR_{object}$ curve, but represent 63.4% of the latency perceived by the user, shown in the $SFAR_{latency}$ curve. Therefore, the percentage of latency that could be reduced taking into account only this factor is 36.6%.

A deep analysis of this result shows that this poor potential reduction of latency is due to the fact that users access most of the servers only once. Figure 12.12 shows how the $SFAR_{latency}$ is distributed through the servers. Each point of the curve represents the ratio of servers that have a $SFAR_{latency}$ lower than its corresponding value in the X axis. As one can observe, more than 60% of servers have all their latency in the first pages. For this reason, the potential latency reduction highly depends on the server workload since, as shown in Figure 12.12, there is a noticeable amount of servers that could benefit from prefetching to a higher extent. For instance, 20% of servers have a $SFAR_{latency}$ lower than 37.4% (i.e., a potential latency reduction higher than 62.6%).

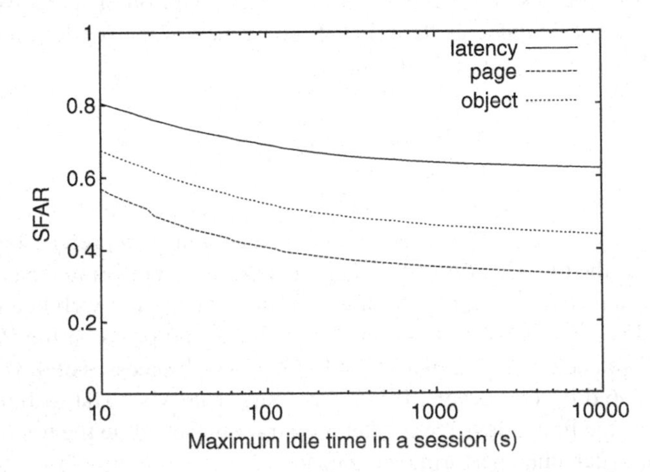

Fig. 12.11 SFAR at the server as a function of the maximum idle time in a session

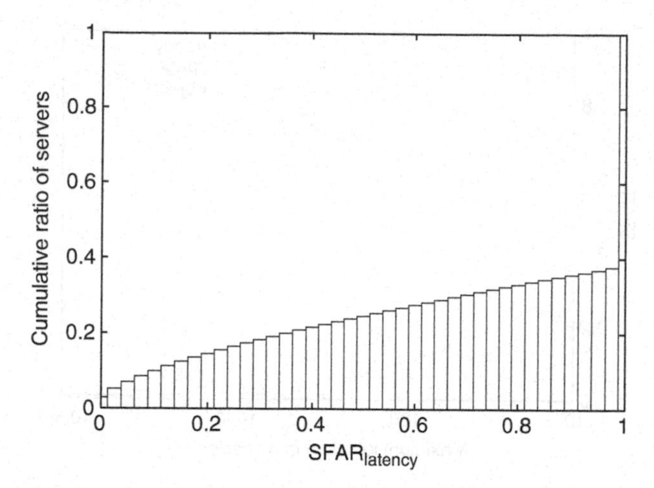

Fig. 12.12 Cumulative relative frequency histogram of SFAR$_{latency}$ at the server with $t = 1{,}800$ s to web servers

Table 12.2 First seen ratio at different servers

Metric	Server				
	A (%)	*B* (%)	*C* (%)	*D* (%)	*E* (%)
FSR$_{object}$	0.5	0.8	0.6	0.4	0.6
FSR$_{latency}$	3.9	9.3	1.7	1.8	1.7

It is not possible to infer a precise FSR value at the servers from a proxy trace because it gathers only a part, in most cases not significant, of the requests received by a server. To obtain an approximate value, we calculated the FSR in the top-5 servers with more load gathered in the proxy. Table 12.2 shows that, among these servers, the FSR$_{object}$ is lower than 0.8%, although the latency of these objects represents 9.3% of the user-perceived latency.

In summary, the main limitation to reduce the latency when the prediction engine is located at the server is thc inability of predicting the first access to the server. This fact makes it impossible to reduce the user-perceived latency beyond 36.6% with this prefetching architecture.

12.4.3 Predicting at the Client

Locating the predictor engine at the client has two main advantages: (i) the client is the element where the SFAR is the lowest since the predictor engine can gather all the user accesses, and (ii) it is possible to gather, and therefore, to predict, cross-server transitions. In contrast, this location has, as main shortcoming, the low amount of accesses that it receives, making FSR high.

Regarding the prediction limit that affects the prediction of objects never seen before, we found a FSR$_{object}$ value of 41.6%, which is the percentage of objects

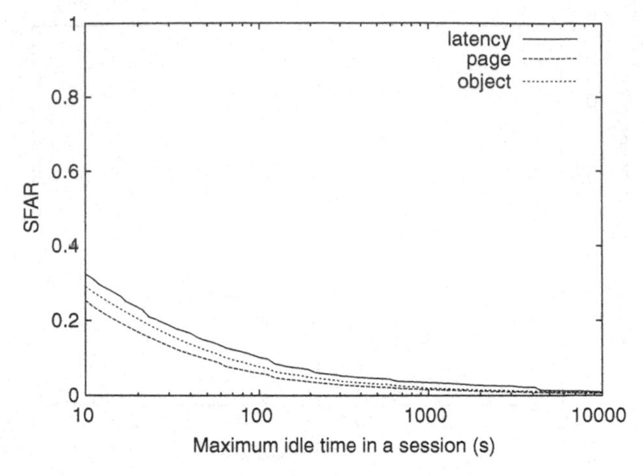

Fig. 12.13 SFAR at the client as a function of the maximum idle time in a session

demanded by a user that has never been accessed by that user before. The latency of these objects represent 45.6% of the total latency perceived by users, which is the latency that cannot be reduced when the predictor is located at the client.

Figure 12.13 shows the prediction limit related to the session first-access factor. As one can observe, the values are noticeably lower than those measured at the server. For a t value of 1,800 s, only 1.3% of pages are the first ones in a session, which represent 2.8% of the user-perceived latency.

In summary, unlike the predictor located at the server, the main limitation of the web prefetching techniques for reducing the user-perceived latency is the access to objects never seen before. This fact means that the maximum latency that this web prefetching architecture could reduce is about 54.4% of the latency perceived by users.

12.4.4 Predicting at the Proxy

Locating the predictor at the proxy has similar advantages to locating it at the client, since both gather all the requests performed by the client. For this reason, their SFAR matches. However, the FSR has a better value in the proxy than in the client, due to the high amount of clients that are usually connected to the proxy. To analyze how the FSR changes depending on the amount of connected clients, experiments were run varying the number of users connected from 40 to 9,359, randomly selected from the original trace. As the results are highly dependent on the selected set of users that are connected to the scaled proxy, results for between 20 and 100 sets of users are averaged to achieve a 95% confidence interval less than 0.04.

Figure 12.14 shows that in a small sized proxy with 40 clients connected, 34% of user accesses were made to an object never requested before. This FSR decreases

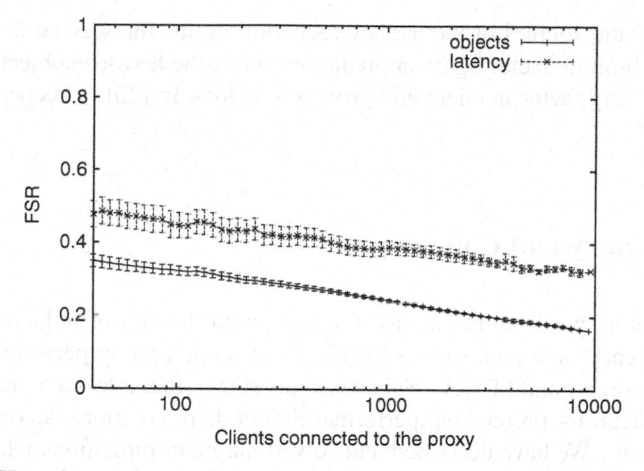

Fig. 12.14 FSR at the proxy with 95% confidence intervals depending on the amount of clients connected to the proxy

Table 12.3 Summary of predicting limits depending on where the prediction engine is located

Factor	Client (%)	Proxy (%)	Server (%)
$SFAR_{object}$	1.3	1.3	45
FSR_{object}	41.6	16.3	0.4
Max. object pred.	58.4	83.7	55
$SFAR_{latency}$	2.8	2.8	63.4
$FSR_{latency}$	45.6	32.6	1.7
Max. lat. reduct.	54.4	67.4	36.6

logarithmically to 16% when considering all the users connected. It occurs in a similar way when looking at the latency curve, which decreases from 47% with a 40-client proxy to 33% with the 10,000-user proxy.

The values obtained for SFAR and FSR let us state that the first seen objects are the main limiting factor of prediction engines in the proxy in order to reduce the user-perceived latency.

12.4.5 Prefetching Limits Summary

Our study summarizes the potential prediction capabilities and the sources of the limits of each prefetching architecture (see Table 12.3). The best element to implement a single predictor is the proxy, which could reduce up to 67% of the perceived latency. The prediction engines located at servers can only reduce 36.6% the overall latency. This is due to the high amount of servers that are accessed only once per session. However, there is a noticeable amount of servers that can benefit from implementing the prediction engine to a higher extent.

Regarding the sources of the latency reduction limits, the session first-accesses are the main limiting factor in server predictors, while the first seen objects represent the main limiting factor in client and proxy predictors. In [26], a deeper study was presented.

12.5 Summary and Conclusions

Web prefetching has been researched for years with the aim of reducing the user-perceived latency; however, studies mainly focus on prediction performance rather than on the user's point of view. This work has shown that a fast and accurate prediction is crucial for prefetching performance, but there are more factors involving the user benefit. We have described and solved the main limitations when evaluating the performance from the user's perspective. To fairly evaluate web prefetching techniques, an experimental framework has been developed. By simulating the whole web architecture, all performance issues were measured.

One of the most important steps in a performance evaluation study is the correct choice of the performance indexes, but we found a wide heterogeneity of metrics in the literature. After its analysis, we clarified indexes definitions and detected the most meaningful metrics for the users. In spite of the focus on metrics used in web prefetching, some of the conclusions obtained for these metrics are common in any web performance evaluation study.

We also proposed an evaluation methodology to consider the cost and the benefit of web prefetching at the same time, although this approach could be used to any technique that implies a tradeoff.

Once the methodological basics were established, the main prediction algorithms were tuned and compared. Results showed that, given an amount of extra requests, the DDG algorithm always reduces the latency more than the other existing algorithms. In addition, these results were achieved without increasing the order of complexity of the existing algorithms.

We analyzed the limits on performance depending on the web prefetching architecture and found that a proxy-located predictor is the one less limited to reduce user-perceived latency since the potential limit of latency reduction is 67%. The prediction engines located at servers can only reduce up to 37% the overall latency. However, there is a noticeable amount of servers that can benefit from implementing this prediction engine to a higher extent.

References

1. Google web accelerator. http://webaccelerator.google.com/
2. Albrecht, D.W., Zukerman, I., Nicholson, A.E.: Pre-sending documents on the WWW: A comparative study. In: Proceedings of the Sixteenth International Joint Conference on Artificial Intelligence. Stockholm, Sweden (1999)

3. Bestavros, A.: Using speculation to reduce server load and service time on the WWW. In: Proceedings of the 4th ACM International Conference on Information and Knowledge Management. Baltimore, USA (1995)
4. Bonino, D., Corno, F., Squillero, G.: A real-time evolutionary algorithm for web prediction. In: Proceedings of the IEEE/WIC International Conference on Web Intelligence. Halifax, Canada (2003)
5. Bouras, C., Konidaris, A., Kostoulas, D.: Efficient reduction of web latency through predictive prefetching on a WAN. In: Proceedings of the 4th International Conference on Advances in Web-Age Information Management, pp. 25–36. Chengdu, China (2003)
6. Bouras, C., Konidaris, A., Kostoulas, D.: Predictive prefetching on the web and its potential impact in the wide area. World Wide Web **7**(2), 143–179 (2004)
7. Cao, P.: Wisconsin web cache simulator. http://www.cs.wisc.edu/ cao
8. Chen, X., Zhang, X.: Popularity-based PPM: An effective web prefetching technique for high accuracy and low storage. In: Proceedings of the International Conference on Parallel Processing. Vancouver, Canada (2002)
9. Chen, X., Zhang, X.: A popularity-based prediction model for web prefetching. IEEE Computer **36**(3), 63–70 (2003)
10. Chen, X., Zhang, X.: Coordinated data prefetching for web contents. Computer Communications **28**, 1947–1958 (2005)
11. Cherkasova, L., Ciardo, G.: Characterizing temporal locality and its impact on web server performance. In: Proceedings of the 9th International Conference on Computer Communication and Networks. Las Vegas, USA (2000)
12. Cohen, E., Kaplan, H.: Prefetching the means for document transfer: a new approach for reducing web latency. Computer Networks **39**(4), 437–455 (2002)
13. Cohen, E., Krishnamurthy, B., Rexford, J.: Efficient algorithms for predicting requests to web servers. In: Proceedings of the IEEE INFOCOM '99 Conference. New York, USA (1999)
14. Cooley, R., Mobasher, B., Srivastava, J.: Data preparation for mining World Wide Web browsing patterns. Knowledge and information systems **1**(1), 5–32 (1999)
15. Crdenas, L.G., Sahuquillo, J., Pont, A., Gil, J.A.: The multikey web cache simulator: A platform for designing proxy cache management techniques. In: Proceedings of the 12th Euromicro Conference on Parallel, Distributed and Network based Processing. La Corua, Spain (2004)
16. Crovella, M., Barford, P.: The network effects of prefetching. In: Proceedings of the IEEE INFOCOM'98 Conference. San Francisco, USA (1998)
17. Davison, B.D.: Assertion: Prefetching with GET is not good. In: Proceedings of the 6th International Workshop on Web Caching and Content Distribution. Boston, USA (2001)
18. Davison, B.D.: NCS: Network and cache simulator – an introduction. Tech. rep., Department of Computer Science, Rutgers University (2001)
19. Davison, B.D.: Learning web request patterns. In: Web Dynamics – Adapting to Change in Content, Size, Topology and Use, pp. 435–460. Springer (2004)
20. Domènech, J., Pont, A., Sahuquillo, J., Gil, J.A.: A user-focused evaluation of web prefetching algorithms. Computer Communications **10**(30), 2213–2224 (2007)
21. Domènech, J., Gil, J.A., Sahuquillo, J., Pont, A.: DDG: An efficient prefetching algorithm for current web generation. In: Proceedings of the 1st IEEE Workshop on Hot Topics in Web Systems and Technologies (HotWeb). Boston, USA (2006)
22. Domènech, J., Gil, J.A., Sahuquillo, J., Pont, A.: Web prefetching performance metrics: A survey. Performance Evaluation **63**(9–10), 988–1004 (2006)
23. Domènech, J., Pont, A., Gil, J.A., Sahuquillo, J.: Guidelines for evaluating and adapting web prefetching techniques. In: Proceedings of the XVII Jornadas de Paralelismo. Albacete, Spain (2006)
24. Domènech, J., Pont, A., Sahuquillo, J., Gil, J.A.: A comparative study of web prefetching techniques focusing on user's perspective. In: Proceedings of the IFIP International Conference on Network and Parallel Computing (NPC 2006). Tokyo, Japan (2006)

25. Domènech, J., Pont, A., Sahuquillo, J., Gil, J.A.: Cost–benefit analysis of web prefetching algorithms from the user's point of view. In: Proceedings of the 5th International IFIP Networking Conference. Coimbra, Portugal (2006)
26. Domènech, J., Sahuquillo, J., Gil, J.A., Pont, A.: The impact of the web prefetching architecture on the limits of reducing user's perceived latency. In: Proceedings of the 2006 IEEE / WIC / ACM International Conference on Web Intelligence. Hong Kong, China (2006)
27. Domènech, J., Sahuquillo, J., Pont, A., Gil, J.A.: How current web generation affects prediction algorithms performance. In: Proceedings of the 13th International Conference on Software, Telecommunications and Computer Networks (SoftCOM). Split, Croatia (2005)
28. Dongshan, X., Junyi, S.: A new Markov model for web access prediction. Computing in Science and Engineering **4**(6), 34–39 (2002). DOI http://dx.doi.org/10.1109/MCISE.2002.1046594
29. Duchamp, D.: Prefetching hyperlinks. In: Proceedings of the 2nd USENIX Symposium on Internet Technologies and Systems. Boulder, USA (1999)
30. Fan, L., Cao, P., Lin, W., Jacobson, Q.: Web prefetching between low-bandwidth clients and proxies: Potential and performance. In: Proceedings of the ACM SIGMETRICS Conference on Measurement and Modeling Of Computer Systems, pp. 178–187. Atlanta, USA (1999)
31. Fielding, R., Gettys, J., Mogul, J., Frystyk, H., Masinter, L., Leach, P., Berners-Lee, T.: Hypertext transfer protocol – HTTP/1.1 (1999)
32. Fisher, D., Saksena, G.: Link prefetching in Mozilla: A server driven approach. In: Proceedings of the 8th International Workshop on Web Content Caching and Distribution (WCW 2003). New York, USA (2003)
33. Group, N.W.: Internet web replication and caching taxonomy. RFC 3040 (2001)
34. Ibrahim, T.I., Xu, C.Z.: Neural nets based predictive prefetching to tolerate WWW latency. In: Proceedings of the 20th IEEE International Conference on Distributed Computing Systems. Taipei, Taiwan (2000)
35. Jiang, Y., Wu, M.Y., Shu, W.: Web prefetching: Costs, benefits and performance. In: Proceedings of the 7th International Workshop on Web Content Caching and Content Distribution. Boulder, USA (2002)
36. Jiang, Z., Kleinrock, L.: Prefetching links on the WWW. In: Proceedings of the IEEE International Conference on Communications. Montreal, Canada (1997)
37. Jiang, Z., Kleinrock, L.: An adaptive network prefetch scheme. IEEE Journal on Selected Areas in Communications **16**(3), 358–368 (1998)
38. Khan, J.I., Tao, Q.: Partial prefetch for faster surfing in composite hypermedia. In: Proceedings of the 3rd USENIX Symposium on Internet Technologies and Systems. San Francisco, USA (2001)
39. Khan, J.I., Tao, Q.: Exploiting webspace organization for accelerating web prefetching. In: Proceedings of the IEEE/WIC International Conference on Web Intelligence. Halifax, Canada (2003)
40. Kim, Y., Kim, J.: Web prefetching using display-based prediction. In: Proceedings of the IEEE/WIC International Conference on Web Intelligence. Halifax, Canada (2003)
41. Kokku, R., Yalagandula, P., Venkataramani, A., Dahlin, M.: NPS: A non-interfering deployable web prefetching system. In: Proceedings of the USENIX Symposium on Internet Technologies and Systems. Palo Alto, USA (2003)
42. Kroeger, T.M., Long, D.D., Mogul, J.C.: Exploring the bounds of web latency reduction from caching and prefetching. In: Proceedings of the 1st USENIX Symposium on Internet Technologies and Systems. Monterey, USA (1997)
43. Lau, K., Ng, Y.K.: A client-based web prefetching management system based on detection theory. In: Proceedings of the Web Content Caching and Distribution: 9th International Workshop (WCW 2004), pp. 129–143. Beijing, China (2004)
44. Lee, H.K., Vageesan, G., Yum, K.H., Kim, E.J.: A proactive request distribution (prord) using web log mining in a cluster-based web server. In: Proceedings of the International Conference on Parallel Processing (ICPP'06). Columbus, USA (2006)

45. Loon, T.S., Bharghavan, V.: Alleviating the latency reduction and bandwidth problems in WWW browsing. In: Proceedings of the 1st USENIX Symposium on Internet Technologies and Systems. Monterey, USA (1997)
46. Markatos, E., Chronaki, C.: A top-10 approach to prefetching on the web. In: Proceedings of the INET' 98. Geneva, Switzerland (1998)
47. Mozilla: Link prefetching faq. http://www.mozilla.org/projects/netlib/LinkPrefetchingFAQ.html
48. Nanopoulos, A., Katsaros, D., Manolopoulos, Y.: Exploiting Web Log Mining for Web Cache Enhancement, vol. 2356, chap. in Lecture Notes in Artificial Intelligence (LNAI), pp. 68–87. Springer-Verlag (2002)
49. Nanopoulos, A., Katsaros, D., Manolopoulos, Y.: A data mining algorithm for generalized web prefetching. IEEE Trans. Knowl. Data Eng. **15**(5), 1155–1169 (2003)
50. Padmanabhan, V.N., Mogul, J.C.: Using predictive prefetching to improve World Wide Web latency. Computer Communication Review **26**(3), 22–36 (1996)
51. Palpanas, T., Mendelzon, A.: Web prefetching using partial match prediction. In: Proceedings of the 4th International Web Caching Workshop. San Diego, USA (1999)
52. Pearson, O.: Squid user's guide. http://squid-docs.sourceforge.net/latest/book-full.html
53. Rabinovich, M., Spatscheck, O.: Web Caching and Replication. Addison-Wesley (2002)
54. Sarukkai, R.: Link prediction and path analysis using Markov chains. Computer Networks **33**(1-6), 377–386 (2000)
55. Schechter, S., Krishnan, M., Smith, M.D.: Using path profiles to predict http requests. In: Proceedings of the 7th International World Wide Web Conference. Brisbane, Australia (1998)
56. Tao, Q.: Impact of webspace organization and user interaction behavior on a prefetching proxy. Ph.D. thesis, Kent State University (2002)
57. Teng, W.G., Chang, C.Y., Chen, M.S.: Integrating web caching and web prefetching in client-side proxies. IEEE Transactions on Parallel and Distributed Systems **16**(5), 444–455 (2005)
58. UCB, LBNL, VINT: Network simulator ns (version 2). http://www.isi.edu/nsnam/ns
59. Vahdat, A., Anderson, T., Dahlin, M., Belani, E., Culler, D., Eastham, P., Yoshikawa, C.: WebOS: Operating System Services for Wide Area Applications. In: Proceedings of the 7th Symposium on High Performance Distributed Computing Systems. Chicago, USA (1998)
60. Venkataramani, A., Yalagandula, P., Kokku, R., Sharif, S., Dahlin, M.: The potential costs and benefits of long-term prefetching for content distribution. Computer Communications **25**, 367–375 (2002)
61. Wu, B., Kshemkalyani, A.D.: Objective-optimal algorithms for long-term web prefetching. IEEE Transactions on Computers **55**(1), 2–17 (2006)
62. Yang, Q., Huang, J.Z., Ng, M.: A data cube model for prediction-based web prefetching. Journal of Intelligent Information Systems **20**(1), 11–30 (2003). DOI http://dx.doi.org/10.1023/A:1020990805004

Chapter 13
Knowledge Management System Based on Web 2.0 Technologies

Guillermo Jimenez and Carlos Barradas

Abstract Most of the research work on knowledge management systems has been addressed to knowledge representation, storage, and retrieval. However, user interaction has suffered from the same limitations faced by most current Web-based systems. Web 2.0 technologies bring completely new elements that make possible designing user interfaces similar to those that could be built in windowing environments of current desktop platforms. These technologies open new possibilities to enhance user experience when working with Web-based applications. This chapter shows how Web 2.0 technologies could be used to design user interaction in a knowledge management system. Details presented could be useful to improve online interaction with Web-based support systems (WSS) in other application domains.

13.1 Introduction

Moving support systems online is an increasing trend in many research domains. In such contexts, Web-based Support Systems (WSS) are an emerging multidisciplinary research area that studies the support of human activities with the Web as the common platform, medium, and interface. One of the goals of building WSS is to extend the human physical limitation of information processing in the information age [18].

A key issue of WSS research is to identify both domain-independent and domain-dependent activities before selecting suitable computer and Web technologies to support them. In this chapter, we explore the use of so-called Web 2.0 technologies

---

G. Jimenez and C. Barradas
Center for Intelligent Computing and Robotics, Tecnolóogico de Monterrey, Av. E. Garza Sada 2501, 64849 Monterrey, NL
e-mail: guillermo.jimenez@itesm.mx; carlos.barradas@itesm.mx

J.T. Yao (ed.), *Web-Based Support Systems*, Advanced Information and Knowledge Processing, DOI 10.1007/978-1-84882-628-1_13,

on enhancing user interaction for WSSs systems. A case study shows how Web 2.0 technologies are used to implement state-of-the-art user interaction to multiple types of information. The information consists of a knowledge base describing how to build a small plane called RV-10. Different teams work in a long-term asynchronous collaboration effort, thus making necessary recording, and management of information to share it among the teams.

The chapter describes the general goals of knowledge management systems, suggests the use of Web-based technology for implementing interfaces that enhance user experience, provides a methodology for Web-based systems development, and describes a case study applying them to the development of a particular knowledge management system. The first three aspects cover several of the most important issues cited by experts to make successful a knowledge management system. The main concern here is how a Web-based knowledge management system could be developed and enhanced by the latest technologies. Specifically, the chapter concentrates on describing how Web 2.0 technologies could be combined in designing and implementing easy-to-use knowledge management systems, for updating and reusing knowledge, thus motivating users to use the system.

13.2 Knowledge Management Systems

Knowledge management could be defined as a systematic process to acquire, classify, and communicate knowledge to people in the organization with the purpose of providing information that could be used to enhance their performance [7]. It is important to distinguish the difference among data, information, and knowledge. The data are a unit of information; information is the unification of several data through a logic relationship; and knowledge is a flow of ideas, information, and experience, which, once evaluated and understood, could generate other ideas or even more knowledge.

Knowledge could be classified as tacit or explicit [8]. Tacit knowledge is what a person only "knows" but is not capable of explaining or sharing with others because it is a constituent part of how he or she perceives the world (e.g., know-how). On the other hand, explicit knowledge can be shared and explained without major difficulties so others could use it (e.g., know-what).

Figure 13.1 shows the information sources and the general process of classification that should be applied in order to make it possible to store information in an easy-to-access personalized repository. How the knowledge should be handled highly depends on the type of knowledge from the organization; it is necessary to filter it at several stages to segment the knowledge in order to simplify its distribution and use.

It is at this point that knowledge management systems come into use, to help in the classification of the explicit knowledge [5]. This makes possible store groups of ideas, living experiences, or even certain solutions to specific situations, with the purpose of recording what happened so that we may know when these situations

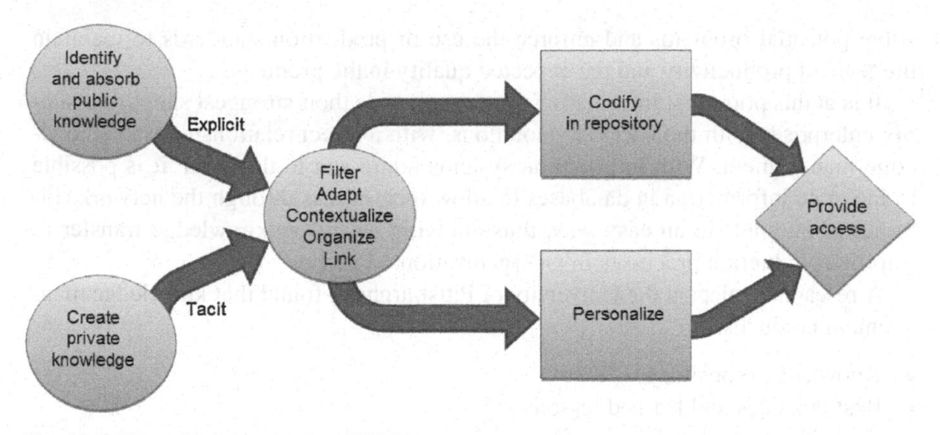

Fig. 13.1 Flow of information in a KBMS

arise again in the organization. This also brings about continuous improvement and enhances work quality, thus helping enterprises in achieving their strategies and goals.

One of the challenges of these systems is balancing the information overload, and thus knowledge refers to potentially useful content [7]. This also means that, to be effective, it is necessary to generate the consciousness of documenting what is done or solved, with the condition that it is relevant and not repetitive. Another challenge is that a knowledge management system will not produce the expected results if no information is stored in it; this requires that users acknowledge that the system should be seen as a support tool for their work, so it could be accepted and its use promoted.

Knowledge management is no more than concentrating technology forces in one single place to help others learn, adapt to the competence, and, above all, survive [16]. The idea is not to create a compilation of information that no one will use, but to store it with the purpose of having a tool that provides solutions which could be applied and adapted when necessary.

Knowledge management implies, among other things [16]:

- Generate new knowledge.
- Allow its access from external sources.
- Transfer the knowledge.
- Employ it on decision making.
- Include it in the processes and, or services.

One of the functions of technology is to provide appropriate tools to perform a great step in the creation of innovative products and high-quality services [6]. Collaboration networks provide adequate media for information sharing among specialized people taking advantage of information technology. It is thus the case that a knowledge management system is a repository not only of ideas, but also of solutions. For instance, it is possible to provide support along all the supply chain process to help solve potential problems and enforce the use of production standards to maintain the level of productivity and the expected quality in the products.

It is at this point that information systems provide their strongest support to supply enterprises with the necessary solutions, with a direct relationship with knowledge management. With information systems addressed to this goal, it is possible to integrate information in databases to allow their access through the network (Internet or intranet) in an easy way, thus allowing necessary knowledge transfer to improve production processes in an organization.

A research project at the University of Pittsburgh [9] found that knowledge management produces:

- Knowledge repository
- Best practices and learned lessons
- Networks of experts
- Communities of practice

For any one of these points it was identified that knowledge management is part of the strategic vision of the enterprise to take competitive advantage and produce correct and adaptable solutions. At the same time, it helps to determine if the processes being applied are adequate; that is, to be sure that things are done right so they can be taken into account as a basis in similar situations.

A special aspect also detected by the University of Pittsburgh's research [9] is that through a system it is possible to capture enterprise knowledge in an organized way using technological tools. In this way, it is possible to construct a solution that in the short term is an investment, and will help in innovation, sustainability, motivation, well-defined tasks and responsibilities, operations standardization, and quality assurance, among others.

Knowledge is not important by itself, but the processes of understanding, learning, and creation that enable the transmission of that knowledge to others [19]. One of the main obstacles in enterprise regarding knowledge and best practices dissemination is the acceptance by people, because generally there are only a few individuals who know all the "secrets" of daily activities and it is very difficult and complicated to share knowledge with others in a simple and correct way. It is at such times that a knowledge management system should provide the following aspects [1]:

- Spontaneous participation. Users should be motivated to participate in the system.
- Common objectives. Identification of users to achieve similar results and improve the community.
- Social relationships. Promote knowledge generation of tacit knowledge and convert it to explicit knowledge to be able to share it.
- Common repertoire. Specify everything in a previously agreed language to enable all personnel to understand and specialized personnel to act accordingly.

As a consequence, the goal is reusing the generated knowledge to take advantage of its potential and produce alterative solutions to organizations [1]. The results are very easy to evaluate: effectiveness and efficiency of the system after users have accessed the knowledge management system. That is, by analyzing how, after they fed or extracted information from the system, users increase their knowledge to make their work better.

A successful knowledge management system should use the better available technologies in its implementation. Technologies should be those that better capture and support requirements. It should be able to provide the best results at lower possible costs. Users should be able to store new knowledge and extract existing knowledge to help them design better solutions to improve the quality in an enterprise. The following section describes and proposes Web 2.0 technologies or concepts as a good alternative to knowledge management systems implementation. We will insist later on in this chapter that it could be also very useful to enhance the current state of the art in developing WSSs by making them easier to interact with.

13.3 Web 2.0

Information technology has been able to establish itself as an important element and is so involved in our daily life that we depend on it increasingly every day [14]. This is no exception with enterprises, which have been able to grow as technology sets the path; from the appearance of electronic commerce and the adoption of brink and mortar enterprises to the dot-com, collaboration systems, the expansion of the Internet, and wireless systems. All of this is a result of the necessity to keep innovating, and with it new tendencies emerge in the applications development field.

Web 2.0 is more a concept than a technology [3], because it relies on previously existing infrastructure and programming languages; the novelty is on how applications are constructed and their intended use. Web 2.0 applications promote collective intelligence, that is, they are used to enable user communities to work collectivity to create concepts and processes.

The basic principle behind Web 2.0 is a culture of participation for sharing knowledge or experience. The concept could be used in enterprises at least in two ways [3]:

- Understand communities where product and services are delivered.
- Improve knowledge management and increase productivity.

In this way, competitive intelligence becomes highly relevant to enterprises by offering communication with customers to show them new products on the market, and additionally to promote practices and propose improvements in their business processes.

Web 2.0 could refer to any of the following characteristics [17]:

- A read–write Web
- Sites that simplify collaboration and collective creation (Figure 13.2)
- Sites with a visually rich and highly interactive user interface, through the use of a new generation of technologies.

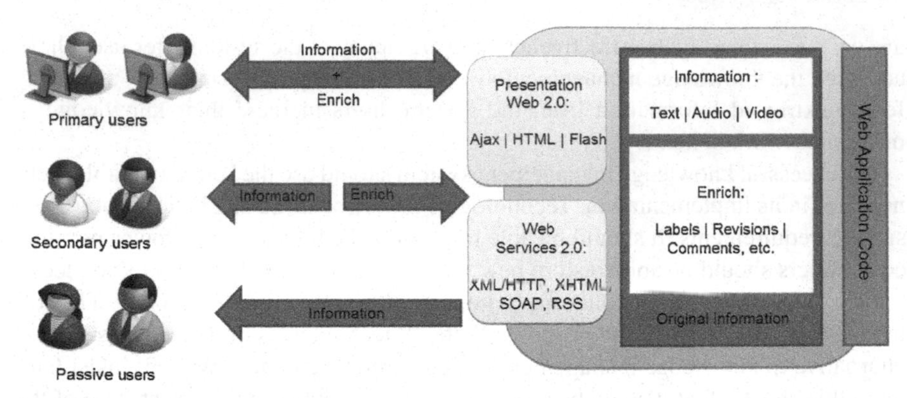

Fig. 13.2 Participation model on the Web 2.0

- The Web as a platform for creating applications or offering Web services to dynamically create pages from others.

Figure 13.2 summarizes research results that identify three roles in the Web 2.0 which define the way to work and interact with applications (Hinchcliffe 2005):

- Primary users are very participative because they store and request information, and enrich it. Making the applications keeps them active and useful overall.
- Secondary users in most of the cases are information consumers, although eventually they could contribute with observations or commentaries. Their role is important because probably they are a larger group than the primary users and some of them could become primary users as they acquire more experience.
- Lastly, passive users only consume resources and normally comprise the larger number of Web users, using information without providing feedback.

Currently there are three main models for application software development [13]:

- Desktop. In this, applications keep the main control of the data model, thus creating a total dependency between the application, the operating system, the data format, and the system's organization. In this model, users are adjusted to the characteristics of the software. For instance: a word processor, a worksheet, etc.
- Web 1.0 or the "old school" of applications. At this stage, functionality relies on the use of electronic forms that convey data to one application server, employing the schema of request–response that depends mainly on data going and returning without connectivity problems, and the use of cookies and user sessions, which, combined with the electronic forms, allow constructing transactional applications in multiple steps.
- RIA (Rich Internet Application). This model pretends to simplify even more the access to robust applications, through the use of a Web browser, where the computer or the operating system being used is worthless. The goal is to achieve more productivity without considering whether the user is at work or home, because data travels using standards such as XML and multimedia contents could be shown without any trouble using plugins, which most Web browsers already include or are downloadable from the Web.

The Internet has changed too much since its conception, which started as a text-based communication technology and at the same time allowed to generate other more powerful means of communication [4]. Currently, Web applications are one of the more important fields of software development, and are growing faster, with many new possibilities of interaction.

Internet applications offer a large number of benefits compared to desktop applications [4]. They are highly accessible; do not require installation; it is possible to update them at any moment; and overall, they allow access to a great amount of information without involving complex networks. At least, these advantages allow less cost of development and support compared to "normal" applications. Although in some cases Internet applications offer lesser usability because they are simpler, they still keep gaining terrain to desktop applications. The Rich Internet Application (RIA) concept refers to Internet applications that seek to fulfill the usability hole mentioned before, through the inclusion of more codes at the Web browser site, offering additional levels of interaction, compared to desktop applications [4].

RIA technology is strongly associated with a new tendency of Internet application development publicly known as Web 2.0 which offers the following advantages [10]:

- Web-based interactive applications, such as desktop applications.
- Application development without economic restrictions or limitations (keeping them independent of large software houses).
- Additional user support through the use of standards, simplifying system interoperability.
- Preserving investment in Web technologies (keeping in mind that they are accessible regardless of the connections).

RIAs allow the creation of WSS with sophisticated and interactive visual elements that provide a more enhanced user experience than was possible with Web 1.0 technologies [3]. Currently, there are many technologies that could be used to develop RIAs. The following section provides a short description of technologies and concentrates on an open-source alternative.

13.4 Rich Internet Applications Architecture

The recent trends in Web software development focus on enhancing the user experience through the concept of RIAs. In an effort to achieve that appear new terms such as AJAX and development frameworks supporting it. One of such frameworks is OpenLaszlo. This section discusses these concepts and introduces OpenLazlo, an open-source-based framework, as a low-cost alternative for RIA development using AJAX. There exist several other AJAX frameworks with similar characteristics (e.g., SmartClient, XUI, ThinWire); the intention here is just to describe one of them.

A new breed of Web applications has appeared in response to the limited user support provided by the 1.0 generation of Web applications [12]. This new proposal is accompanied by the RIA concept and the set of technologies known as AJAX.

AJAX is an acronym of Asynchronous JavaScript And XML. It is a set of not necessarily new technologies for Web-based software development that include presentation standards based on XHTML (Extensible HTML), cascade style sheets (CSS), many newer display options, user interaction, data manipulation and exchange, asynchronous information recovery, and the possibility of linking all these elements through JavaScript [12]. The software components that integrate AJAX were developed and standardized in the last decade and were improved recently, making them more adequate for enterprise use [15]. Following is a summarization of their functionality:

- DHTML. Helps designers to develop more interactive Web pages, for instance, when the mouse pointer is above an element in a Web page, the color or the text size could be changed. Users are also able to drag and drop images to different places on the Web page.
- XML. Used to code and transfer data between a server and a Web browser. The fundamental goal of XML is to allow platform interoperability for data interchange in the Internet.
- CSS (Cascading style sheets). Provide developers more control on how browsers display Web pages to give a design and specific appearance to every element forming the user interface.
- JavaScript. Interacts with the XHTML code to make AJAX pages more interactive, making server calls to request new data from the application container while the user continues interacting with the page.

Figure 13.3 shows architectures for a traditional (Web 1.0) application (a) and a Web 2.0 (AJAX) application (b). In the Web 1.0 application, the user issues an HTTP request to the Web server, who processes it and returns a Web page to the user; every subsequent information request blocks the application until the system updates the page. In an AJAX application, an AJAX engine running on the browser intercepts data from the user, displays requested data, and handles interaction on the user side. In case information is needed from the server, the request is performed in the background while still allowing the user to interact with the application [20].

In this way, users operating in an RIA-based application could be more productive, because in many cases they will not need to wait for requests: their Web browser will not be blocked.

13.5 Rich Internet Application Frameworks

As explained before, AJAX is a generic concept for problem solving. It is not a programming language, product, or specific toolset downloadable from a provider. It includes a variety of concepts, software platforms, and architectures. Currently there exist many RIA frameworks: XUI, ThinWire, ARP, SmartClient, Silverlight, just to name a few. In this section, we describe OpenLaszlo, an open-source framework for rich Internet applications. Many of the available RIA frameworks offer similar

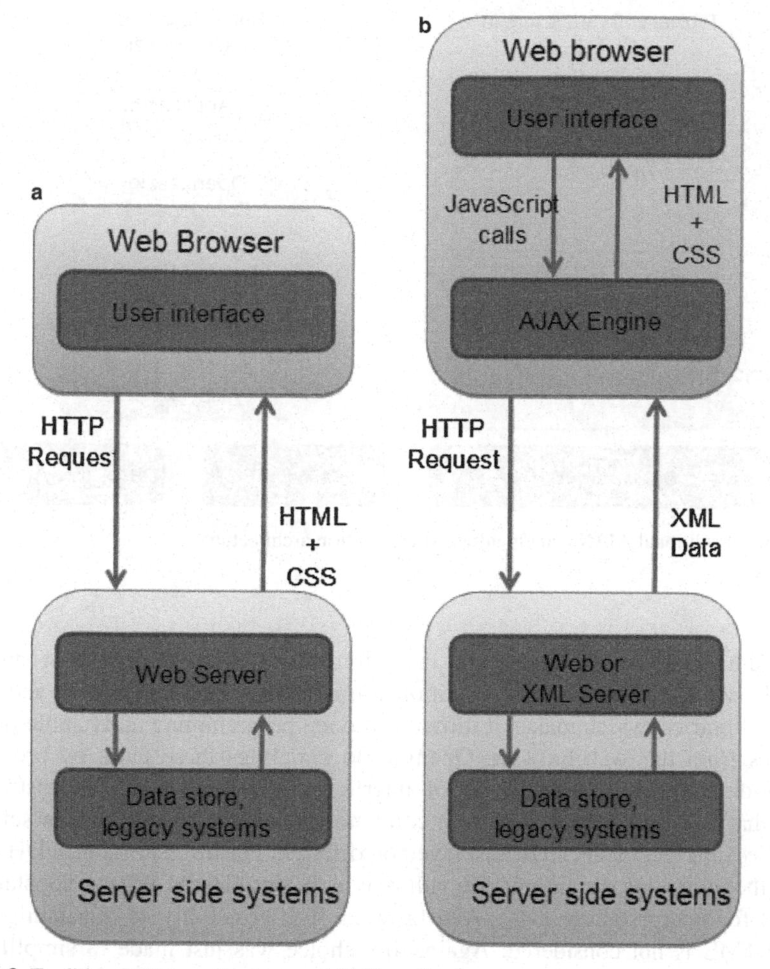

Fig. 13.3 Traditional Web application vs. AJAX application

support for application development. Our choice was based in that, at the time we started analyzing RIA development, OpenLaszlo was one of the most stable tools.

OpenLaszlo was developed by Laszlo Systems, a San Mateo, California-based company [2]. It was modeled as a development suite for AJAX applications using a programming language called LZX, which is based on XML and JavaScript, thus simplifying code syntax and interpretation, even for new programmers. LZX is an object-oriented and case-sensitive programming language. JavaScript is used for event control.

OpenLaszlo includes a compiler that is able to translate LZX source files to two types of formats: Flash or Dynamic HTML (DHTML), at the developer's choice. The compiler is written in Java, but it is not necessary to understand Java to write an application in LZX. However, it is necessary to install the Java JDK and the OpenLaszlo suite in the same place to be able to execute the programs written in LZX.

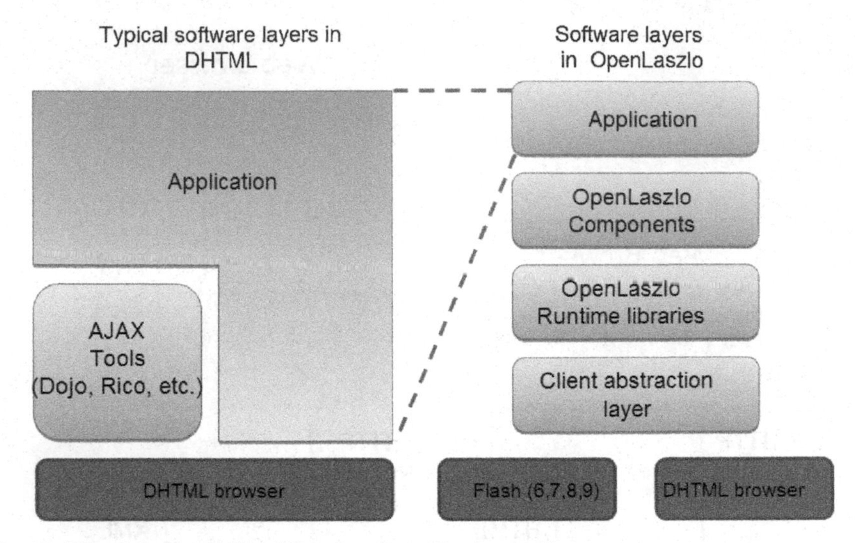

Fig. 13.4 Traditional AJAX and OpenLaszlo application architectures

Figure 13.4 shows the layers in a traditional AJAX application and an OpenLaszlo-based application. Traditional applications lack frameworks and component libraries, which makes it difficult to code programming and handle dependencies from the Web browser. OpenLaszlo simplifies these tasks by providing execution libraries and a rich set of interface components, with an abstraction layer that isolates the developer from concrete characteristics of Web browsers and provides interfaces that can be displayed on different Flash versions, or in DHTML.

In the remaining discussions we will only consider Flash (SWF) as the standard output for interpretation in the Web browser. The possibility of translating LZX to DHTML is not considered. Again, this choice was just made to simplify the presentation.

The OpenLaszlo architecture combines power and usability from a client/server design with the administrative facilities that Web applications provide [11]. Some of them are faster, lighter, and more robust software, with the added benefits of service availability and possibility of remote access.

The OpenLaszlo server is a Java application that can interact with other back-end servers and data sources using a variety of protocols. Applications written in LZX code are compiled into a format that can be interpreted by a Web browser, to build the front end. In the OpenLaszlo context, the client is an LZX application that is executed in the Web browser. The server is the OpenLaszlo server. The LZX client and the OpenLaszlo server communicate using HTTP; the server sends the machine code (bytecode) and the application sends XML.

Figure 13.5 shows layers of OpenLaszlo runtime architecture. The Web browser is on the user side; that is where the Flash Player resides and contains the Laszlo application, once it is requested from the Web server. Communication is

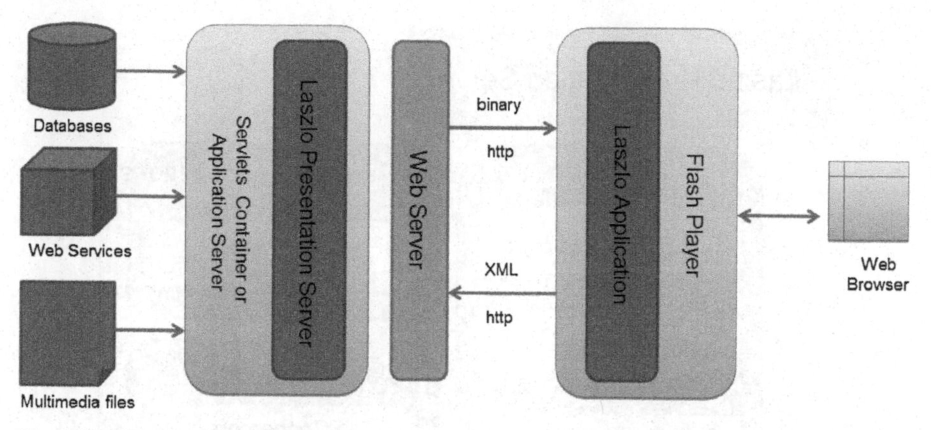

Fig. 13.5 OpenLaszlo runtime architecture

through HTTP and thanks to this it is possible to access different types of content: multimedia, Web services, and databases which are coordinated by the Laszlo Presentation Server.

The server runs in a servlet container (e.g., Apache Tomcat, WebSphere, Weblogic, Jetty, etc.), which runs in a Java Runtime Environment (JRE) version 1.4 o superior. The server consists of five main subsystems which are shown in Figure 13.6 and described below.

- The compiler interprets LZX labels and JavaScript and translates them to executable code, which will be sent to the client environment.
- The media encoder translates a wide range of multimedia data types to deliver them to the Flash player.
- The data manager translates all data into a format recognizable by the OpenLaszlo applications, allowing the retrieval of XML information via HTTP.
- The persistent connections manager handles user authentication and real-time communication with the required systems, which are prepared for that.
- The cache contains the most recently compiled version of any OpenLaszlo application. The first time it is requested, the application is compiled and the resulting file (SWF or DHTML) is sent to the client. A copy is maintained in the server and thus future requests do not need to be compiled again.

Figure 13.6 also shows that the Laszlo Presentation Manager (LPM) works on top of the operating system and Java runtime layers. HTTP requests go to the LPS and it first associates the request to a destination by interacting with the persistent connections manager. After that, the compiler detects the LZX code to be interpreted, establishes connection to the data sources, and generates in the cache the resultant contents to return and deliver to the client. In this way, internal communication among the LPS components is performed internally in that layer.

The client architecture consists of an OpenLaszlo Runtime Library (ORL) which is compiled in every OpenLaszlo application. This library provides services for

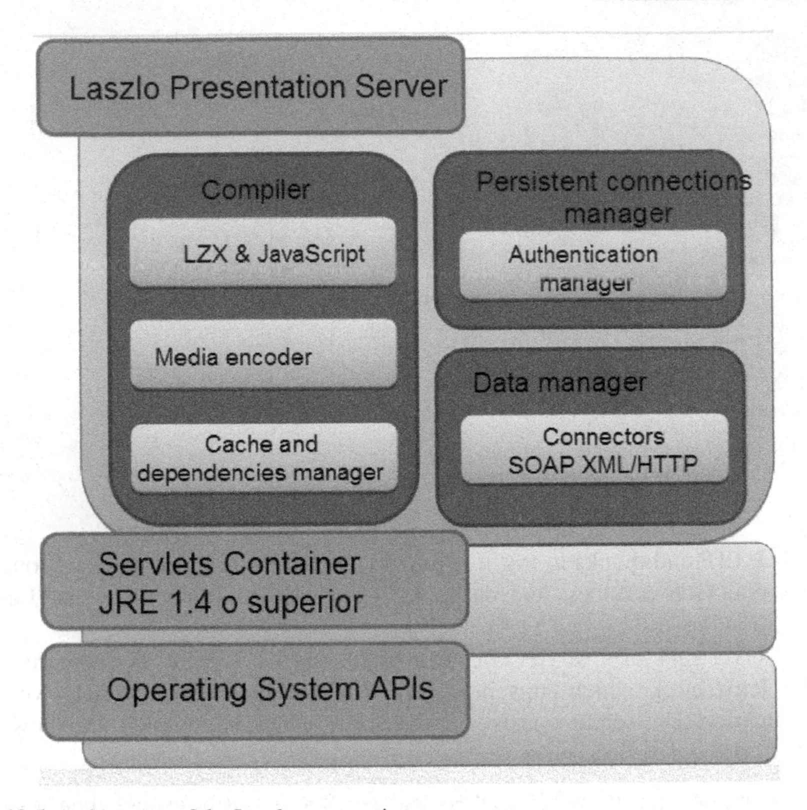

Fig. 13.6 Architecture of the Laszlo presentation server

graphic and sound presentation. None of these components depends on the Flash object model; it is open to deliver contents in DHTML and there exists the possibility to extend it to support other future formats.

When an OpenLaszlo application is launched, all libraries necessary to run an LZX application are downloaded to the client at the beginning. The connection with the server is maintained during the entire session.

The four main components at the client side (ORL) are:

- The events system recognizes and handles application events (such as mouse clicks or data requests to the server).
- The data loader/linker serves as traffic manager accepting connections from the network to the server and links them to its visual components for the correct display of the data.
- The design and animation system provides an interface between programming and application's visual elements, controlling states, animations, position, and times on screen.
- The OpenLaszlo services system includes time handling, sound, dialog panes, and informative texts from the application.

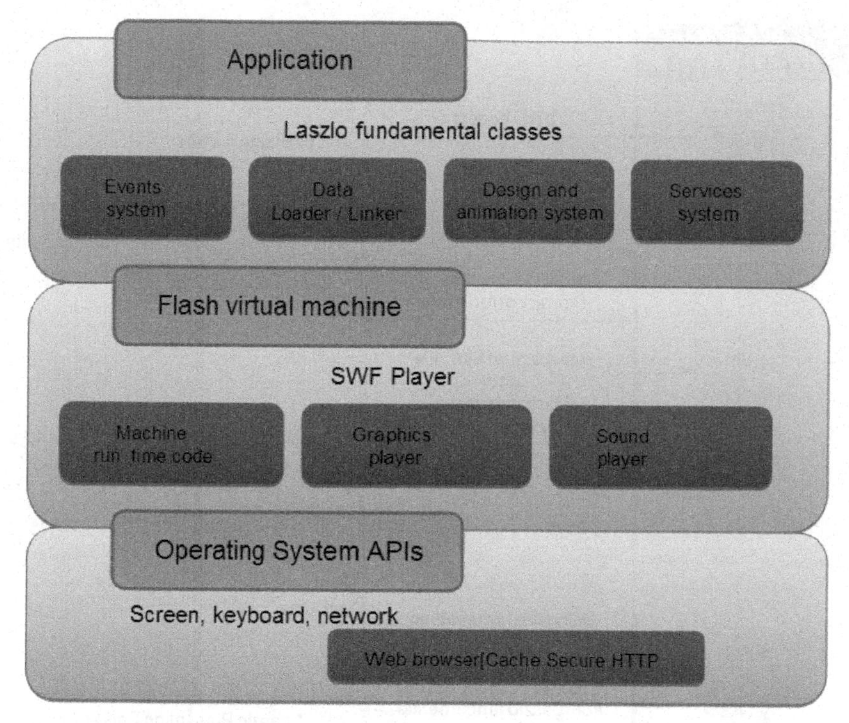

Fig. 13.7 OpenLaszlo client side architecture

The basic classes at the client side of an OpenLaszlo application are shown in Figure 13.7. At the client side, the communication paths and components are established. All of this is embedded in the Flash engine for multimedia playing and correct display of the data in the Web browser.

The client architecture simplifies the association of actions concerning OpenLaszlo exclusively at the application layer. Event handling for visual effects is sent and interpreted by the Flash virtual machine. Figure 13.7 shows such separation and every component handles its own functionality, but keeps direct communication to send events to the corresponding layer, which results in a well-structured relationship.

An OpenLaszlo application for database access works in the following way:

- The user introduces a URL in the Web browser.
- The server localizes the resource requested by the client, ordinarily an LZX file.
- The resources (graphs, sources, sounds, and other applications components) are compiled only once and translated to bytecode.
- The OpenLaszlo server delivers an executable file (SWF) to the Web browser, including Laszlo fundamental classes for its correct interpretation.
- The Web browser receives the SWF file and invokes the Flash Player to display the application.
- The user interacts with the application requesting data access.

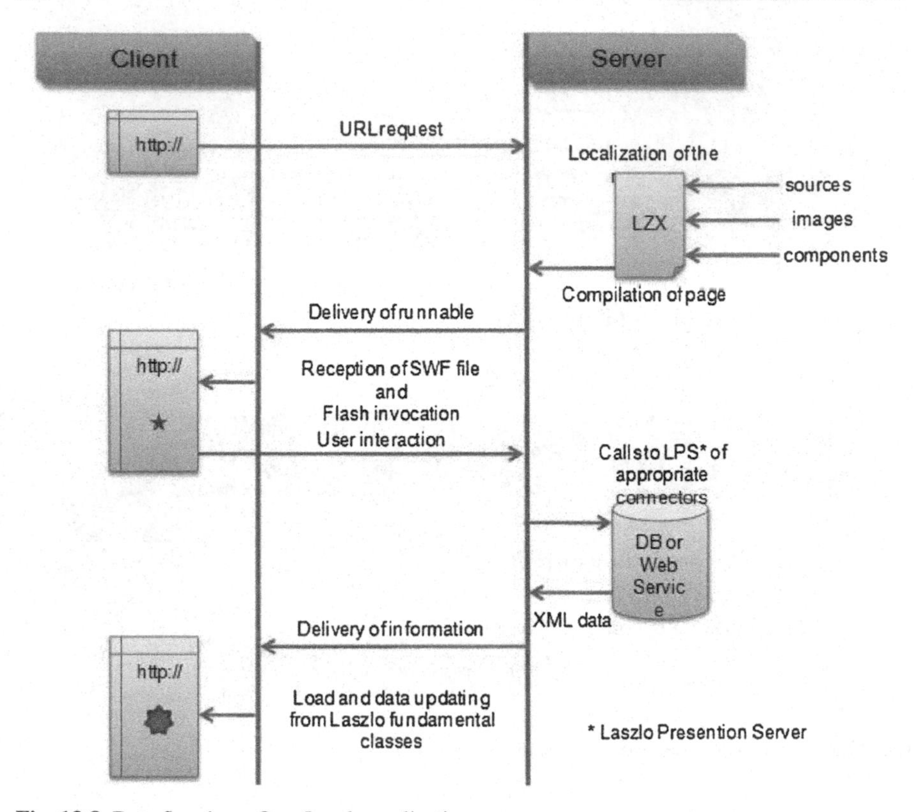

Fig. 13.8 Data flow in an OpenLaszlo application

- The OpenLaszlo server (LPS) calls the appropriate connector to retrieve the requested information (XML, SOAP, Web services).
- If necessary, the LPS compiles the data and sends the information back to the client.
- The Laszlo fundamental classes at the client side link the data to the interface appropriate objects and the screen elements are updated with the information just received.

Figure 13.8 summarizes the process of how data flow in an OpenLaszlo application, from the starting of a URL request until the contents are dynamically updated at the user screen.

In OpenLaszlo applications the presentation logic is separated from the business logic. The server sends information in binary and compressed format, instead of sending text only, thus reducing the amount of data transmitted against HTML-based applications and other Web-based applications. The cache both at the client and server sides eliminates unnecessary code execution and data transmission.

The programming cycle in OpenLaszlo consists of seven steps:

(a) Start the OpenLaszlo server.
(b) Write source code using a text editor and saving it with the .lzx file extension.

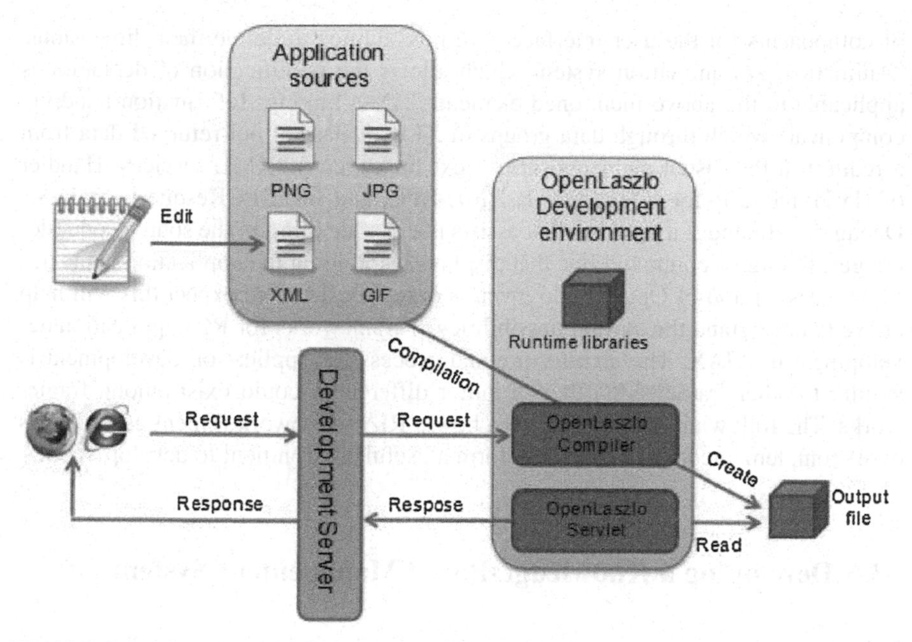

Fig. 13.9 Compilation process in OpenLaszlo

(c) Save the file in the appropriate folder on the server.
(d) Compile the application.
(e) Debug and modify the program.
(f) Repeat steps 2 to 5 until the program is correct.
(g) Deploy the application.

Application compilation could be done in two ways: the application can be loaded from the Web browser, thus forcing a pre-execution of the application and displaying errors, if any, at the Web page; or the compilation can be in a command window, depending on the operating system.

Figure 13.9 shows the process involved in developing an OpenLaszlo application. The process starts with writing the source code using a text editor, inclusion of data sources (e.g., multimedia files and database connections) and all of them specified in source LZX expressed in the XML standard. At compilation time, the OpenLaszlo framework loads libraries, links components for visual elements, and generates an output file, which is the application to be displayed by the Web browser in the client. In the same way, the compilation process could be fired from the Web browser, storing the application in the servlets container to be able to access the output file.

The runtime libraries or Laszlo Fundamental Classes (LFC) contain components for the user interface, data sources, and network services. Fundamental classes in Laszlo support the following characteristics: "Components. Library of many user interface elements including Web forms, configurable grid like structures to show information in a spreadsheet style and tree like (hierarchical) views." Design. Avariety of components for the user interface with possibilities to define their dimensions. "Animation. An animation system which allows the specification of declarations applicable to the above mentioned elements." Data linking. Information-handling components which through data groups in XML (datasets) put returned data from a request in the visual elements (grids, text boxes, etc.) "XML services. Handler of HTTP requests for XML, SOAP, XML-RPC, and JavaRPCResquest services." Debugging. Includes a program that assists in error handling in the source code (debugger) through a command line that displays warning on the application context.

The description of OpenLaszlo given is quite detailed. We expect this will help above to understand the current possibilities of frameworks for RIA application development in AJAX. The architecture and process for application development is similar to other frameworks, though minor differences could exist among frameworks. The following section describes how a RIA framework (OpenLaszlo in this case) complemented with other tools form a useful environment to develop WSS.

13.6 Developing a Knowledge-Based Management System

This section describes three technological perspectives of a knowledge-based management system (KBMS) using a RIA development framework. The first perspective is an architecture showing the system's three fundamental layers. The second perspective shows the relationships among the tools used to develop the system. The third perspective shows the integration of technologies for the runtime environment.

As depicted in Figure 13.10, the architecture of a KBMS consists of three layers. The three specific layers in the architecture are:

- Web browser. Handles user requests and information presentation.
- Application server (Web server). Contains the system's logic and communication with external and lower layers.
- Knowledge repository. Stores structured records of information and complementary files in different formats such as video, audio, and pictures.

One goal of the architecture is keeping a single repository of relevant information of the knowledge base. Such information may consist of all kinds of knowledge generated by the organization: processes, techniques, standards, and operational models. The information stored in the repository could be of many types, but it should be organized in a specific structure to help knowledge users update the information. The structure is defined by a data model (more detail on this is presented on a case study later in the chapter).

The benefit of implementing the KBMS on the Web is allowing users access the system using a Web browser. To start interacting with the KBMS, users only need the URL or Web address where the system is hosted. Once the system is started, interaction is performed by clicking on buttons, links, or options to navigate in the system.

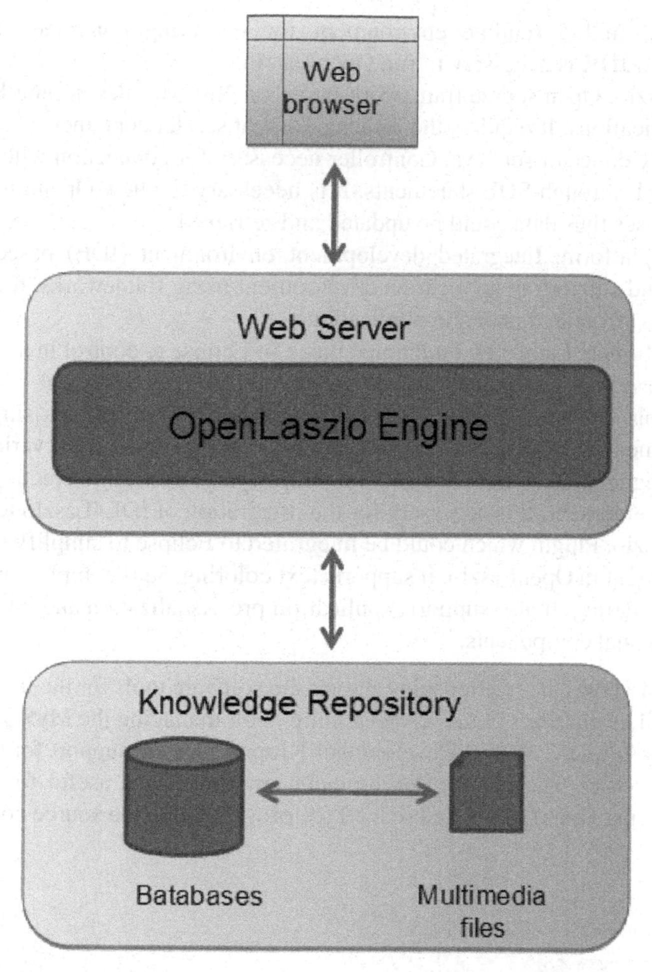

Fig. 13.10 Architecture of a KBMS

To implement innovative systems, it is necessary the use of recent technological tools. Such is the case of RIA. The following set of tools could be appropriate to achieve the goals of creating a flexible environment for RIA application development:

- XAMPP. An open source and platform independent software suite which includes: MySQL relational database, Apache Web server, interpreters for PHP and Perl scripting languages, phpMyAdmin which is a management tool for MySQL, FileZilla FTP server, and the Mercury mail server. The name of this application suite derives from the main tools it includes: Apache, MySQL, PHP, Perl; X refers to the possibility of running it on different operating systems.
- Java Development Kit (JDK). A Java based environment for developers, which includes a compiler, archive generator component, document generator,

debugger, and a runtime environment for executing compiled Java based programs. JDK is necessary to run OpenLaszlo.

- OpenLaszlo. Open source framework based on XML for developing Rich Internet Applications. It requires the Apache Tomcat servlet container.
- MySQL Connector for Java. Controller necessary for connection with databases in MySQL through SQL statements. It is necessary to allow OpenLaszlo access to databases thus data could be updated and retrieved.
- Eclipse plattform. Integrated development environment (IDE) based on open source and supporting application development using frameworks, tools, and libraries to create and maintain applications.
- Eclipse Tomcat Launcher. Functional piece in Eclipse to control in a simple way the Tomcat server and integrate it to the development environment.
- Web Tools Platform (WTP). An extended toolset for Eclipse that simplifies the development of Web applications. It includes graph editors for a variety of programming languages and assistants for simplifying application creation and software development. It is necessary for the integration of IDE4Laszlo to Eclipse.
- IDE4Laszlo. Plugin which could be integrated to Eclipse to simplify application development in OpenLaszlo. It supports text coloring, auto-completion, and LZX code formatting. It also supports application pre-visualization and integrated design of visual components.

Figure 13.11 shows the relationships among the software tools for the integrated environment. The main use of the XAMPP suite is for managing the MySQL database with phpMyAdmin. The Java Development Kit provides the support for the Tomcat application server. The Eclipse development environment is useful for two tasks; first for running OpenLaszlo, and second for programming the source code.

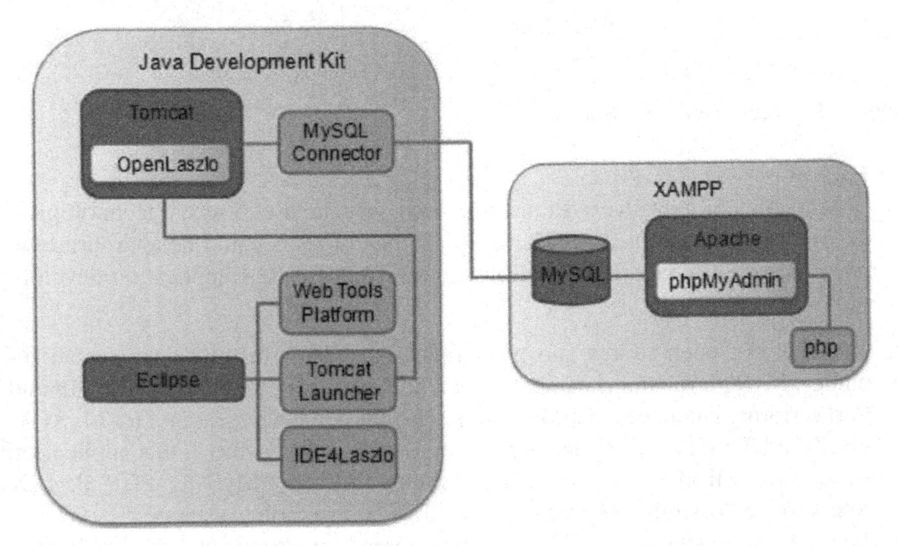

Fig. 13.11 Relationships among development tools

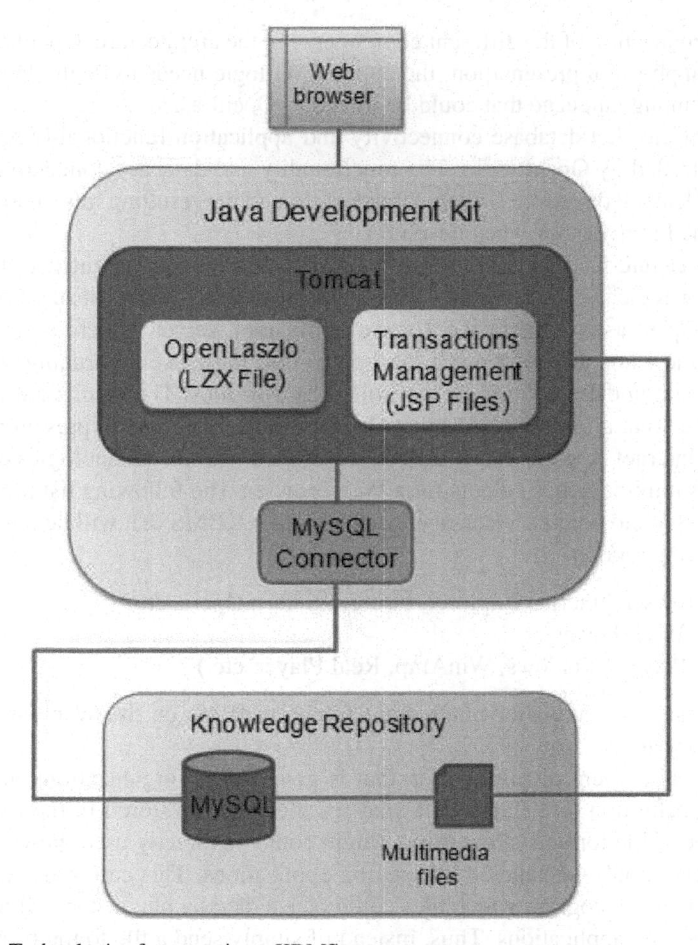

Fig. 13.12 Technologies for operating a KBMS system

The generic architecture of a KBMS was shown on Figure 13.10. Figure 13.12 provides a detailed view of the integration of technologies used to support the operation of a KBMS system. As Figure 13.12 shows, the heart for KBMS operation is the Java Environment where the Tomcat application server is executed. OpenLaszlo handles user requests to the KBMS with the help of Transactions Management component.

The application logic for connection management to the database and information recovery is handled by the MySQL connector. JSP files are executed and handled exclusively by Tomcat. Access to multimedia files is controlled by the JSPs, just leaving the presentation logic to OpenLaszlo which return request results to clients.

The above description of the architecture is highly simplified and does not reflect problems that have to be solved at application development time. Several problems may arise while writing the source code. These problems are ordinarily related to the interconnection of the different components in the architecture. OpenLaszlo just handles application presentation; the application logic needs to be implemented in a programming language that could be linked to OpenLaszlo.

This means that database connectivity and application functionality are not totally controlled by OpenLaszlo. The functionality and database connectivity needs to be implemented separately from OpenLaszlo and the resulting information delivered to the Laszlo server when needed.

The user interface for a Rich Internet Application is implemented in LZX, the application logic is implemented in several JSP files. JSP is the most adequate option because OpenLaszlo runs in the Tomcat application server, which is also able to run JSP files. For this reason, all the application logic (i.e., searching, insertion, deletion, and updating) could be controlled by JSP files. The results are delivered to OpenLazlo in a format (XML) that OpenLaszlo can process to present results in the Rich Internet Application. A KBMS developed with the technologies described above has minimal requirements for a Web browser. The following list presents the Web browser and plugins necessary for the specific KBMS that will be described in the following section:

- Web browser (Internet Explorer, Firefox, Safari, Opera, etc.)
- Adobe Flash Player
- Media Player (Windows, WinAmp, Real Player, etc.)

All of these applications or plugins are freely available on the Web for different operating systems.

The great amount of information that is generated in organizations makes difficult the definition of a standard format for files that are stored in the knowledge base. Several file formats allow that a plugin could be directly used; however, other formats could only be handled by specific applications. This case requires the application server recognize which files could be handled by plugins and which others require specific applications. Thus, instead of simply send a file to the browser, the application should decide if particular files need specific applications. In this last case, a link to the file should be provided to the application for its correct treatment. Most of the multimedia file formats in current use could be handled by a plugin (e.g., Mp3, Wav, Avi, Mpg, Wmv, Png, Jpg, Gif, Bmp). These files could be directly displayed in the Web browser, by playing video or audio, showing the image or photo. However, for other file formats such as Word, or Excel, the application should be executed.

It is important to mention that as more plugins and applications are installed on the client, the different types of knowledge that can be handled on a system increases. For instance, currently there exists several 3D modeling tools whose models could only be displayed using a specific plugin or application.

This section presented the general architecture and specific tools to build KBMS. The section that follows shows code segments of a knowledge management system. The reader not interested on implementation details at the moment, but just get a general understanding on how Web based support systems are developed under Web 2.0 technologies, can skip the following section.

13.7 Implementing a Knowledge Management System

This section shows code fragments describing how components of a KBMS' architecture could be implemented. For readers not familiar with JSP and/or XML the section could be skipped without loss of understanding of the general approach of developing a KBMS with the technologies described in the previous section. The interested reader could refer to JSP or XML sources prior to try to understand the code fragments.

The first code segment is depicted in Figure 13.13. It is part of a JSP source file to connect MySQL, which holds the database that stores the knowledge we can be interested in extracting / updating. Information from the database is extracted / updated by SQL statements and obtained result is used to generate a XML file. The XML file will have several elements grouped under a "knowledge" object, these elements are: item, details, general, and concrete.

This code will be run by the Tomcat application server (in our particular case), at reception of the particular request from the client/browser. The browser will display or use the XML file according to the application.

A segment of on example of XML file generated by the JSP from Figure 13.13 is shown on Figure 13.14. It may be important to understand that the labels used correspond to a previous agreement between client and server on how the result will be represented. The structure of the XML file should be consistent with the agreed protocol defined for the implementation.

Figure 13.15 shows a code segment of the LZX source file to display the XML object, from Figure 13.14, in a grid on the Web browser page. Without paying too much attention to the code syntax, the elements on the XML could be identified in the LZX source code.

```
. . .
    try {
        Class forName("com mysql jdbc Driver");
        connection =
Drivermanager.getConnection("jdbc:mysql://localhost:3306/laszlo?user=root
&password=admin");
        Statement stmt = connection createStatement();
        ResultSet rs = stmt executeQuery("select * from knowledge");
        while (rs.next()) {
    %>
    <knowledge id="<%= rs.getString("id")%>"
                item="<%= rs getString("item")%>"
        details="<%= rs.getString("details")%>"
        general="<%= rs.getString("general")%>"
        concrete="<%= rs.getString("concrete")%>"/>
    }
. . .
```

Fig. 13.13 JSP connection to MySQL and XML generation

```
<xml_data>
 <knowlege id="226"item="Manufacturing topics" details="List of manufacturing topics"
general="Product development" concrete="Automatization"/>
 <knowledge id="227"item="RV10" details="General information of the RV10"
general="Images of the RV10" concrete="Dimensions, images, etc"/>
</xml_data>
```

Fig. 13.14 XML object generated by the JSP component

```
...
<mygrid datapath="DSrodosregs:/xml_data/" shownitems="7"
            showhlines="true" contentdatapath="knowledge">

            <gridtext width="50" datapath="@id" editable="false"
                        sortable="true"resizable="false">
                        ID
            </gridtext>
            <gridtext width="${(canvas width / 5) - 10}" datapath="@item"
                        editable="false"sortable="true"resizable="false">
                        Item
            </gridtext>
            <gridtext width="${(canvas.width / 5) - 10}"
                        datapath="@details"editable="false"sortable="true"
                        resizable="false">
                        Description
            </gridtext>
```

Fig. 13.15 LZX code fragment to interpret an XML file

Although very brief, previous code segments could help to get a general understanding of how one goes from concept to facts while developing Web based support systems based on the architecture presented in Figure 13.12. The following section describes a specific KBMS system developed using the architecture and technologies described.

13.8 Case Study: The RV10 Project

This section describes a project aimed at assembling a small plane. The main goal of the project is that engineering students acquire a set of different competencies on plane construction and use of software technologies to support the process. Many people participate at different times in the so called RV10 project. One problem they are facing is how to communicate among teams' members, and recording knowledge for project's future reference. An aim of the project is that experience obtained could be useful in future projects.

Besides assembling the plane, participants are modeling in a 3D software (CATIA) all pieces. Videos describing the correct way to assemble complex components are being recorded. Documents on recommendations and lessons learned are being generated. Other documents are digitalized plans and charts. The project requires that all participants have access to all the knowledge generated on the process and search for different types of suggestions from experts.

This section describes a specific KBMS implemented to assist participants in the RV10 project. The KBMS was developed based on architectures and technologies described in earlier sections of this chapter. Several screenshots show the user interface to support the functionality required by the project.

Figure 13.16 shows a sketch of the plane being built. The detailed documentation of the project consists from any single screw and bolt, to subsections of the plane.

One of the main goals in the RV10 project is keep a knowledge base of direct experiences derived from the practice in assembling the plane. Techniques, methods and assembly procedures, design, manufacturing, control and instrumentation of plane construction should be recorded, managed, and made available for searching.

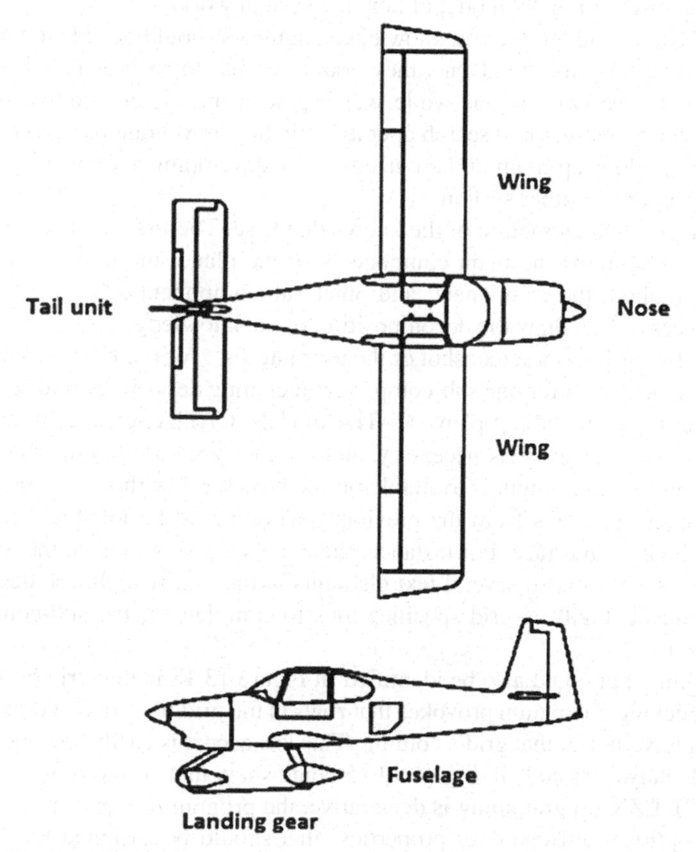

Fig. 13.16 Sketch of the RV10 plane

Components of the plane are modeled in a 3D design tool (CATIA/DELMIA). The 3D tool has among other possibilities, functionality for rendering images for their dynamic presentation. Users are able to see different perspectives of components by rotating, resizing, and shading the images. A viewing tool could be used to analyze designs, without requiring the complete tool installed on every user site, this is an advantage because the complete tool requires significant hardware resources in the workstation.

The teams participating in the project are following the Aerospace Standard AS9100B for documentation, which is highly compliant (90%) to ISO 9001:2000. This is important to guaranty that the learning curve of any newcomer to the project is not a problem. It is also important because participants (many of them engineering students) will become familiar with the AS9100B standard.

To simplify understanding of the general structure of the knowledge base, a structure that resembles the main components of the plane was designed. Figure 13.17 shows the structure of the KBMS for the RV10 project. Only one component is detailed (flap) in Figure13.17, the other components have a similar structure. The structure is of relevance to the knowledge engineers (or domain experts) and users. It will support their mental model of how the system works.

To make the system effective, knowledge engineers should be able to understand how knowledge is structured thus they could be able to record their knowledge. Users would have two options while working with the system: follow the structure or perform unstructured search operations in the knowledge base. The software developer should keep in mind these needs while developing a knowledge management system, or any other system.

Note the tree-like structure of the knowledge base. The first level is the root; the second level contains the main components of the plane; the third level contains information about the component; and other sub-components. The tree can go as deep as necessary to show the decomposition of the knowledge base.

Figure 13.18 shows a screenshot of the user interface for the KBMS. The screenshot shows the details for one sub-component including description and several files of different types including: plans, CATIA models, CATIA screenshot, and so on. The list of files can grow as necessary, and the file types are unimportant as long as the corresponding plugin is available on the browser. For those interested in details, the code segments from the previous section could be identified in the elements in the user interface. For instance, there are several buttons at the top (of the dark area), there are also several text elements in the first rows: these describe the sub-component; finally a grid specifies files to complement the sub-component's information.

Something that could also be identified in Figure 13.18 is that grid headings are buttons: clicking one button provokes that rows in the grid be sorted and redisplayed according to values in that grid's column. This last aspect is easily specified in LZX (a careful analysis to code in Figure 13.15 could show that columns are declared as "sortable"). LZX programming is declarative, the programmer just specify mandatory and optional attributes (or properties) that should be assigned to screen elements. LZX contains lots of properties that could be assigned to LZX components.

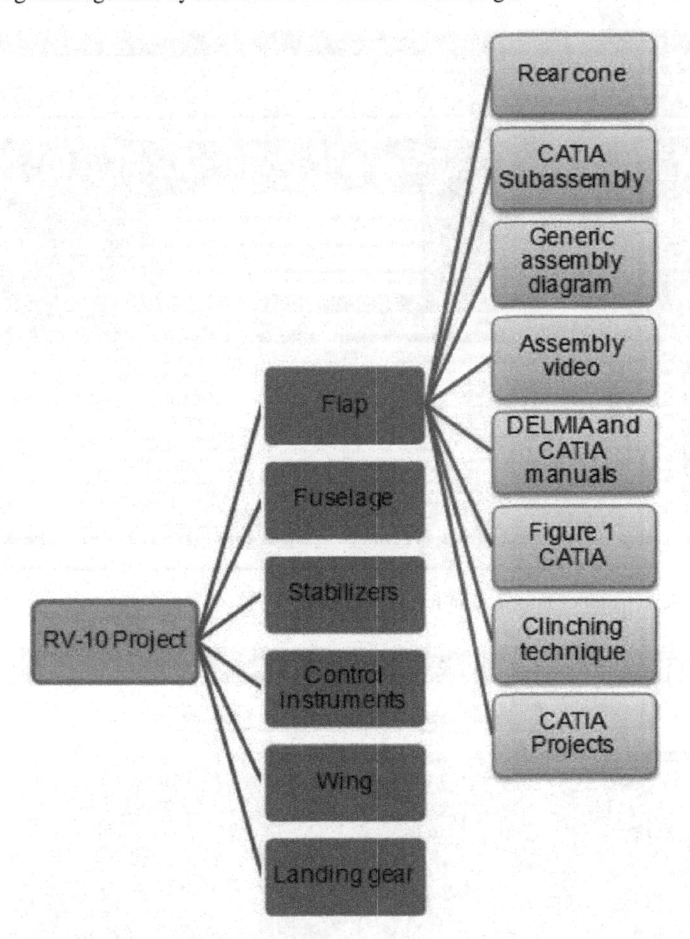

Fig. 13.17 KBMS structure for the RV10 project

Another screenshot is shown on Figure 13.19. In this case, the browser uses the picture viewer utility already installed in the operating system to show the picture.

Lastly, Figure 13.20 shows a screenshot of a video describing how several components should be assembled. A video player is used to play the video file. Many different types of video files could be used, as long as the video player is installed in the client. Some video formats and/or players could offer several advantages above others. For instance, a video could be played (forward or backward), paused/continued, skip in some sections, etc. The video designer should be aware of these characteristics, which is not part of the Web based support system itself, but from tools available for video reproduction.

It is also important here take into account the platforms of potential users. For several video formats a video player could be available, the video player has some cost, or any other thing that precludes the clients of having installed or install the video player. Similar decisions should be made to choose the most appropriate picture and sound formats.

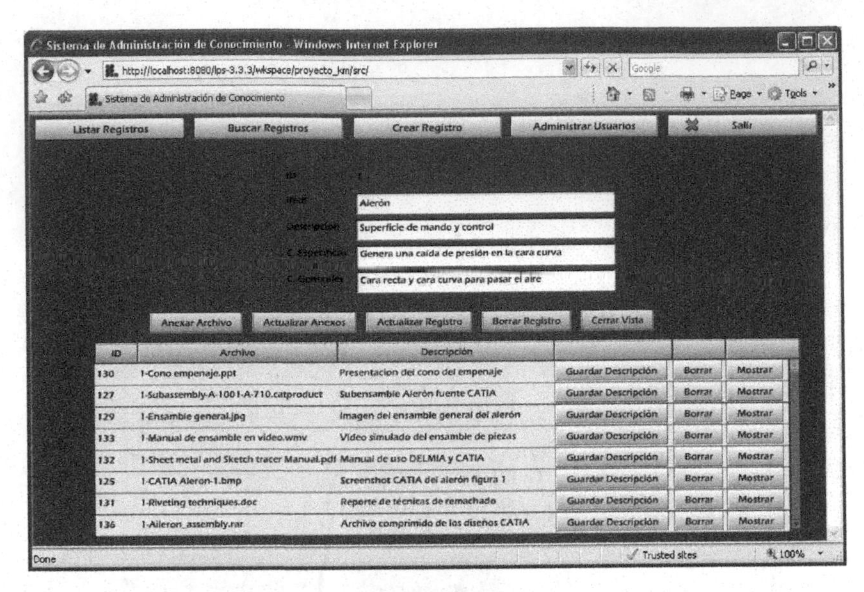

Fig. 13.18 Screenshot of information for one component

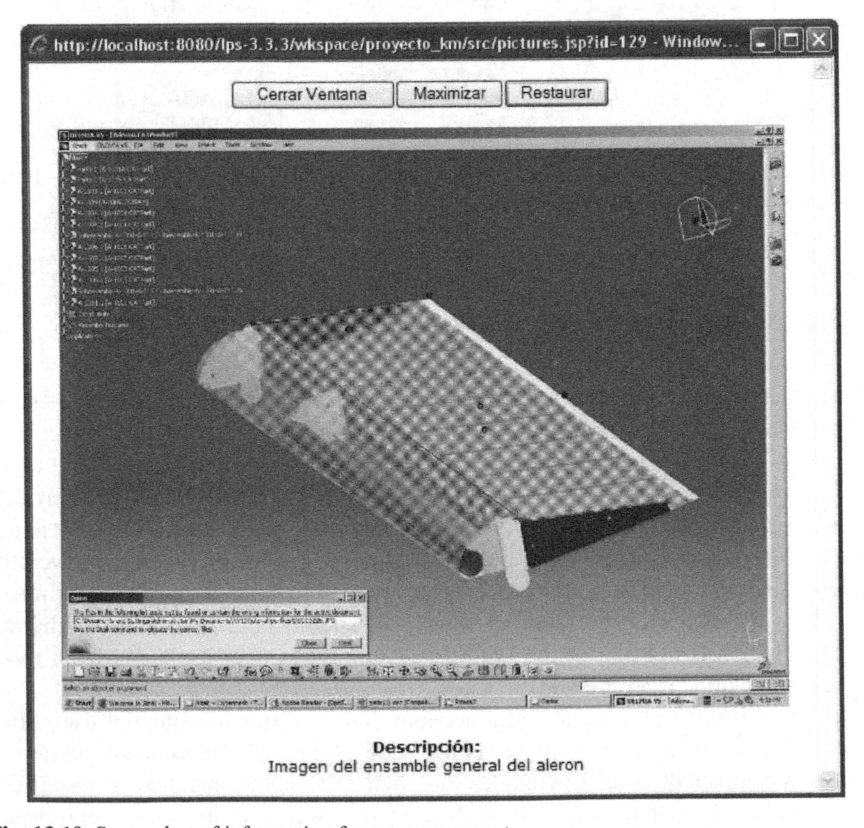

Fig. 13.19 Screenshot of information for one component

Fig. 13.20 Screenshot for a video

These last comments could help to understand that a careful analysis of the possibilities of a Web based support system go beyond programming, or being able to implement specific functionality. Web based systems are aimed at running in different platforms thus the availability of players for file types should be considered.

13.9 Conclusions

The continuous development of concepts, approaches, and technologies allow that Web-based support systems with many new capabilities could be developed. This chapter described how currently available concepts and technologies, could be complemented by a methodology for a special type of Web-based support systems. Knowledge based management systems share a fundamental and similar requirement of many application domains: users should be able to utilize a rich interface to simplify interaction and enhance their experience when using systems. By understanding the architectures and technologies involved in rich Internet applications, developers will be in possibility of creating systems that were just a dream on the past.

Software developers need to know a great number of technologies for improving the user experience in Web-based support systems. The chapter presented a set of technologies that complement each other to develop the Web-based support systems that will be in common use in the near future. The details provided could help developers create environments which will simplify their tasks.

Although the method applied here was not emphasized, one could grasp that it consists in three main stages: domain requirement elicitation, architectural design, technological architecture definition, and implementation. In the chapter, requirements where those imposed by knowledge based management systems; architecture was client/server, the technological architecture was described in detail; implementation was clarified by including code segments of the main architectural components.

Web-based support systems for other domains, could use a similar method to the one presented here.

Acknowledgements The authors acknowledge the support received from Tecnolóico de Monterrey and CEMEX through grant number CAT030 to carry out the research reported in this document.

References

1. A. Agostini, S. Albolino, G. De Michelis, F. De Paoli and R. Dondi, Stimulating knowledge discovery and sharing, Proceedings of the 2003 international ACM SIGGROUP conference on Supporting group work, ACM Press, 2003.
2. C. Coremans, AJAX and Flash Development with OpenLaszlo: A Tutorial, Brainy Software, 2006.
3. G. Vossen and S. Hagemann, Unleashing Web 2.0: From Concepts to Creativity, Morgan Kaufmann, 2007.
4. J. Eichorn, Understanding AJAX: Using JavaScript to Create Rich Internet Applications, Prentice Hall, 2007.
5. P. Gottschalk, Knowledge Management Systems: Value Shop Creation, Idea Group Publishing, London, 2007.
6. A. Gregory, Collaborate to accumulate, Manufacturing Computer Solutions, 6(3): p. 7, 2000.
7. J. Hahn, MR Subramani, A framework of knowledge management systems: issues and challenges for theory and practice, Proceedings of the twenty first international conference on Information systems, Association for Information Systems, 2000.
8. A. Jashapara, Knowledge Management: An Integrated Approach, Prentice Hall, 2005.
9. WR King, PV Marks, S. McCoy, The most important issues in knowledge management, Communications of the ACM2, 45(9):93-97, 2002.
10. N. Klein, M. Carlson, G. MacEwen, Laszlo in Action. Manning Publications, 2007.

11. Laszlo Systems, OpenLaszlo, The premier open-source platform for rich internet applications, http://www.openlaszlo.org/, Accessed 2007.
12. A. Mesbah and A. van Deursen, An Architectural Style for Ajax, in Conference on Software Architecture, 2007.
13. D. Moore, R. Budd, E. Benson, Professional Rich Internet Applications: AJAX and Beyond, Wiley Publishing, Inc, Indianapolis, USA, 2007.
14. T. O'Reilly, What Is Web 2.0., http://www.oreillynet.com/pub/a/oreilly/tim/news/2005/09/30/what-is-web-20.html?page=1. Accessed 2005.
15. LD Paulson, Building Rich Web Applications with Ajax, IEEE Computer, p. 4, 2005.
16. J. Rowley, What is knowledge management? Library Management, 20(8):416-419, 1999.
17. LD Soto, Web 2.x and Web 3.x, in Software Guru, Conocimiento en práctica, 2007.
18. J. T. Yao, Web-based Support Systems, Tutorial at the 2007 IEEE/WIC/ACM International Conference on Web Intelligence, Silicon Valley, 2007.
19. G. Winch, Knowledge management and competitive manufacturing, Engineering Management Journal, p. 130–134, 2000.
20. E. van der Vlist, D. Ayers, E. Bruchez, J. Fawcett, and A. Vernet, Professional Web 2.0 Programming, Wiley Publishing, Inc, Indianapolis, USA, 2007.

Part III
Design and Development of Web-Based Support Systems

Chapter 14
A Web-Based System for Managing Software Architectural Knowledge

Muhammad Ali Babar

Abstract The Management of architectural knowledge is vital for improving an organization's architectural capabilities. Despite the recognition of the importance of capturing and reusing architectural knowledge, there is no suitable support mechanism. We have developed a conceptual framework for capturing and using architectural knowledge to support the software architecture process. A knowledge management system is envisioned to support the proposed framework. Such a system should facilitate architectural knowledge sharing among stakeholders, who may be collocated or geographically distributed. This chapter presents and discusses the design, implementation, and deployment details of a web-based architectural knowledge management system, called PAKME, to support the software architecture process. This chapter also discusses different usages of PAKME in the context of the software architecture process. The chapter concludes with a brief description of the use of PAKME in an industrial setting.

14.1 Introduction

The software architecture (SA) process consists of several activities (such as design, documentation, and evaluation), which involve complex knowledge-intensive tasks [52, 57]. The knowledge that is required to make suitable architectural choices and to rigorously assess those design decisions is broad, complex, and evolving. Such knowledge is often beyond the capabilities of any single architect. The software architecture community has developed several methods (such as a general model of software architecture design [35], the Architecture Tradeoff Analysis Method (ATAM) [24], and architecture-based development [16]) to support a disciplined

M. Ali Babar
Software Development Group IT University of Copenhagen, Rued Langaards, Vej 7, DK-2300, Copenhagen, S. Denmark
e-mail: malibaba@itu.dk

J.T. Yao (ed.), *Web-Based Support Systems*, Advanced Information and Knowledge Processing, DOI 10.1007/978-1-84882-628-1_14,

architecture process. Although these approaches help manage complexity by using systematic approaches, they give little support to provide or manage the knowledge required or generated during the software architecture process.

The requisite knowledge can be technical (such as patterns, tactics, and quality attribute analysis models) or contextual (such as design options considered, trade-offs made, assumptions, and design reasoning) [5]. The former type of knowledge is required to identify, assess, and select suitable design options for design decisions. The latter is required to provide the answers about a particular design option or the process followed to select that design option [27, 31]. If not documented, knowledge concerning the domain analysis, patterns used, design options evaluated, and decisions made is lost, and hence is unavailable to support subsequent decisions in development lifecycle [20, 48, 58].

Recently, various researchers [38, 44] have proposed different ways to capture contextual knowledge underpinning design decisions. An essential requirement of all these approaches is to describe software architecture in terms of design decisions and the knowledge surrounding them usually referred to as design rationale. However, architecture design decisions and contextual knowledge are seldom documented in a rigorous manner. Our research has identified that there is lack of suitable methods, techniques, and tools for capturing and managing architecture design decisions, and knowledge underpinning them [7, 55].

In order to provide an infrastructure for managing architectural knowledge [3, 5], we have developed a framework to provide theoretical underpinning and conceptual guidance for designing and implementing a Web-based knowledge management system [3]. This framework has been applied to design and implement a Web-based architectural knowledge management system called PAKME (Process-based Architecture Knowledge Management Environment).

This chapter describes the theoretical concepts and practical issues involved in managing architectural knowledge for supporting the software architecture process. It elaborates on the process and logistics involved in designing and implementing a Web-based architectural knowledge management system like PAKME. The chapter discusses the requirements and architectural aspects of PAKME and explains the objective and usage of its features. The chapter also describes a study carried out to assess PAKME in an industrial setting to support the architecture evaluation process.

14.2 Background and Motivation

This section presents the main concepts underpinning software architecture evaluation. It is aimed at setting the context for the rest of the chapter by mentioning the importance of incorporating software architecture evaluation into the software development process and discussing the relationships between software architecture, quality attributes, and scenarios.

14.2.1 Architecture-Based Software Development

Software architecture embodies some of the earliest design decisions, which are hard and expensive to change if found flawed during downstream development activities. Since the quality attributes (such as maintainability, and reliability) of complex software systems largely depend on the overall software architecture of such systems [15], a systematic and integrated approach is required to address architectural issues throughout the software development lifecycle; such an approach is called architecture-based development [16]. One of the main characteristics of architecture-based development is the role of quality attributes and architecture styles and patterns, which provide the basis for the design and evaluation of architectural decisions in this development approach [46]. Figure 14.1 shows a high-level process model of architecture-based development that consists of six steps, each having several activities and tasks.

Architectural requirements are those that have broad cross-functional implications. Such requirements are usually elicited and specified using quality-sensitive

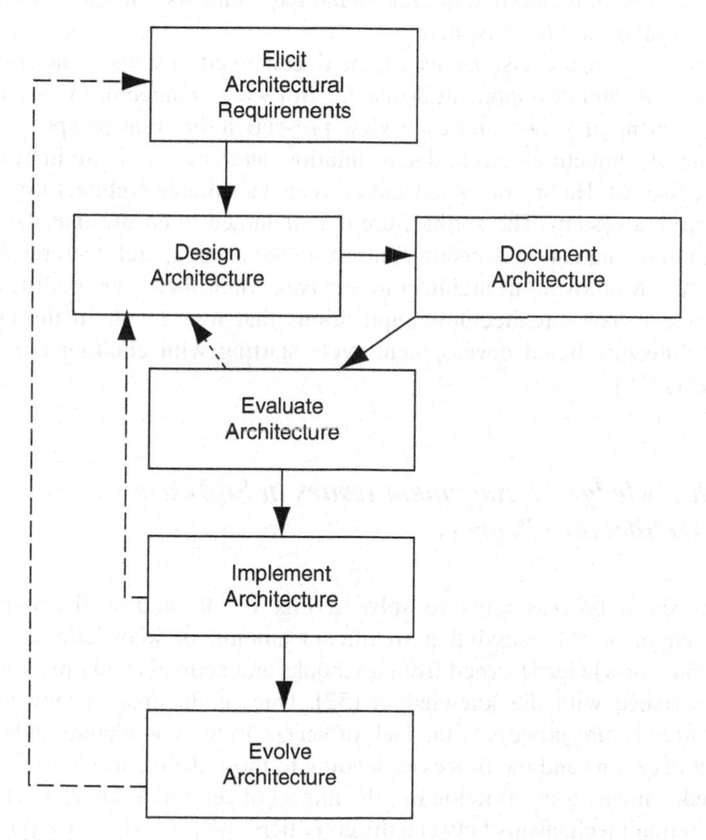

Fig. 14.1 A high level process model of architecture-based development

scenarios [15]. Scenarios have been used for a long time in several areas of different disciplines (military and business strategy, decision making,). The software engineering community has started using scenarios in user-interface engineering, requirements elicitation, performance modeling, and more recently in SA evaluation [41]. Scenarios are quite effective for specifying architectural requirements because they are very flexible. Scenarios can also be used to characterize most of the architectural requirements. For example, we can use scenarios that represent failure to examine availability and reliability, scenarios that represent change requests to analyze modifiability, scenarios that represent threats to analyze security, or scenarios that represent ease of use to analyze usability. Moreover, scenarios are normally concrete, enabling the user to understand their detailed effect [45].

Architecture design is an iterative process, making incremental decisions to satisfy functional and architectural requirements. Architecture design decisions are mainly motivated by architectural requirements, and provide the criteria used to reason about and justify architectural choices [16]. Architects usually enlist several design options, which may have the potential of satisfying different nonfunctional requirements. Then, a selection is made from the available design options in order to satisfy all or most of the desired nonfunctional requirements. This selection process involves several trade-off decisions.

Architecture design decisions need to be documented to support the subsequent design, analysis, and development decisions. Software architecture is usually documented in terms of views, and each view presents a different perspective of the architecture. Architecture design, documentation, and evaluation are iterative steps in the process [16]. Having designed and evaluated a suitable architecture, it is possible to create a system. The architecture is maintained to ensure that the detailed design and implementation decision conform to the original architectural decisions and rationale. Moreover, an architecture evolves when there are modification requests that can have architectural implications that may result in the continuation of architecture-based development cycle starting with eliciting architectural requirements [16].

14.2.2 Knowledge Management Issues in Software Architecture Process

The architecture process aims to solve a mix of ill- and well-defined problems, which involve processing a significant amount of knowledge. Architects require topic knowledge (learned from textbooks and courses) and episodic knowledge (experience with the knowledge) [52]. One of the main problems in the software architecture process is the lack of access to the knowledge underpinning the design decisions and the processes leading to those decisions [5, 20]. This type of knowledge involves information like the impact of certain middleware choices on communication mechanisms between different tiers, why an API is used instead of a wrapper, and who to contact to discuss the performance of different architectural choices.

Much of this knowledge is episodic and is usually not documented [57]. The absence of a disciplined approach to capture and maintain architectural knowledge has many downstream consequences. These include:

- The evolution of the system becomes complex and cumbersome, resulting in violation of the fundamental design decisions
- Inability to identify design errors
- Inadequate clarification of arguments and information sharing about the design artifacts and process

The SA community has developed several methods (such as ATAM [24] and PASA [59]) to support a disciplined approach to architectural practices. Some of these do emphasize the need for knowledge management to improve reusability and grow organizational capabilities in the architecture domain. Except for [23], there is no approach that explicitly states what type of knowledge needs to be managed and how, when, where, or by whom. Also, none of the current approaches provides any conceptual framework to design, develop, and maintain a suitable repository of architecture knowledge. Hence we posit that the lack of suitable techniques, tools, and guidance is why architecture design knowledge is not captured.

The software engineering community has been discovering and documenting architecture knowledge accumulated by experienced researchers and practitioners in the form of architecture or design patterns [21, 30]. These patterns attempt to codify implicit knowledge. However, we have found that the amount of information provided and the level of abstraction used may not be appropriate for the architecture stage – too much detail is counterproductive as expert designers usually follow breadth-first approach [52]. Moreover, we have found that the existing formats of pattern documentation are not appropriate for explicating the schemas of the relationships among scenario, quality attributes, and patterns in a way that makes this knowledge readily reusable. This results in little use or reuse of the architectural artifacts (such as scenarios, quality attributes, and tactics) informally described in patterns' documentation [1, 6].

14.2.3 Support for Architectural Knowledge Management

To address the above-mentioned problems caused by the lack of suitable knowledge management technologies (e.g., methods, techniques, tools) to support the software architecture process, we have proposed a framework for managing architectural knowledge throughout the lifecycle of a software architecture as described in Section 14.2.1. This framework is aimed at improving the quality of the software architecture process by providing effective knowledge management (KM) structures to facilitate the management of implicit architecture knowledge generated during the architecture process, informally described in sources such as [11, 15, 21, 30]. This framework consists of techniques for capturing architectural knowledge (both technical and contextual), an approach to distill and document architecturally significant information from patterns, and a data model to characterize architectural constructs, their attributes, and relationships. The theoretical concepts underpinning this framework have been derived from KM [50,53], experience factories [13, 14], and pattern-mining [6, 60] paradigms. A detailed description of the framework and the theoretical concepts underpinning it can be found in [5].

One of the central objectives of this framework is to provide a set of generic guidelines for developing and maintaining a repository of architectural knowledge. The novelty of this framework resides in its ability to incorporate all the components into an integrated approach, which has been implemented as a Web-based architectural KM system, whose details have been reported in Section 14.3.

14.3 Tool Support for Managing Architectural Knowledge

In this section, we provide a detailed description of a knowledge management system that we have developed to support our framework for capturing and managing architectural knowledge. The process-based architecture knowledge management environment (PAKME) is a Web-based system that is aimed at providing knowledge management support for different activities of the software architecture process described in Section 14.2.1. It has been built on top of an open-source groupware platform, Hipergate [34]. This platform provides various collaborative features including contact management, project management, online collaboration tools, and others that can be exploited to build a groupware support environment. This environment incorporates architecture knowledge management features for geographically distributed software development teams in the context of global software development.

14.3.1 The Architecture of PAKME

The architecture of PAKME is shown in Figure 14.2. This logical view of the architecture of PAKME provides an abstract description of the gross structure of the system, its components, their organization, and interaction with each other. The component-based architecture is aimed at supporting a flexible and extensible system that can be implemented either from scratch, built on the top of an existing framework, or integrated with various tools that are aimed at supporting software architecture-related activities such as requirements management and modeling.

The major components of the PAKME architecture are the following:

User interface component – The component provides a mechanism for a user to interact with PAKME in order to use any of the architectural knowledge management services. This component has several kinds of functions that help manage a user's interaction with PAKME. For example, user registration and authentication function enables a user to register with PAKME and then login onto the system to

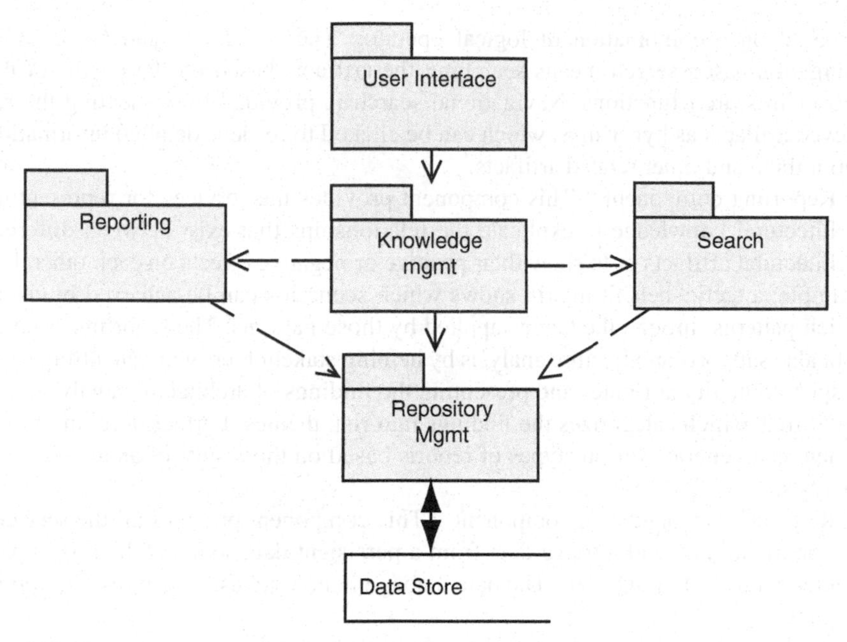

Fig. 14.2 The main components included in PAKME's architecture

use its different features based on the privileges assigned to a particular user. The user interface component also provides various forms and editing tools to enter new or update existing architectural knowledge. The knowledge acquisition forms are designed in such a way as to capture architecture knowledge according to the requirements of the abstract architecture design knowledge template and the concrete architecture design knowledge template described in [5].

Knowledge management component – This component provides common services to store, retrieve, and update artifacts that make up architectural knowledge such as patterns, scenarios, quality attributes, and others. PAKME's knowledge base is logically divided into organizational knowledge (generic), and project-specific knowledge (concrete). This component also provides services to instantiate generic architectural artifacts into concrete artifacts, for example, creating concrete scenarios to characterize quality attributes for a certain project based on the general scenarios stored in architecture knowledge repository. Moreover, this component can also store and manage artifacts created outside PAKME but attached to different artifacts created within PAKME, for example, design models or requirement documents attached to scenarios. This component uses the services of a data management component to store, maintain, and retrieve architectural artifacts.

Search component – This component helps search through the knowledge repository for desired artifacts. This component incorporates three types of search functions: keyword-based search, advanced search, and navigation-based search. Keyword-based search facility explores the repository for a desired artifact utilizing the key words that are attached as meta-data to each artifact. Advanced search is based on a combination of logical operators (i.e., And, Or, and Not), while navigation-based search means searching the artifacts based on the results of the two main search functions. Navigational search is provided by presenting the retrieved artifacts as hyperlinks, which can be clicked to retrieve detailed information about them and other related artifacts.

Reporting component – This component provides the services for representing architectural knowledge to explicate the relationships that exist between different architectural artifacts or to show their positive or negative effects on each other. For example, a tactic–benefit matrix shows which scenarios can be achieved by using which patterns through the tactics applied by those patterns. The reporting component also supports architecture analysis by helping stakeholders generate utility trees to specify quality attributes and presenting the findings of architecture analysis as a result tree, which categorizes the findings into risk themes. Furthermore, this component can generate various types of reports based on the results of an architecture analysis.

Repository management component – This component provides all the services to store, maintain, and retrieve data from a persistent data source, which is implemented with PostgreSQL 8.0. The data management logic uses Postgres's scripting language.

14.3.2 The Data Model of PAKME

PAKME is a repository-driven knowledge management system. Hence, the data model underpinning PAKME is one of the key elements of this system. A well-thoughtout and suitable data model is one of the most important artifacts needed for the development of an automated system for storing and accessing the data that characterize architectural knowledge. Figure 14.2 presents a partial data model that identifies the main architectural constructs and their relationships. PAKME's data model logically divides architectural knowledge into organizational (generic), and project-specific (concrete). Access to a repository of generic knowledge enables designers to use accumulated "wisdom" from different projects when devising or analyzing architectural decisions. The project-specific repository captures and consolidates knowledge and rationale specific to a project such as design history, analysis findings, and architectural views for stakeholders.

To develop the data model for PAKME presented in Figure 14.3, we used several approaches to arrive at an appropriate set of architectural constructs, establishing relationships and cardinalities among them, and identifying their attributes and constraints:

- We read several textbooks on software architecture (e.g., [15, 19, 21, 24, 30]) and analyzed their examples and case studies to support our exploration of relevant architectural constructs and their attributes.
- Our work on comparison and classification of SA evaluation methods [8] was an important means of identifying relevant literature.

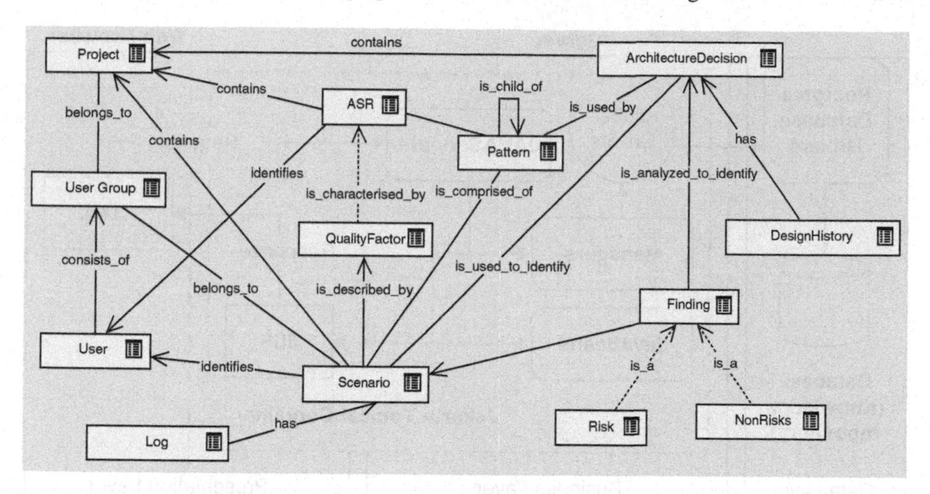

Fig. 14.3 A partial data model characterizing architectural knowledge

- We reviewed a selected set of gray literature (such as PhD theses and technical reports) in software architecture (e.g., [17,18,47]) and standards for documenting architectures [36].
- We reviewed the literature in other engineering disciplines and human–computer interaction (e.g., [25, 31, 49, 51]) to discover appropriate constructs, which describe DR.

Having identified the entities, relationships among them, and their respective attributes to characterize architectural knowledge, we assessed the data model by reference to the literature, in particular [36, 47], which provide reference models for describing and evaluating software architecture.

14.3.3 Implementation

We have built PAKME on top of an open-source groupware platform, Hipergate (http://www.hipergate.org). Hipergate provides a suite of collaborative applications, which can easily be extended to support various collaborative activities. We have described some of the main reasons for our decision to build PAKME using a groupware platform like Hipergate in Section 14.3. Some of the other reasons for choosing Hipergate for developing PAKME include:

- It is completely free and the source code is available for modifications.
- It has the large number of features available to support project collaboration and management.
- Its APIs are well documented and it supports a wide range of packages and functionalities.
- It supports multiple databases and operating systems.

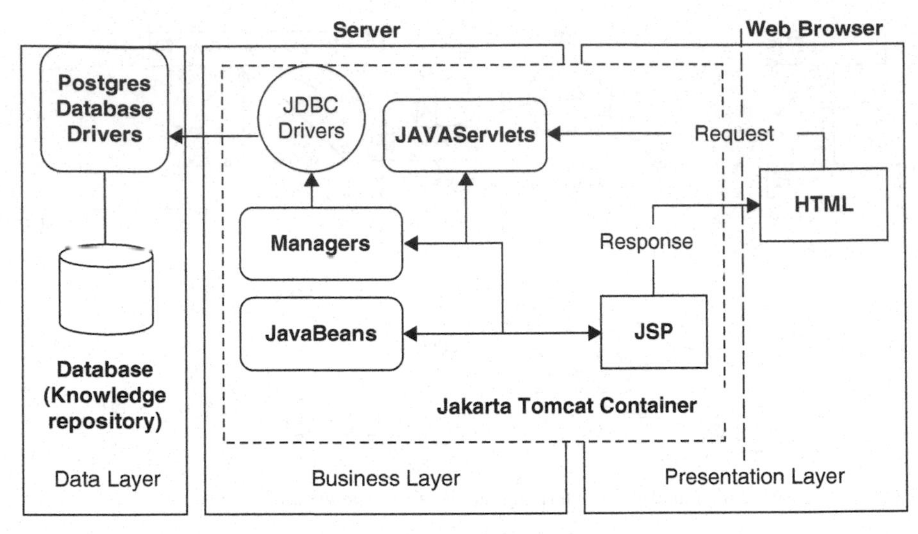

Fig. 14.4 A layered model of implementing PAKME's architecture

Hence, the implementation architecture of PAKME conforms to the constraints placed by Hipergate. PAKME has been implemented using J2EE framework and related technologies. The choice of J2EE platform was made for several reasons: Hipergate has been developed using J2EE framework, a large number of reusable features are available, well-documented APIs, and platform independence. The implemented architecture consists of three layers as shown in Figure 14.4. These three layers are presentation tier, business tier, and data tier.

Presentation Tier This layer is responsible for capturing and presenting architectural knowledge managed by PAKME using HTML and Java Server Pages (JSPs) technologies. The presentation layer is composed of several dozens of forms designed, based on templates provided by our architectural knowledge management framework for capturing and/or presenting architectural knowledge. HTML-based forms gather the requests or data from a user and send these as requests to the servlet in the business layer. When a reply is received, the Java Server Pages are responsible to render the HTML back to the user through a Web browser. The Java Server Pages technology helps dynamically create Web pages required by a knowledge management system like PAKME. JSPs in the presentation layer also interact with the business tier, which has been implemented using servlets technology. JSPs are embedded in the forms to perform data validation at the client's side and to perform some functions such as opening a new window and refreshing data and controlling a session using cookies.

Business Layer This layer represents the business logic required by PAKME as well as Hipergate. This layer has been implemented using Java Servelets. Once a servlet receives a request from the Web client, it directs relevant request managers to handle the request. The JDBC then acts as an interface to a suitable database driver in order to extract information from the database to perform the required operations. As shown in Figure 14.4, the business tier is deployed on Tomcat of Apache Web Server. Apart from being one of the most reliable Web application deployment platforms, Tomcat was chosen because it is supported by Hipergate.

Data Layer This layer consists of a database to store and manage architectural knowledge. This layer implements the data model described in Section 14.3.2. using an open-source relational database management system, PostgresSQL. We decided to use PostgresSQL as it has very good reputation for reliability, data integrity, and correctness. It provides powerful functionalities to build database views and functions. PostgresSQL is one of the database management systems being supported by Hipergate. We have implemented PAKME's data model on PostgresSQL by modifying Hipergate's data model. There are more than three dozen tables related to PAKME. The Data management logic for PAKME has been implemented using functions and triggers written in PostgresSQL's scripting language. By implementing data management logic using PostgresSQL scripting language, it is expected that performance gains can be achieved and the architectural knowledge managed by PAKME's repository can easily be made available to other tools for various purposes. Moreover, PAKME can easily be integrated into another platform.

14.4 Managing Architectural Knowledge with PAKME

This section describes different ways of managing architectural knowledge using PAKME. Most of the approaches to managing knowledge can be categorized into codification and personalization [32]. Codification concentrates on identifying, eliciting, and storing knowledge as information in repositories, which are expected to support the high-quality, reliable, and rapid reuse of knowledge. Personalization resorts to fostering interaction among knowledge workers for explicating and sharing knowledge. Although this chapter focuses on those features of PAKME as a Web-based knowledge management system that support codification, PAKME also supports personalization as it not only provides access to architectural knowledge but also identifies the source of knowledge. Hence, PAKME supports a hybrid strategy for managing knowledge by using a Web-based system [26]. Here we briefly discuss the four main services of PAKME:

- The knowledge acquisition service provides various forms and editing tools to enter new generic or project-specific knowledge into the repository. The knowledge capturing forms are based on various templates that we have designed to help maintain consistency during knowledge elicitation and structuring processes.
- The knowledge maintenance service provides different functions to modify, delete and instantiate the artifacts stored in the knowledge repository. Moreover, this service also implements the constraints on the modifications of different artifacts based on the requirements of a particular domain.

- The knowledge retrieval service helps a user to locate and retrieve the desired artifacts along with the information about the artifacts associated with them. PAKME provides three types of search mechanisms. A basic search can be performed within a single artifact based on the values of its attributes or keywords. An advanced search string is built using a combination of logical operators within a single or multiple artifacts. Navigational search is supported by presenting the retrieved artifacts and their relationships with other artifacts as hyperlinks.
- The knowledge presentation service presents knowledge in a structured manner at a suitable abstraction level by using templates (such as provided in [4]) and representation mechanisms like utility and results trees described in [15].

These services not only satisfy the requirements identified by us to provide knowledge management for different methods (such as reported in [35, 42]) developed to support the software architecture process, but also support many of the use cases proposed in [44] for managing architectural knowledge.

14.4.1 Capturing and Presenting Knowledge

There are two main strategies to elicit and codify knowledge:

- Appoint a knowledge engineer to elicit and codify knowledge from individuals or teams [54, 57].
- Provide a tool to encode the knowledge into the system as part of the knowledge creation process.

The latter is called contextualized knowledge acquisition [33]. Each strategy has its strengths and weaknesses. To take the advantage of the strengths of both strategies, PAKME helps elicit and codify architectural knowledge using either of these strategies. We have been using PAKME by embedding it into knowledge creation processes. Its repository has been populated by capturing knowledge from several J2EE [9] patterns and architecture patterns [21], case studies described in [15, 24], and design primitives [17]. PAKME provides several forms based on different templates to help users elicit and structure knowledge before storing it into the repository. Templates are aimed at keeping the process consistent across users [42]. Figure 14.6 shows a form for capturing a general scenario, which can be elicited from a stakeholder or extracted from a pattern. Each scenario can have several attributes attached to it, including source documents, revision history, and a set of keywords. PAKME's repository contains hundreds of general scenarios (Figure 14.5 shows some of them).

Figure 14.7 shows a template for capturing and presenting patterns. A pattern may be composed of other patterns and each pattern may have several tactics attached to it. To support reusability at the design decision level, PAKME's repository contains design options, which are design decisions that can be considered and/or evaluated to satisfy one or more functional or nonfunctional requirements. For example, Java RMI or publish-scribe design options can be used for event notification

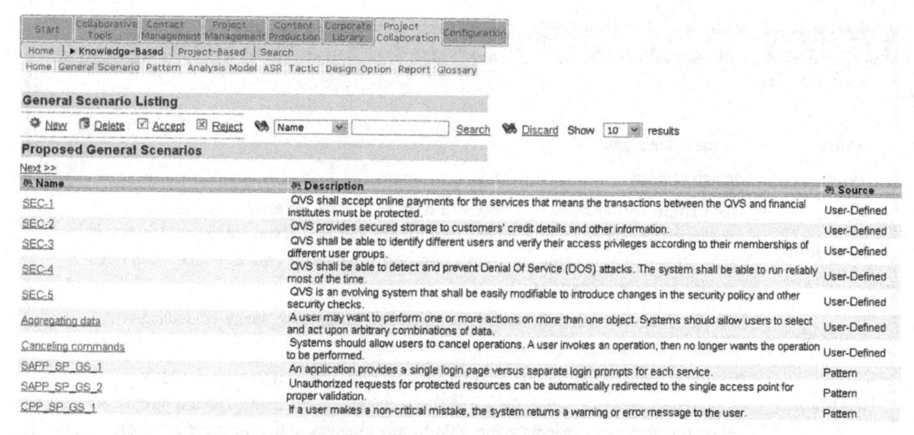

Name	Description	Source
SEC-1	QVS shall accept online payments for the services that means the transactions between the QVS and financial institutes must be protected.	User-Defined
SEC-2	QVS provides secured storage to customers' credit details and other information.	User-Defined
SEC-3	QVS shall be able to identify different users and verify their access privileges according to their memberships of different user groups.	User-Defined
SEC-4	QVS shall be able to detect and prevent Denial Of Service (DOS) attacks. The system shall be able to run reliably most of the time.	User-Defined
SEC-5	QVS is an evolving system that shall be easily modifiable to introduce changes in the security policy and other security checks.	User-Defined
Aggregating data	A user may want to perform one or more actions on more than one object. Systems should allow users to select and act upon arbitrary combinations of data.	User-Defined
Canceling commands	Systems should allow users to cancel operations. A user invokes an operation, then no longer wants the operation to be performed.	User-Defined
SAPP_SP_GS_1	An application provides a single login page versus separate login prompts for each service.	Pattern
SAPP_SP_GS_2	Unauthorized requests for protected resources can be automatically redirected to the single access point for proper validation.	Pattern
CPP_SP_GS_1	If a user makes a non-critical mistake, the system returns a warning or error message to the user.	Pattern

Fig. 14.5 General scenarios captured in PAKME's repository

hipergate :: Edit General Scenario - Microsoft Internet Explorer provided by D...
Edit General Scenario
Name Canceling commands *
Description Systems should allow users to cancel operations. A user invokes an operation, then no longer wants the operation to be performed. ***
Source User-Defined
Pattern User-Defined Pattern * (if source is *Pattern*)
Documents (Please add documents.) Add Documents
Keywords (Please add keywords.) Add Keywords
Logs (it is created only if *Revision Reason* is not empty)
Revision Reason

Fig. 14.6 The interface to capture a general scenario

purposes. Each design option is composed of one of more architectural and/or design patterns and each of them is composed of one or more tactics. For example, the publish–subscribe design option applies the publish-on-demand design pattern.

PAKME captures design options as contextualized cases from literature or previous projects. A design option case consists of problem and solution statements, patterns and tactics used, rationale, and related design options. Rationale for each design option are captured in a separate template, which is designed based on

hipergate :: View Pattern - Business Delegate - Microsoft Internet Explorer provided b...

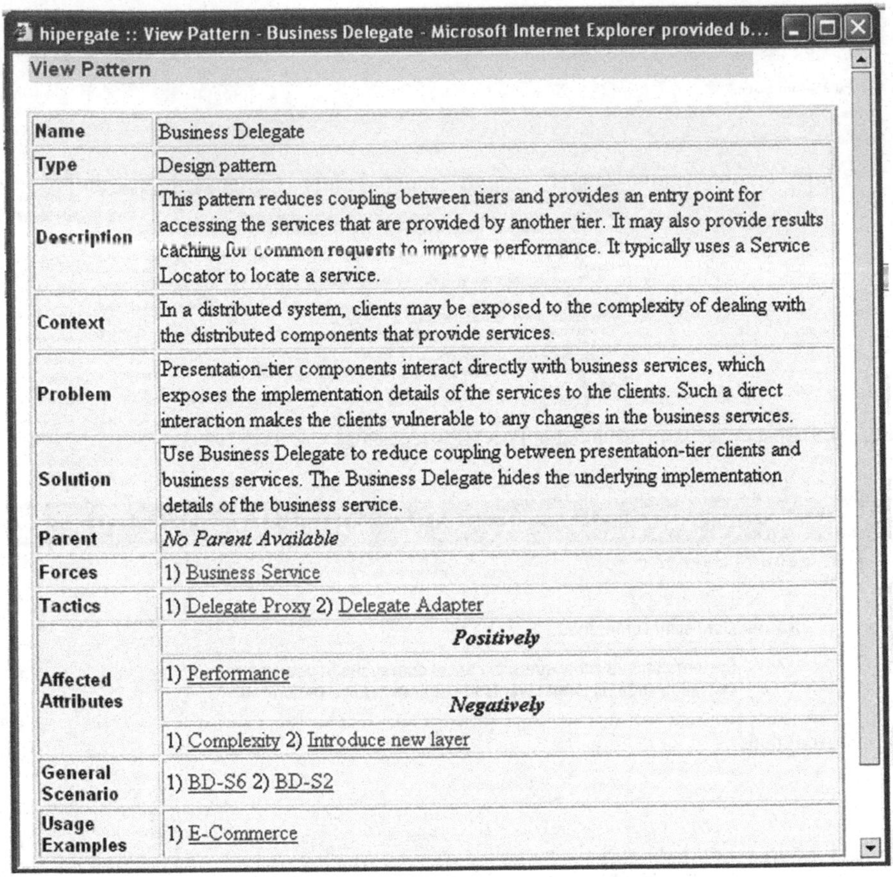

View Pattern

Name	Business Delegate
Type	Design pattern
Description	This pattern reduces coupling between tiers and provides an entry point for accessing the services that are provided by another tier. It may also provide results caching for common requests to improve performance. It typically uses a Service Locator to locate a service.
Context	In a distributed system, clients may be exposed to the complexity of dealing with the distributed components that provide services.
Problem	Presentation-tier components interact directly with business services, which exposes the implementation details of the services to the clients. Such a direct interaction makes the clients vulnerable to any changes in the business services.
Solution	Use Business Delegate to reduce coupling between presentation-tier clients and business services. The Business Delegate hides the underlying implementation details of the business service.
Parent	*No Parent Available*
Forces	1) Business Service
Tactics	1) Delegate Proxy 2) Delegate Adapter
Affected Attributes	***Positively***
	1) Performance
	Negatively
	1) Complexity 2) Introduce new layer
General Scenario	1) BD-S6 2) BD-S2
Usage Examples	1) E-Commerce

Fig. 14.7 Template to capture and present patterns

practitioners' opinions about rationale reported in [55] and templates proposed in [23, 58]. Figure 14.8 shows a partial design option. By capturing design options as cases, PAKME enables architects to reason about design decisions using a case-based approach [43].

Recently, there has been an increased emphasis on describing software architecture as a set of design decisions [20, 39]. Kruchten et al. have proposed a taxonomy of architectural decisions, their properties, and relationships among them [44].

Figures 14.9 shows that PAKME can capture many of the attributes and relationships of architectural decisions as described in [44] using templates proposed in [58]. In PAKME, architectural decision can be described at different levels of granularity. An architectural decision is a selected design option, which can be composed of an architectural pattern, a design pattern, or a design tactic. Like a design option, each architectural decision also captures rationale using a template. The rationale describes the reasons for an architectural decision, justification for it, trade-offs made, and argumentation leading to the design decision. Hence, PAKME captures

hipergate :: View Design Option - Database Server - Microsoft Internet Explorer

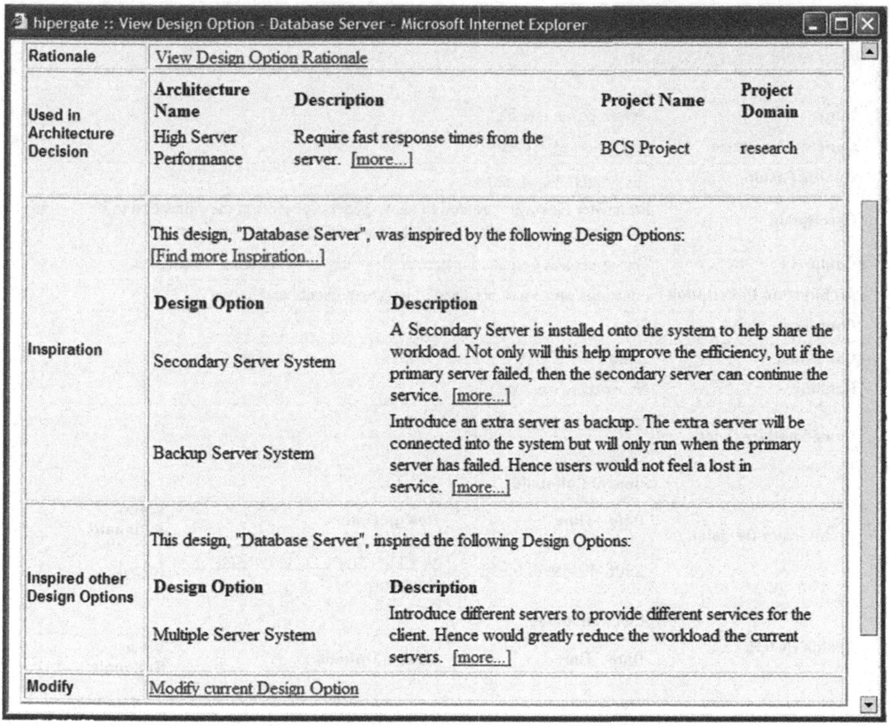

Rationale	View Design Option Rationale
Used in Architecture Decision	**Architecture Name** / **Description** / **Project Name** / **Project Domain** High Server Performance / Require fast response times from the server. [more...] / BCS Project / research
Inspiration	This design, "Database Server", was inspired by the following Design Options: [Find more Inspiration...] **Design Option** / **Description** Secondary Server System / A Secondary Server is installed onto the system to help share the workload. Not only will this help improve the efficiency, but if the primary server failed, then the secondary server can continue the service. [more...] Backup Server System / Introduce an extra server as backup. The extra server will be connected into the system but will only run when the primary server has failed. Hence users would not feel a lost in service. [more...]
Inspired other Design Options	This design, "Database Server", inspired the following Design Options: **Design Option** / **Description** Multiple Server System / Introduce different servers to provide different services for the client. Hence would greatly reduce the workload the current servers. [more...]
Modify	Modify current Design Option

Fig. 14.8 A partial view of a design option case

rationale for design options as well as for architectural design decisions, which are made by selecting one or more suitable design options from a set of considered or assessed design options.

Moreover, traceability is also provided. Each architectural design decision describes the design options considered but rejected, concrete scenarios to be satisfied, and a model of architectural decision attached as design artifacts (shown in Figure 14.9). Revisions of design decisions and reasons are logged for later review. Design decisions are time stamped and annotated with the decision maker's details, which can be used to seek explanation for that design decision. Hence, PAKME supports the description of an architectural design decision in ways suggested in [39,58] with the attributes and relationships proposed in [44].

14.4.2 Supporting Knowledge Use/Reuse

This section describes various ways in which PAKME facilitates architectural knowledge use/reuse. Let us first consider how PAKME supports the reuse of design options in making architectural decisions. We have developed a four-step process for

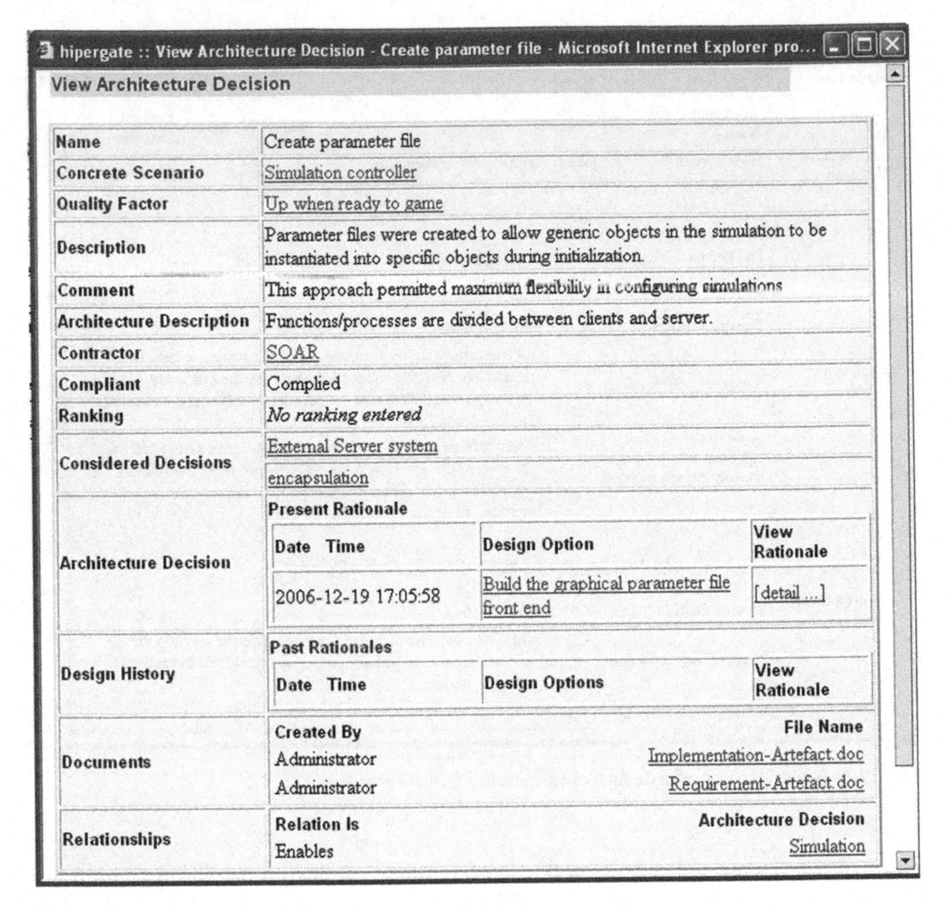

Fig. 14.9 An architecture decision captured in PAKME

reusing design options, which are captured as contextualized cases of architectural design decisions. The process starts when a user has a new requirement that needs architectural support. This requirement would characterize a quality goal and would have been specified using concrete scenario. In order to satisfy that requirement, an architect needs to make a new design decision. To address that requirement, the architect would then have two options:

- Search and retrieve a previous design option from the knowledge repository.
- Create a new design option to solve the given problem.

If the architect decides to search through the knowledge repository for cases of design options, they can perform a search to retrieve a list of design options. Figure 14.10 shows that a user can build a complex search string based on various attributes of a design options stored in PAKME's repository. After reviewing the retrieved list of design options, the architect can either reuse an existing design option in its original form or modify it according to the current context. Figure 14.11

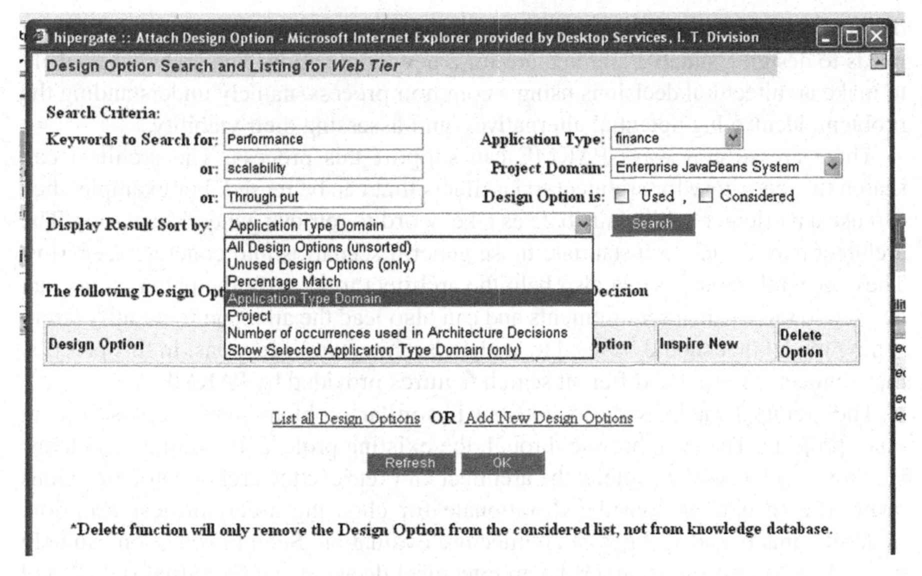

Fig. 14.10 PAKME's interface for searching design option cases

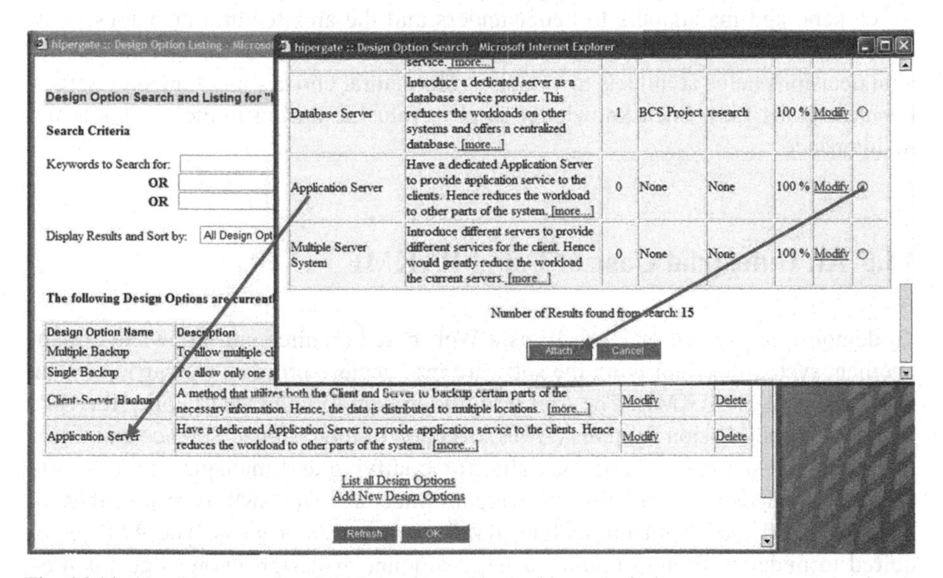

Fig. 14.11 Attaching a retrieved design option to an architecture design

shows that a retrieved design option can be used by attaching it to an architecture design decision. If a design option is modified, it is considered a new design option but it is linked to the original design option for traceability. This new design option can be chosen as an architecture design decision and it is stored in the repository as a design option, which is an architectural decision. To demonstrate the other ways of reusing architecture knowledge with PAKME, let us consider that an architect needs to design a suitable architecture for a new application. The architect is likely to make architectural decisions using a common process, namely understanding the problem, identifying potential alternatives, and assessing their viability.

There are several ways PAKME can support this process. The architect can search the repository for architectural artifacts that can be reused. For example, they can use a particular quality attribute as a keyword to retrieve general scenarios. The architect may decide to instantiate those general scenarios into concrete scenarios. These general scenarios can also help the architect to identify the patterns that can be used to satisfy their requirements and can also lead the architect to identify a reasoning model that should be used to analyze architectural decisions. In this process, the architect can use the different search features provided by PAKME.

The architect may decide to find out if similar problems have been solved in other projects. They can browse through the existing projects for similar problems. Having found a similar project, the architect can retrieve the architectural decisions taken, design options considered, rationale for choosing a certain design option, trade-offs made, and findings of architecture evaluation. Such information can help the architect to decide whether the architectural decision can be reused or not, and how much tailoring is required. Project-specific knowledge can also help designers, developers, and maintainers to better understand the architectural decisions, their constraints, and reasoning behind it. Availability of the rationale behind the design decisions helps architects to explain architectural choices and how they satisfy business goals [58]. Such knowledge is also valuable during implementation and maintenance.

14.5 An Industrial Case of Using PAKME

To demonstrate the use of PAKME as a Web-based architectural knowledge management system for improving the software architecture process, we carried out an industrial trial of PAKME. The trial was undertaken to tailor and deploy PAKME in the Airborne Mission Systems (AMS) division of the Defence Science and Technology Organisation (DSTO), Australia, for codifying and managing process and domain knowledge of evaluating software architecture. The AMS is responsible for evaluating software architectures for aircrafts acquisition projects. The AMS is required to understand and organize a large amount of design knowledge for mission systems' architectures to support the evaluation process. Recently, AMS has decided to improve its architectural evaluation practices by codifying and reusing an architecture evaluation process, architecture design knowledge, and contextual knowledge using a Web-based knowledge management system like PAKME, which is expected to help AMS to capture and manage architecture knowledge. It is expected that the use of an architecture knowledge management tool will systemize the process and help organize the architecture design knowledge and contextual information required or generated during a software architecture evaluation process.

This objective is expected to be achieved by embedding PAKME in the software architecture evaluation process.

Having tailored and deployed PAKME in AMS's environment, a study was carried out to assess PAKME's support for AMS's architecture evaluation process. This study involved using PAKME for capturing and managing architectural knowledge to support architecture evaluation of an aircraft system. It was a postmortem analysis of the architecture evaluation conducted without using PAKME. The study was also aimed at investigating how the introduction of PAKME could help capture and manage architectural knowledge, and whether or not the evaluation process is improved by using PAKME.

For this study, a number of quality factors were chosen as measures for the mission system architecture evaluation. These quality factors were growth, security, adaptability, and reconfigurability as some of them are shown in Figure 14.12. The evaluation process involved AMS's evaluators comparing alternative design decisions from multiple hypothetical tenders, to simulate the type of evaluation completed during the real evaluation of an aircraft acquisition project. The evaluation was performed by measuring each scenario against the quality attributes, as well as assigning a measure of risk to each of the proposed architectural design decisions.

View Architecture Decision

<table>
<tr><td>Name</td><td colspan="3">Compliance with the High Level Architecture (HLA)</td></tr>
<tr><td>Concrete Scenario</td><td colspan="3">Reconfigure Wargame 2000</td></tr>
<tr><td>Quality Factor</td><td colspan="3">Reconfigurability</td></tr>
<tr><td>Description</td><td colspan="3">The HLA was developed under the leadership of the Defense Modeling and Simulation Office (DMSO) to support reuse and interoperability across the large numbers of different types of simulations developed and maintained by the DoD.</td></tr>
<tr><td>Comment</td><td colspan="3">HLA compliance in Wargame 2000 is achieved via an HLA gateway to interface with other systems that are HLA compliant.</td></tr>
<tr><td>Architecture Description</td><td colspan="3">Functions/processes are divided between clients and server.</td></tr>
<tr><td>Contractor</td><td colspan="3">NICTA</td></tr>
<tr><td>Compliant</td><td colspan="3">Complied</td></tr>
<tr><td>Ranking</td><td colspan="3">No ranking entered</td></tr>
<tr><td>Considered Decisions</td><td colspan="3">No Alternative Decisions</td></tr>
<tr><td rowspan="3">Architecture Decision</td><td colspan="3">Present Rationale</td></tr>
<tr><td>Date Time</td><td>Design Option</td><td>View Rationale</td></tr>
<tr><td>2006-12-19 17:12:24</td><td>Provide a virtual gaming site</td><td>[detail ...]</td></tr>
<tr><td rowspan="2">Design History</td><td colspan="3">Past Rationales</td></tr>
<tr><td>Date Time</td><td>Design Options</td><td>View Rationale</td></tr>
<tr><td>Documents</td><td colspan="3">No Documents Associated</td></tr>
<tr><td>Relationships</td><td colspan="3">No Relations Associated</td></tr>
</table>

Fig. 14.12 An architectural design decision example

14.5.1 Use of PAKME's Knowledge Base

PAKME's knowledge base was populated with the AMS domain knowledge consisting of quality attributes and its quality factors that characterize the quality model being used by DSTO to evaluate software architecture of avionics systems. The process of transferring AMS's quality model into PAKME was guided by the ISO 9126 quality model [37], the quality attributes, and the general scenarios provided by SEI [12], and the experience of avionics domain experts gathered in a workshop aimed at eliciting and capturing knowledge underpinning AMS's quality model.

This quality model was created in PAKME's repository and made available to all the AMS's evaluators. PAKME was also populated with general scenarios for characterizing each of the quality attributes included in the quality model as shown in Figure 14.5. Apart from the scenarios characterizing the AMS quality model, several hundred general scenarios were stored in PAKME to support architecture evaluation. The general scenarios stored in PAKME are used to generate concrete scenarios for different evaluation projects.

Domain-specific general scenarios captured in PAKME's repository can provide assurance that the breadth of topics applicable to a system has been analyzed, and deriving concrete scenarios from those general scenarios allows analysis of system behaviour under task-specific criteria. Subsets of general scenarios can be assembled into meta-projects that are applicable to a class of aircraft systems such as helicopters and fast jets. The general scenarios from the meta-projects are imported into project-specific repository to support architecture evaluation of a system. Figure 14.3 shows a user-defined scenario for a generic mission system architecture captured under meta-projects in PAKME.

PAKME's repository was also populated with avionics-specific general design options. These design options were captured from the architecture solutions proposed for the system reviewed for this case study, AMS domain experts, and case studies on avionics systems reported in sources such as [15]. Each design option was captured as a design decision case as shown in Figure 14.4. These generic design options were used as input to design decision making or evaluation processes.

14.5.2 Use of PAKME's Project Base

For this study, the AMS team created a new project in PAKME and populated its project base with the project quality model to specify quality factors with concrete scenarios based on the scenarios used to characterize the AMS's quality model built during the above-mentioned workshop.

Each architecture decision proposed by different contractors for satisfying required scenarios of the project was identified and entered into PAKME. Each decision was also linked to the concrete scenarios satisfied by that architecture decision. An example of a design decision affecting architectural quality is the use of a layered architecture including an isolation layer to reduce the impact of change, and thus improving flexibility, technology refreshment, and growth capability. This architecture design decision was stored in PAKME along with the rationale.

Each architecture decision of this project was also captured as a design option in the generic knowledge base of PAKME. The evaluation team also captured several design options based on their domain knowledge. Figure 14.12 shows one of the architecture design decisions evaluated using PAKME for this case study. During architecture evaluation, each architecture design decision was assessed with respect to the design options, which are expected to satisfy the same concrete scenario. Having populated PAKME with the project-specific architecture knowledge, the evaluation team evaluated the architecture design decisions proposed for an aircraft system by several contractors. For this evaluation, the team used PAKME for accessing the architectural knowledge required for the evaluation and capturing the findings and rationale for evaluation.

The architecture evaluation process involved determining whether or not a concrete scenario is satisfied by the proposed architecture decision. If there were more than one proposed architecture decision for a scenario, architecture decisions were assigned a ranking based on an evaluator's opinion about each architecture decision's capability of achieving a certain level of required quality factor. Figure 14.12 shows how an architectural decision is captured along with rationale using a template provided by PAKME. Evaluators recorded their findings using the provided templates. Each finding describes whether or not a certain architecture decision complied with the relevant requirement, its ranking, and rationale/justification for the findings. Based on the evaluation findings, architecture decisions were categorized as risks or non-risks. Risks were further categorized under various risk themes as suggested in [24].

At the end of evaluation, different reports were generated for project managers. Figure 14.13 shows one view of the Web-based evaluation report generated by PAKME. At the top of Figure 14.13 is finding matrix (showing concrete scenario, architecture decision, and respective findings) based on the architecture evaluation carried out using PAKME for this case study. Web-based report presents main elements of findings as hyperlinks, which can be browsed for detailed information as is shown in Figure 14.13 for concrete scenario and architecture decisions. PAKME also generates PDF reports for evaluation teams and management based on specified criteria.

14.5.3 Observations from the Study

Based on their experiences of evaluating architectures with and without using a knowledge management tool like PAKME, AMS staff involved in this study reported that the use of PAKME proved quite encouraging. During the simulated architecture evaluation project, the AMS's evaluators used PAKME as a

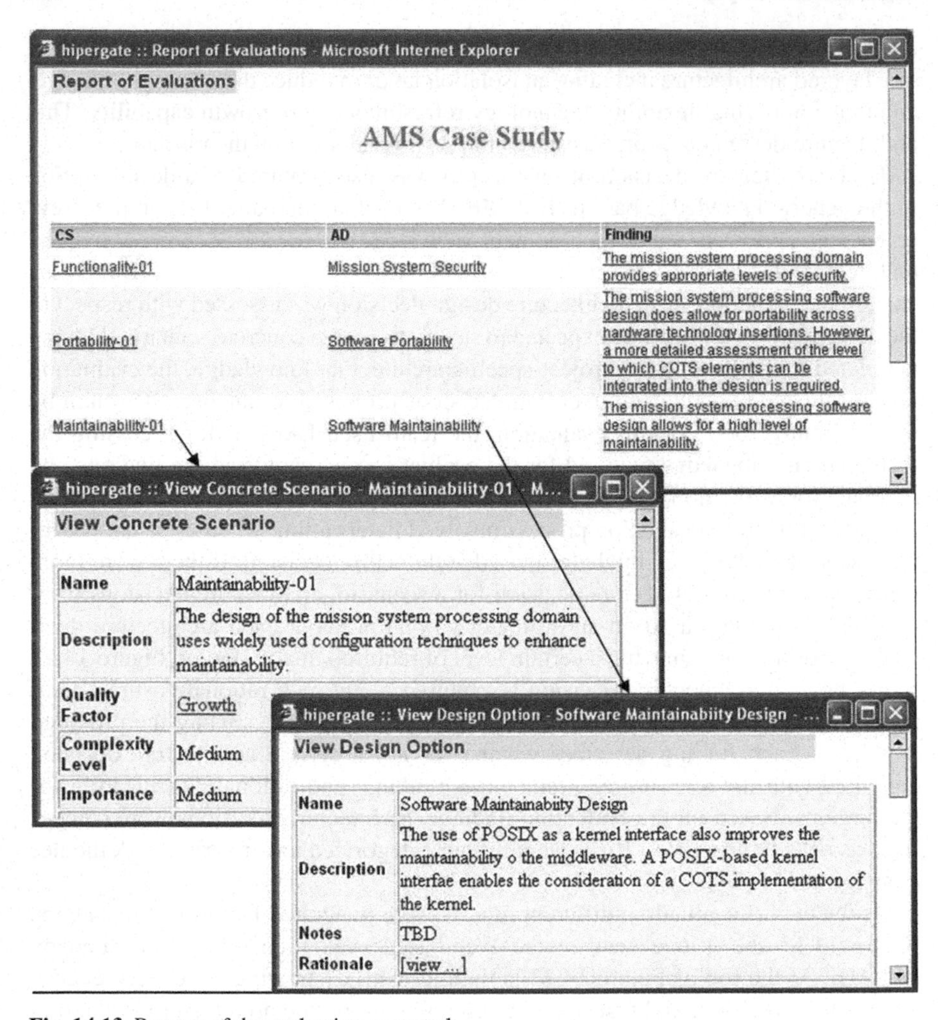

Fig. 14.13 Reports of the evaluation case study

communication and knowledge-sharing mechanism. They reported that the general scenarios and design options captured in the knowledge base helped them in generating concrete scenarios and understanding proposed solutions. Having access to a codified quality model through a Web-based system provided all the evaluators with the same understanding of the quality requirements. Moreover, the evaluators found PAKME's templates to capture justification for evaluation decision are very useful. Overall, the evaluators and subject matter experts found that the use of an evaluation framework and knowledge management tool brought added rigor to the evaluation process. It is anticipated that the management of evaluation decisions and their justification using PAKME would minimize the need for contacting the evaluator of past projects for explanation. Hence, the AMS management were now more convinced that an architectural knowledge management tool like PAKME would provide them with several benefits and help them to institutionalize a disciplined software architecture evaluation process.

Customizing and trialing PAKME for the AMS architecture evaluation process, the research team faced several organizational, technical, and logistical challenges that needed to be overcome in order to complete this project. We believe that some of them are general enough for any research–industry collaboration aimed at tailoring and deploying a Web-based tool like PAKME in the defence domain.

One of the key organizational challenges was that the research team did not have clearance to access the documentation of the software architectures being evaluated. Nor did they have domain expertise. Hence, they had to gain a certain level of domain understanding in order to help the AMS build their domain-specific quality model using scenarios, identify and capture design options and patterns from architectural descriptions of the systems being evaluated, and determine requirements for customizing PAKME. The AMS's management helped researchers overcome this challenge by providing continuous access to their domain experts, who helped researchers to understand the unique technical and organizational requirements of the process of evaluating software architecture and integrating a Web-based knowledge management tool like PAKME in that process.

The research team also found that certain requirements were unique to the AMS domain and difficult to implement without making significant changes in existing user interface and data model. For example, the AMS data is classified based on different levels of security clearance for the data user. There are four different levels of security clearance (top secret, secret, classified, and unclassified). Moreover, each quality attribute scenario can have more than one architecture decision proposed by different contractors. The AMS required that each combination of a concrete scenario and proposed solution needed to have its own findings attached and any relevant artifacts in a matrix shown in Figure 14.13.

Based on our experiences during these modifications and AMS's feedback on the configuration issues with PAKME's templates, we believe that a Web-based architectural knowledge management system like PAKME needs to be designed to be heavily customizable at deployment time, and that a rigid tool is unlikely to be widely successful. We also realized early on that the key to making PAKME useful is to incorporate it into the process already being used for evaluating a software architecture by AMS team. It was also realized that PAKME need not be perceived as a replacement to the existing tools rather a complementary tool, which can eventually be integrated with the existing tools.

Overall, the research team felt that the modified version of PAKME provided AMS with an effective and efficient mechanism to organize and understand large amounts of architectural knowledge. The current version of PAKME is suitable for capturing and managing several types of architectural knowledge and artifacts of an airborne mission system, and supporting a rigorous and systematic architecture evaluation process using well-known methods like ATAM [24].

14.6 Related Work

In response to the increasing realization of the importance of providing suitable tooling support for capturing and sharing architectural knowledge (AK), several researchers have developed various tools [2, 22, 39, 56] for managing architectural knowledge and others have identified requirements with the intention of providing such tools [29]. In this section, we briefly discuss some of the architectural knowledge management tools.

One of the earliest tools for managing architectural knowledge is Archium, which models design decisions and their relationships with resulting components [38]. Archium is a Java tool that integrates requirements, decisions, architectures, and implementation models, and provides traceability among a wide range of concepts. Archium uses an architecture description language (ADL) to describe the architectures from a component and connector view, and stores and visualizes design decisions and its rationale. However, Archium is not a Web-based tool; rather it is a desktop-based tool which cannot be used in distributed arrangement. Architecture Rationale and Element Linkage (AREL) is a UML-based tool that aims to assist architects to create and document architectural design with a focus on architectural decisions and design rationale [56]. AREL is implemented by a UML tool called Enterprise Architect (EA) that is based on the Microsoft-operating platforms. It uses standard UML notation to represent design and decision entities. Any customized architectural knowledge is captured by applying predefined tagged template of a stereotype. Import tools are available to extract information from Word-based requirement specifications, and plug-in tools are available to enable graphical tracing and analysis of the information. This tool has been tested on an industrial electronic payment system specification. The Knowledge architect is another recently developed toolset for managing architectural knowledge [40]. The toolset comprises a repository, a server, and a number of clients. The knowledge architect clients capture and manage architectural knowledge in different formats and contexts. One client is the Word plug-in, which allows architects to capture and manage architectural knowledge in MS Word documents. A second client, also a plug-in, captures and manages architectural knowledge of quantitative analysis models expressed in MS Excel. A third client, the knowledge architect Explorer, is a visualization tool that supports the analysis of the captured architectural knowledge. This tool allows for the exploration of the AK by searching and navigating through the web of traceability links among the knowledge entities. It has been reported that this toolset has been applied in two industrial case studies, one involving the Word plug-in, the other one the Excel plug-in.

EAGLE [28] is an architectural knowledge management portal that claims to implement best practices from knowledge management for improving architectural knowledge sharing. The main features of EAGLE include integrated support for both codified and personalized architectural knowledge support for stakeholder-specific content, subscription, and notification. EAGLE focuses on stakeholder collaboration during the architecture process by enabling them to connect to colleagues or other involved stakeholders by retrieving "who is doing what" and "who knows what." Additionally, codified architectural knowledge in a document repository or best practice repository can also be easily accessed using advanced search mechanisms. The Architecture Design Decision Support System (ADDSS) is a Web-based tool developed for managing architectural decisions [28]. ADDSS captures both architectures and decisions following an iterative process and simulates the way in which software architects build their architectures as a set of successive refinements. ADDSS also provides links between requirements and architectures for forward and backward traceability. During the reasoning process, ADDSS users can reuse general knowledge in the form of patterns and architectural styles. A query system provides valuable information to the architect of related requirements, decisions, and architectures stored in the tool. Moreover, ADDSS also enable the users to generate PDF reports about the detailed description of the architectures and their decisions. Different features and applications of ADDSS have been demonstrated in several cases in research and development environments. Hence, ADDSS has the closet similarity with PAKME in terms of the process support and objectives.

14.7 Summary

The software architecture community has increasingly been emphasizing the importance of systematically capturing and managing architectural knowledge to improve the software architecture process. We have developed a framework for capturing and sharing architectural knowledge to support the software architecture process. This framework consists of techniques for capturing design decisions and contextual information, an approach to distilling and documenting architectural knowledge from patterns, and a data model to characterize architectural constructs, their attributes and relationships [4, 5]. In order to support this framework, we have developed a Web-based architectural knowledge management called PAKME.

This chapter has provided an elaborated narration of the process and logistics involved in designing and implementing a Web-based architectural knowledge management system like PAKME. It also has discussed the requirements and architectural aspects of PAKME and explained various features provided by PAKME. It has also briefly described a study carried out to assess PAKME in an industrial setting to support the architecture evaluation process. The description of the study is aimed at showing how PAKME has been tailored and deployed in an industrial setting to serve as a repository of an organization's architecture knowledge analogous to an engineer's handbooks, which consolidate knowledge about best practices in a certain domain [10]. During the trial, PAKME has proven to be adaptable and useful to complex domains like defense. It is expected that this chapter can provide a frame of reference for designing, developing, and deploying a Web-based architectural knowledge management system like PAKME to help organizations to build their architectural competencies by systematically capturing and widely sharing architectural knowledge generated or required in the software architecture process.

Acknowledgements The author acknowledges the contributions from the AMS's team, in particular Andrew Northway, on the reported study. This work is based on the research carried out when the author was working with the National ICT Australia. Several people contributed to the theoretical, implementation, and evaluation aspects of PAKME, in particular, Ian Gorton, Andrew Northway, Lingzhi Xu, Janice Wing Yin Leung and Louis Wong. A part of the manuscript was prepared while the author was working with Lero, University of Limerick, Ireland.

References

1. Ali-Babar, M.: Scenarios, quality attributes, and patterns: Capturing and using their synergistic relationships for product line architectures. In: International Workshop on Adopting Product Line Software Engineering. Busan, South Korea (2004)
2. Ali-Babar, M., Gorton, I.: A tool for managing software architecture knowledge. In: Proceedings of the 2nd Workshop on SHAring and Reusing architectural knowledge – Architecture, rationale, and Design Intent (SHARK/ADI 2007), Collocated with ICSE 2007. IEEE Computer Society, Minneapolis (2007)
3. Ali-Babar, M., Gorton, I., Jeffery, R.: Capturing and using software architecture knowledge for architecture-based software development. In: 5th International Conference on Quality Software. IEEE Computer Press, Melbourne, Australia (2005)
4. Ali-Babar, M., Gorton, I., Jeffery, R.: Toward a framework for capturing and using architecture design knowledge. Tech. Rep. TR-0513, University of New South Wales, Australia (2005)
5. Ali-Babar, M., Gorton, I., Kitchenham, B.: A framework for supporting architecture knowledge and rationale management. In: A.H. Dutoit, R. McCall, I. Mistrik, B. Paech (eds.) Rationale Management in Software Engineering, pp. 237–254. Springer (2006)
6. Ali-Babar, M., Kitchenham, B., Maheshwari, P., Jeffery, R.: Mining patterns for improving architecting activities – a research program and preliminary assessment. In: Proceedings of the 9th International conference on Empirical Assessment in Software Engineering. Keele, UK (2005)
7. Ali-Babar, M., Tang, A., Gorton, I., Han, J.: Industrial perspective on the usefulness of design rationale for software maintenance: A survey. In: Proceedings of the 6th International Conference on Quality Software (2005)
8. Ali-Babar, M., Zhu, L., Jeffery, R.: A framework for classifying and comparing software architecture evaluation methods. In: Proceedings of the 15th Australian Software Engineering Conference, pp. 309–319. Melbourne, Australia (2004)
9. Alur, D., Crupi, J., Malks, D.: Core J2EE Patterns: Best Practices and Design Strategies, 2nd edn. Sun Microsystem Press (2003)
10. Arango, G., Schoen, E., Pettengill, R.: A process for consolidating and reusing design knowledge. In: Proc. 15th Int'l. Conf. of Software Eng., pp. 231–242. Baltimore, Maryland, USA (1993)
11. Bachmann, F., Bass, L., Klein, M.: Deriving architectural tactics: A step toward methodical architectural design. Tech. Rep. CMU/SEI-2003-TR-004, SEI, Carnegie Mellon University, USA (2003)
12. Barbacci, M.R., Klein, M.H., Weinstock, C.B.: Principles for evaluating the quality attributes of a software architecture. Tech. Rep. CMU/SEI-96-TR-036, SEI, Carnegie Mellon University (1996)
13. Basili, V.R., Caldiera, G.: Improving software quality reusing knowledge and experience. Sloan Management Review **37**(1), 55–64 (1995)
14. Basili, V.R., Caldiera, G., Rombach, H.D.: The experience factory. In: J.J. Marciniak (ed.) Encyclopedia of Software Engineering. John Wiley & Sons (2001)
15. Bass, L., Clements, P., Kazman, R.: Software Architecture in Practice, 2 edn. Addison-Wesley (2003)

16. Bass, L., Kazman, R.: Architecture-based development. Tech. Rep. CMU/SEI-99-TR-007, Software Engineering Institute (SEI), Carnegie Mellon University, Pittsburgh, USA (1999)
17. Bass, L., Klein, M., Bachmann, F.: Quality attribute design primitives. Technical Report CMU/SEI-2000-TN-017, SEI, Carnegie Mellon University, USA (2000). URL http://www.cse.unsw.edu.au/~malibaba/research-papers/softwareArchitectures/00tn017.pdf
18. Bengtsson, P.: Architecture-level modifiability analysis. Ph.d. thesis, Blekinge Institute of Technology, Sweden (2002)
19. Bosch, J.: Design & Use of Software Architectures: Adopting and evolving a product-line approach. Addison-Wesley (2000)
20. Bosch, J.: Software architecture: The next step. In: European Workshop on Software Architecture (2004)
21. Buschmann, F., Meunier, R., Rohnert, H., Sommerlad, P., Stal, M.: Pattern-Oriented Software Architecture: A System of Patterns. John Wiley & Sons (1996)
22. Capilla, R., Nava, F., Perez, S., Duenas, J.D.: A web-based tool for managing architectural design decisions. In: Proceedings of the 1st Workshop on Sharing and Reusing Architectural Knowledge (2006)
23. Clements, P., Bachmann, F., Bass, L., Garlan, D., james Ivers, Little, R., Nord, R., Stafford, J.: Documenting Software Architectures: Views and Beyond. Addison-Wesley (2002)
24. Clements, P., Kazman, R., Klein, M.: Evaluating Software Architectures: Methods and Case Studies. Addison-Wesley (2002)
25. Conklin, J., Burgess-Yahkemovic, K.C.: A process-oriented approach to design rationale. Human-Computer Interaction **6**(3-4), 357–391 (1991)
26. Desouza, K.C., Evaristo, J.R.: Managing knowledge in distributed projects. Communication of the ACM **47**(4), 87–91 (2004)
27. Dutoit, A.H., Paech, B.: Rationale management in software engineering. In: S.K. Chang (ed.) Handbook of Software Engineering and Knowledge Engineering, vol. 1. World Scientific Publishing, Singapore (2001)
28. Farenhorst, R., Izaks, R., Lago, P., Vliet, H.: A just-in-time architectural knowledge sharing portal. In: Proceedings of the 7th Working IEEE/IFIP Conference on Software Architecture (WICSA). Vancouver, Canada (2008)
29. Farenhorst, R., Lago, P., van Vliet, H.: Effective tool support for architectural knowledge sharing. In: Proceedings of the First European Conference on Software Architecture (2007)
30. Gamma, E., Helm, R., Johnson, R., Vlissides, J.: Design Patterns-Elements of Reusable Object-Oriented Software. Addison-Wesley, Reading, MA (1995)
31. Gruber, T.R., Russell, D.M.: Design knowledge and design rationale: A framework for representing, capture, and use. Tech. Rep. KSL 90-45, Knowledge Systems Laboratory, Standford University, California, USA, California (1991)
32. Hansen, M.T., Nohria, N., Tierney, T.: What's your strategy for managing knowledge? Harvard Business Review pp. 106–116 (March-April 1999)
33. Henninger, S.: Tool support for experience-based software development methodologies. Advances in Computers **59**, 29–82 (2003)
34. Hipergate: An open source crm and groupware system. URL http://www.hipergate.org/
35. Hofmeister, C., Kruchten, P., Nord, R.L., Obbink, H., Ran, A., America, P.: A general model of software architecture design derived from five industrial approaches. Journal of System and Software **80**(1), 106–126 (2007)
36. IEEE (ed.): Recommended Practices for Architecture Description of Software-Intensive Systems. IEEE Std 1471-2000 (2000)
37. ISO/IEC: Information technology – Software product quality: Quality model (ISO/IEC FDIS 9126-1:2000(E))
38. Jansen, A., Bosch, J.: Software architecture as a set of architectural design decisions. In: Proceedings of the 5th Working IEEE/IFIP Conference on Software Architecture (2005)
39. Jansen, A., van der Ven, J., Avgeriou, P., Hammer, D.: Tool support for architectural decisions. In: Proceedings of the 6th working IEEE/IFIP Conference on Software Architecture, Mumbai, India (2007)

40. Jansen, A., Vries, T., Avgeriou, P., Veelen, M.: Sharing the architectural knowledge of quantitative analysis. In: Proceedings of the Quality of Software-Architectures (QoSA 2008) (2008)
41. Kazman, R., Abowd, G., Bass, L., Clements, P.: Scenario-based analysis of software architecture. IEEE Software Engineering **13**(6), 47–55 (1996)
42. Kazman, R., Bass, L., Klein, M.: The essential components of software architecture design and analysis. Journal of Systems and Software **79**(8), 1207–1216 (2006)
43. Kolodner, J.L.: Improving human decision making through case-based decision aiding. AI Magazine **12**(2), 52–68 (1991)
44. Kruchten, P., Lago, P., Vliet, H.V.: Building up and reasoning about architecture knowledge. In: Proceedings of the 2nd International Conference on Quality of Software Architectures (2006)
45. Lassing, N., Rijsenbrij, D., van Vliet, H.: How well can we predict changes at architecture design time? Journal of Systems and Software **65**(2), 141–153 (2003)
46. Niemela, E., Kalaoja, J., Lago, P.: Toward an architectural knowledge base for wireless service engineering. IEEE Transactions of Software Engineering **31**(5), 361–379 (2005)
47. Obbink, H., Kruchten, P., Kozaczynski, W., Postema, H., Ran, A., Dominick, L., Kazman, R., Hilliard, R., Tracz, W., Kahane, E.: Software architecture review and assessment (sara) report. Tech. rep., SARA W.G., (2001)
48. Pena-Mora, F., Vadhavkar, S.: Augmenting design patterns with design rationale. Artificial Intelligence for Engineering Design, Analysis and Manufacturing **11**, 93–108 (1997)
49. Potts, C.: Supporting software design: Integrating design methods and design rationale. In: J.M. Carroll (ed.) Design Rationale: Concepts, Techniques, and Use, pp. 295–321. Lawrence Erlbaum Associates, Hillsdale, NJ (1995)
50. Probst, G.J.B.: Practical knowledge management: A model that works. Prism, Arthur D. Little (1998). URL http://know.unige.ch/publications/Prismartikel.PDF
51. Regli, W.C., Hu, X., Atwood, M., Sun, W.: A survey of design rationale systems: Approaches, representation, capture and retrieval. Engineering with computers **16**, 209–235 (2002)
52. Robillard, P.N.: The role of knowledge in software development. Communication of the ACM **42**(1), 87–92 (1999)
53. Rus, I., Lindvall, M.: Knowledge management in software engineering. IEEE Software **19**(3), 26–38 (2002)
54. Skuce, B.: Knowledge management in software design: a tool and a trial. Software Engineering Journal **September**, 183–193 (1995)
55. Tang, A., Ali-Babar, M., Gorton, I., Han, J.: A survey of architecture design rationale. Journal of Systems and Software **79**(12), 1792–1804 (2006)
56. Tang, A., Yin, Y., Han, J.: A rationale-based architecture model for design traceability and reasoning. Journal of Systems and Software **80**(6), 918–934 (2007)
57. Terveen, L.G., Selfridge, P.G., Long, M.D.: Living design memory: Framework, implementation, lessons learned. Human-Computer Interaction **10**(1), 1–37 (1995)
58. Tyree, J., Akerman, A.: Architecture decisions: Demystifying architecture. IEEE Software **22**(2), 19–27 (2005)
59. Williams, L.G., Smith, C.U.: Pasa: An architectural approach to fixing software performance problems. In: Proceedings of the International Conference of the Computer Measurement Group. Reno, USA (2002)
60. Zhu, L., Ali-Babar, M., Jeffery, R.: Mining patterns to support software architecture evaluation. In: Proceedings of the 4th Working IEEE/IFIP Conference on Software Architecture (2004)

Chapter 15
CoP Sensing Framework on Web-Based Environment

S.M.F.D. Syed Mustapha

Abstract The Web technologies and Web applications have shown similar high growth rate in terms of daily usages and user acceptance. The Web applications have not only penetrated in the traditional domains such as education and business but have also encroached into areas such as politics, social, lifestyle, and culture. The emergence of Web technologies has enabled Web access even to the person on the move through PDAs or mobile phones that are connected using Wi-Fi, HSDPA, or other communication protocols. These two phenomena are the inducement factors toward the need of building Web-based systems as the supporting tools in fulfilling many mundane activities. In doing this, one of the many focuses in research has been to look at the implementation challenges in building Web-based support systems in different types of environment. This chapter describes the implementation issues in building the community learning framework that can be supported on the Web-based platform. The Community of Practice (CoP) has been chosen as the community learning theory to be the case study and analysis as it challenges the creativity of the architectural design of the Web system in order to capture the presence of learning activities. The details of this chapter describe the characteristics of the CoP to understand the inherent intricacies in modeling in the Web-based environment, the evidences of CoP that need to be traced automatically in a slick manner such that the evidence-capturing process is unobtrusive, and the technologies needed to embrace a full adoption of Web-based support system for the community learning framework.

15.1 Introduction

Community of Practice (CoP) introduced by Etienne Wenger as published in his book titled *Community of Practice: Learning, Identity and Meaning*. The essence

S.M.F.D. Syed Mustapha
Faculty of Information Technology, Universiti Tun Abdul Razak, 16-5 Jalan SS6/12 Kelana Jaya, 47301 Petaling Jaya, Selangor, Malaysia
e-mail: Syedmalek@unitar.edu.my

J.T. Yao (ed.), *Web-Based Support Systems*, Advanced Information and Knowledge Processing, DOI 10.1007/978-1-84882-628-1_15,

of COP is concurrent with the theories of social constructivism [10] and social learning [19]. Social constructivism describes the evolvement of social constructs in terms of concepts and practice that are peculiar to a specific group or society. Social learning theory emphasizes that the learning process takes place through an observation of other people's behaviors and actions. CoP is a marriage of the duo for two reasons. First, the formation of CoP members is a natural process of reacting to similar interest and is not designated or deliberately established. For example, technicians at the Xerox machine company develop a group who share their experiences and tips or even jokes during their coffee breaks that eventually become clich, specifically to their members. They share social constructs such as stories, advice, best practices, and understanding from the continuous and exclusive relationships in a way that they are foreign to outsiders. Second, the CoP weights the learning of doing things from colleagues as the major contribution to learning in comparison to the traditional classroom training. A fresh graduate even with a high-class degree will need guidance and understanding of local practice in order to perform his tasks. This can only be achieved through observing the predecessors or colleagues who provide the training through informal supervision and discussion. This concept is in accordance to the social learning theory where learning through observing is the tenet of the learning process.

The theory of CoP has been well accepted in major corporations worldwide as the mechanism of stimulating knowledge sharing and social engagement within the communities [3, 18, 21]. CoP experts are consulted to determine the presence of the CoP in the organization where observations, interviews, meetings, and questionnaires are performed to elicit the implicit values of the CoP practices. However, the CoP experts are rare in supply, not available at all times, and costly. The values of CoP practices in an organization can change through the years and there is a need of continuous monitoring of its presence. Considering the high demand of CoP experts and consultants and shortages in assessing them, it is imperative to build a community learning framework running on the Web-based platform where the community activities take place. The community learning framework supports the knowledge sharing, knowledge transfer, and community activities that are prescribed by CoP such that its presence can be detected and analyzed automatically. The Web-based community learning has the potential to function in a similar manner as the CoP experts in detecting and analyzing the CoP values. A web-based system is known to be able to support multifarious community activities (i.e., learning, knowledge sharing, and knowledge management) such as net meeting, e-mails, instant messenger, P2P applications, social networking, VoIP, and others in a way that these Web technologies can be tiered as a layer on the community learning framework [4]. There has been an active research area looking at the Web-based design considering multidisciplinary issues [26–28]. Sustainability of communities is essential in CoP and the discussion on utilizing a Web-based information system as supporting tool has been addressed [2, 30]. Nevertheless, building the Web-based community learning framework is not a straightforward task since community learning involves detecting personal interest, group motivation, social trend, social cohesiveness, resource sharing, vision sharing, and others which may be implicit and not directly derived [31]. It is the interest of research to address these challenging issues as to whether Web-based systems can fully support in deriving CoP values that are allusive. The following sections will discuss further CoP and its characteristics, the community learning framework, the Web-based approach in building the community learning framework, and the integrated schema of the entire system.

15.2 Community of Practice (CoP) Characteristics

Readers are encouraged to refer to particular books for excellent reviews on CoP [13, 25]. In this section, the main CoP characteristics are summarized and the features relevant to building the Web-based system as the supporting tools are discussed. There are eight main CoP characteristics, which are the community structure, learning through reification and participation, negotiation of meaning, learning as temporal process, boundary objects, participation (mutual engagement, joint enterprise and shared repertoire), and identity.

Community structure in CoP is informal and its formation is inadvertent. The people in this structure gather based on common interests, work problems, goals, attractions, or visions. The members are not assigned from the top organizational structure and they cut across various divisions in the organization or inter-organization. The group types are not predetermined as they emerge in a serendipitous manner. The members are not fixed to a group and fluid movement between groups is possible. The members build their own contact and social network beyond the organizational formal and traditional business circles. An engineer from a consultant company in Cape Town may engage in making a technical reference with his counterpart in an other engineering firm in Bangalore who had similar experience in dealing with a similar problem. Both of them are not in the same organization or formal business alliance or same task assignment, but they both share common problems. The community members build strong ties between themselves from frequent visits or interactions. CoP extends the definition of the community member to include unanimated objects that play role in situated learning. For example, a teacher uses teaching aids such as interactive courseware to teach his students or an engineer refers to a design plan to communicate with his team members.

Learning through participation and reification fortifies the community social bonding as the underpinning requisite for learning [15]. The community gets together into a deep relationship and intensive discussion such that the participation reveals a reification outcome. This excludes those social group members who frequently gather for long hours for leisure and amusement. The community generates an understanding within itself from the repetitive discussions and meetings so that the ideas, thought, and agreements are reified with some useful objects. For example, a group of technicians who share an emergent problem that is not prescribed in the troubleshooting manuals have identified a list of irregular problems and the solution steps. A group of experts who convene to find the best practice and remedial solution to a common epidemic problem can be categorized under learning through participation and reification. The process of writing rules and procedures, designing plans and layouts building prototypes and models is the process of reification.

Negotiation of meaning is a repetitive process of redefining a subject matter through the community's activities [22]. For example, a group of engineers revisit their design plan and make alterations at each visit due to renewal of personal judgment such as interpretation, understanding, emotion, thinking, impression, belief, strategy, experience, and opinion. The design plan is the focused object of the subject matter for the engineering community in deriving meaning and negotiating through personal judgment among the members. The reformation of the focused object over a series of negotiations by the communities brings about the community's own regeneration. The focused object of the subject matter can be in a tangible form such as documents, architectural model, network design, guidelines or an intangible form such as terminologies, definitions, concepts, ideologies, or arguments. The relationship between two community members can be redefined and reestablished over several interactions and meetings. For example, the understanding and impression toward someone's colleagues can be improved or degenerated over the time of social engagement.

Learning as temporal describes the evolution of the community learning process of the learning object for a given period of time. The community impresses its own cognitive input through criticism, suggestion, recommendation, amendment, modification, planning, redesigning or upgradation that enforces the content development of the learning object. The community stays together over a period of time in engaging the learning process that is in accordance to the historical account of the learning object. For example, in an organization, the policies and regulations are established after the process of brainstorming, discussions, and meetings. During these sessions, there are developments of ideas, thoughts, and opinions that elaborate certain arguments of the possibilities/impossibilities, pros/cons, and boons/banes. The period of learning can be definite or infinite but both leave the traces of the community learning process. The meeting minutes are traceable learning objects as it consists of the attendees, decisions, motions, and the time series of the meetings.

Boundary objects are the community-focused objects that the community interprets and construes during the community's engagements [23]. The objects can appear in concrete or abstract forms and the meanings are shaped by the practice and adoption of certain communities or individuals. Each community establishes a common ground in understanding that is specific to the community. For example, the CSI officer may conceive phone bills as the source of information in understanding a person's contact list while the phone bills are perceived as the source that contains the valid address of a person by the loan institution. The community with a stronghold relationship meets to a certain converging point of translating and interpreting the boundary objects in a consistent manner.

Boundary encounters are generations of ideas, concepts, practices, objects, and artifacts that emerge from members of different communities. In an organization, different members of communities could be the representatives of departments or unit functions that work together on similar projects. New insights may crop up due to multiplicity and diversity in perspective, beliefs, practices, and constraints. For example, establishing the policy for staff leave for operation and office workers will require the consideration of two distinct office cultures.

Participation in the community is characterized in three forms – mutual engagement, joint enterprise, and shared repertoire. Mutual engagement is when the community is engaged in an agreed community activity. The members in the community settle in a mutual form of activities and these can happen in the cases of collaborative design, problem solving, treasure hunt, brainstorming, or game playing where each of them plays the role of his or her own identity without bias. That means there is no indication of domination by certain quarter using one's authority or position in the organization. The participation at the same level and is mutually treated with the same capacity of interest and involvement. For example, each member of an engineering team who collaboratively designs a car has an equal significant role and indispensability regardless of the ranking. Joint enterprise is another participatory characteristic in CoP where the members work together as an enterprise with jointly consensual accountability. It has a strong link with mutual engagement, where sense-making, coordination, cooperative, responsibility, ownership, and interactivity take place. Sense-making is necessary in working relationships to abridge the emotional misinterpretation and to enhance good sense in the community objectivity. Coordination ensures the harmonious relationship especially in dealing with differences and confrontation. Cooperative efforts conjoin the members of CoP into an autonomous solidarity while performing a task in achieving toward a specific goal or mission. Responsibility and ownership are interdependent community behaviors where the former ensures trustworthiness and commitment within oneself in delivering the community's demand while the latter takes possession of the responsibility endowed by the community's force. Joint enterprise requires a strong community link through interaction that results in a continuous social bonding among the members. Shared repertoire is the community's unique belonging of social products that emerge from the social practices and establishment. The social products can be common jokes, clichs, habits, working patterns, community preferences, and affective understanding among the community that are shared exclusively among. For example, the shared repertoire may differ by age group, clan, tribe, geographical location, organization, rank, social position, occupation, religion or race.

Identity is another important characteristic in CoP that describes his or her how an individual sets attributes in the community in terms of role, contribution, expertise, influence, interest, status, and disposition. The individual has to be identifiable by the community by these attributes and they are the traits that fit the community's practices and beliefs. In other words, the individual may portray attributes that are attuned to the social environment he or she is involved in.

In summary, the CoP involves the people in the community learning, the relationships that are established among them, the properties of the learning process, and the social products generated from the community learning process. The objective of the research work is to establish a framework running on Web-based system support that is able to sense the community learning process in an organization. In order to do that, there is a need to determine the CoP objects that are traceable in the framework which will be discussed in the following section [5].

15.3 CoP Objects in the Social Learning Framework

Social learning process as prescribed by CoP is not designed solely to be operated in a computational model. Therefore, the process may take place in a computing or non-computing environment. For example, Allen's serendipitous visit to Joanne's office for a brief greeting may not be captured in a computational form. However, Allen's e-mail messages apologizing for not attending Joanne's party can be detected to conclude that there is a relationship between the two. Depending on the level of the organization's adoption of the ICT, the social learning process can be implemented on ICT platform in two possible ways, namely as supporting tools to facilitate the execution of CoP process and as monitoring tool to trace the execution of the CoP process by detecting the existence of CoP objects [1, 5, 9]). The discussion on methodologies and approaches in capturing community online activities will be the basics to capturing CoP values [6, 17]. Since there is a proliferation of CoP initiative in the company and considering the organizational setting of companies in a globalized market, ICT tools are eminently used in the communication and collaborative work [8, 12]. For example, multinational companies (MNCs) that have workers globally rely on Internet-based applications for online meeting, video conferencing, document management system, task force scheduling, technical support, and resource management. Since the ICT trend has permeated many organizational cultures and practices, it is possible to look at building a computational model for the learning process.

In this section, we revisit the CoP characteristics and identify the CoP objects that are associated with the characteristics. The objects are evidences that are traceable to identify the CoP characteristics. Table 15.1 lists the CoP characteristics and the possible objects that can be used as determining factors.

The social learning framework being proposed assumes that the CoP community operates mainly on the computing environment that is supported by Web-based technology. Even though the current implementation of CoP may not be completely in a computational environment, the future undertaking will need a Web-based social learning framework as information technology has pervaded people's lifestyle and working culture. For that reason, it is significant to consider the novelty of building a Web-based social learning framework specifically to sense the existence of the CoP community based on the CoP objects. The following section discusses the Web-based technology and the framework that can be built for sensing the social learning.

15.4 Web-Based System for Sensing Social Learning Framework

The notion of building a full-fledged Web-based system for social learning is an enterprise move as some of the elements and characteristics of CoP are non-computable and have to be redefined as an alternative in order to be sensed

Table 15.1 CoP characteristics and CoP objects

CoP characteristics	CoP objects
Community structure	⇒ the community structure is volatile where the membership is not fixed ⇒ the hierarchy is not rigidly established (i.e., no member dominates the others in consistent order) ⇒ the members are formed due to common interest ⇒ there is a fluid movement between members of different groups ⇒ members are beyond the formally defined structure ⇒ strong ties between members
Learning through participation and reification	⇒ repetitive participation and deep relationship ⇒ reification as the result of community participatory ⇒ formation of unprecedented solutions or practices
Negotiation of meaning	⇒ evolutionary of community artifacts in the content due to repetitive visit ⇒ evolutionary of community cognitive process over a subject matter
Learning as temporal	⇒ process of learning that takes place over a period of time ⇒ process of learning involves a shift in the knowledge content
Boundary objects	⇒ learning objects that are shared by different group of communities ⇒ multiplicity in how the learning objects are used, understood and interpreted
Boundary encounters	⇒ occasional event where there are distinctive groups encounter ⇒ issuance of new learning objects from the group encounters
Mutual engagement	⇒ agreed community activity ⇒ same level of participation ⇒ mutual significance in contribution
Joint enterprise	⇒ sense-making, coordination, cooperative, responsibility, ownership and interactivity strong relationship and link among members
Shared repertoire	⇒ identifiable social products that are shareable within a community ⇒ identifiable unique social products of certain community group
Identity	⇒ community's recognition on an individual's identity ⇒ identification of an individual's identity

computationally. As mentioned in the earlier example, Allen's visit to Joanne's office can be computationally difficult to be sensed without further enhancement of the Web-based architecture with sensors network or location-based system.

Therefore, a full-fledged Web-based implementation has to introduce the alternative practices that are not common to the traditional CoP. The following subsections describe the architecture that can sense the characteristics of the CoP elements.

15.4.1 Community Structure

Social presence in the Web-based environment has been explored in the past decade through various modes of interaction [7, 11, 14, 24]. The community in CoP emphasizes that the structure should possess the following properties: volatility of the membership in the community, temporal domination in the community participation hierarchy, existence of common interest of certain problems and issues, and fluid movement between members of different groups.

15.4.1.1 Volatility of the Membership

The CoP community comprises of many independent subgroups and they are differentiated by the common interest of each group. Unlike in the formal organizational setting where staffs are appointed in a fixed manner to the department or units based on their job description, the members in the CoP subgroups are volatile, so they are able to crossover to participate in the other subgroups or move back and forth comfortably and freely. Figure 15.1 shows a few members (M1 ⋯ M4) in the CoP community who traverse in different community group at different time points (T1 ⋯ T3). The member's profile comprises attributes that represent his or her unique identity. The profiles are essential for tracing the movement of the members in traversing throughout the available community subgroups. The profiles store

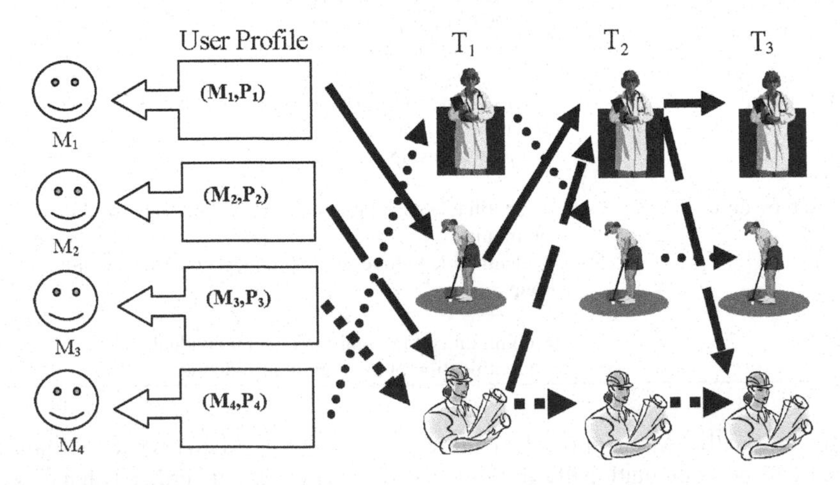

Fig. 15.1 Membership volatility

the information about the user's interest that can be used to recognize the group similarities based on the common interest [20].

15.4.1.2 Temporal Domination in the Community Participation Hierarchy

The Web capabilities determine the experiences of the online community and the real-life community. The Web-based online community experiences have grown from purely text-based communication to graphics, video, audio, and 3D virtual reality. Web-based social network tools and collaborative work applications are auxiliary add-ons to support the community's participatory activities. The complex experiences pose a greater challenge to the Web-based technology in determining the participating behavior of the community. Therefore, the Web-based architecture in identifying the community participation has to envisage the potential incoming technological advances for the Web-based community.

The CoP object for temporal domination in the community participation hierarchy requires us to examine the domination of certain individuals against the interval between time points. The domination should be temporal, in such manner that no one stays in the top hierarchy at all times. The domination in the community is a complex issue such that it is being perceived as domination not only in the discourse but also in the thinking, psychology, culture, and religion. The domination can be categorized in two phases, which are the process of domination and the result of domination. The former measures the activities that take place during the process of domination, while the latter measures the end result of the activities. The aspect of the end result that can be measured in the Web-based environment is not within the scope of CoP object being discussed here. The end result requires a determination of the subject matter of a propaganda that has been inflicted onto a certain group and the effectiveness of the infliction. In the context of CoP, the interest is to look at the domination pattern throughout the participatory activities of the communities, specifically that the domination period is transitional to different groups. The domination of the CoP framework focuses on the community discourse and they are analyzed based on the following aspects:

(a) Number of participation – the membership is measured with the number of times an individual participates in the group. For example, the number of postings in the community forum.
(b) Frequency of participation – the frequency is the ratio between an individual's number of participation against the total number of participation of the community. The frequency is measured on the local community performance.
(c) Focus topics in dominancy – the individual is considered to be dominant if the participation focuses on a certain topic throughout the dominancy.
(d) Temporal dominancy – the period of the dominancy is ephemeral.
(e) Referrals proficiency – an individual who is frequently made as point of reference for a particular subject.

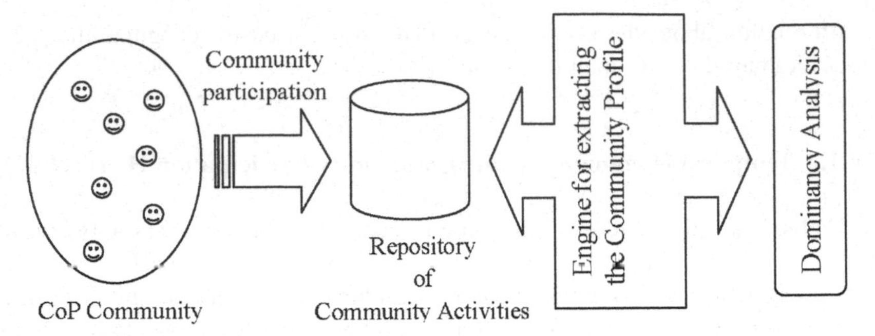

Fig. 15.2 Dominancy analysis process

Figure 15.2 shows the process of analyzing the dominancy from the repository of the community activities. In the text-based communication activity, the repository consists of the logging record, the posted messages, the responses from other members in the community toward a posted message, and the timestamp for each discussion. Some works on modeling the shared community knowledge on the Web space have been discussed [16]. The engine for extracting this information runs periodically to produce the analysis of the dominancy.

15.4.1.3 Existence of Common Interest

A CoP is formed from the de facto members within or beyond the organizational boundary. The essence of the CoP members is the common interest that exists among the members as that is the originating cause of the formation. In respect to this view, it is believed that the common interest of a subgroup in a CoP is unique to each other and at the same time it allows multiple membership of each individual. The common interest is an explicit expression from the members and appears in several forms: activity, communication, and relationship. The activity reflects mutual involvement by the common members over a period of time in a similar type of engagement. Some examples where common interest is exhibited in activities are playing online games or participating in online collaborative design software [29]. The communication indicates the common interest by the topics, issues, language styles, and terminologies used for a particular domain. The relationship describes the establishment of interconnectivity among members over a period of time.

15.4.1.4 Common Interest – Activity

In a control environment, the Web-based activities of the community can be identified as shown in Figure 15.3. The activity of the user is categorized by the Web-based applications provided by the organization. In a nonorganizational environment, the identification process is more challenging since the network is

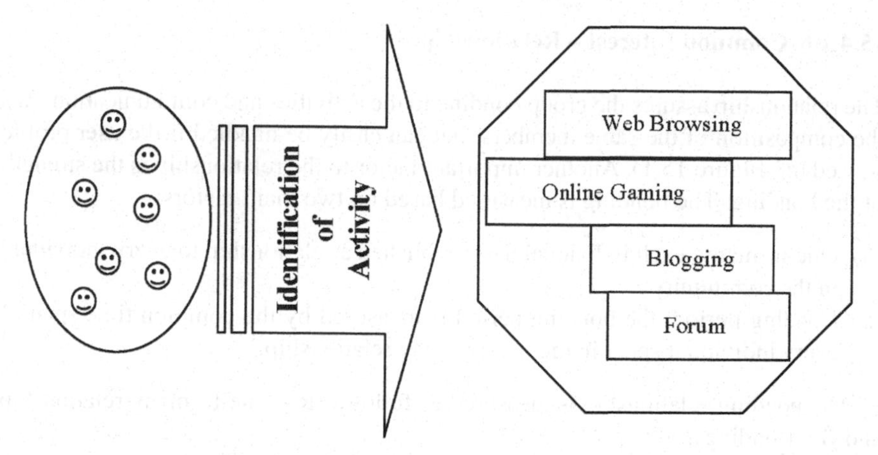

Fig. 15.3 Identification of Web-based activities

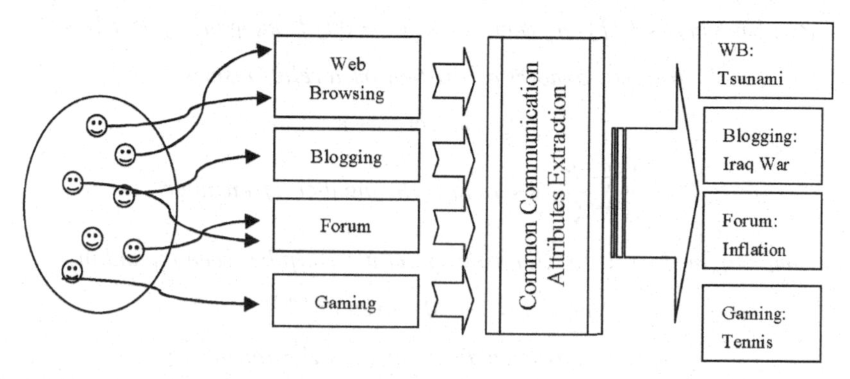

Fig. 15.4 Extracting the common communication attributes

privately protected. A user may be involved in more than one activity at different time durations. The next step to identification of activity is the communication.

15.4.1.5 Common Interest – Communication

Communication is the subsequent step to the activity identification. The communication has attributes that describe the context of the interaction. For example, the Web-browsing activity visits a specific Web portal that reveals the main issue narrated by the text description where the main subject can be extracted using information retrieval (IR) techniques. The IR techniques for extracting the communication attributes are possible for other social media such as blogs, forums, instant messaging, wikis or chat room. The output to the extraction is shown in the right-most column in Figure 15.4.

15.4.1.6 Common Interest – Relationship

The relationship assures the group bonding to the activities and communications has the composition of the same members that can easily be handled if the user profile is used (cf. Figure 15.1). Another important factor to the relationship is the strength of the bonding. The bonding is measured based on two main factors:

(a) One-to-many: each individual has established a relationship to every individual in the community.
(b) Bonding period: the bonding period is measured by the common time span of every individual spent in the one-to-many relationship.

The bonding relationship is measured as follows (R – one-to-many relationship and B – Bonding):

$$\textit{Given } m \textit{ to be the community of } M = \{m_1 \cdots m_M\}$$

$$P_i = \{m_i X m_j \rightarrow 1 \; iff \;\; m_i \; \varphi \; m_j \,|\, m_i X m_j \rightarrow 0 \; iff \;\; m_i \; \overline{\varphi} \; m_j \;\; for \; all \; m_i\}$$

$$\textit{where } \varphi \textit{ denotes an existent of a relationship}$$

$$Q_i = m_i X m_j \;\; \rightarrow 1$$

$$R = \frac{1}{M} \sum_{i}^{m} \frac{P_i}{Q_i} \quad \textit{where } M \textit{ is the number of member}$$

$$m_i^t = t_i(m_i, m_j) \;\; \textit{is the duration of relationship between } m_i \textit{ and } m_j$$

$$\forall m_i \in i = 1 \cdots M, \; \textit{given } m_k^{t=\max}$$

$$\textit{is the maximum duration for all relationship}$$

$$B = \sum_{j=1}^{M} \sum_{i=1}^{m} \frac{m_i^t}{m_k^{t=\max}}$$

15.4.1.7 Fluid Movement Between Groups

The membership that is fixed by the individual interest rather than the formal task assignment creates fluidity in the movement of members between the groups. This factor is essential to differentiate the traditional static and formal membership against the CoP membership.

The members are recognized by the user profile (cf. Figure 15.1) and the groups are distinct community channels (of different interests) that separate the members and at the same time allow them to switch channel. The community channel requires a genuine membership profile to be able to trace the movement accurately. Figure 15.5 shows the movement of the community members at three different time points (T1, T2, T3).

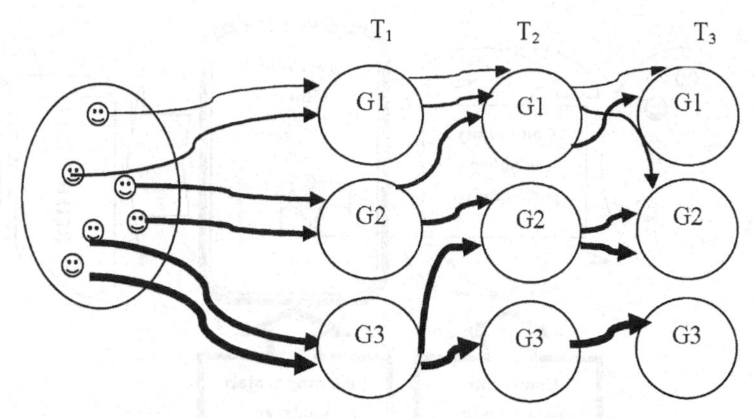

Fig. 15.5 Fluid Movement of Members Between Groups

15.4.2 Learning Through Participation and Reification

The community members in CoP involve in repetitive participation and jive into deep relationships. Repetitive participation affirms the involvement and interest of the member, while deep relationship ensures severity of the involvement. Despite having said that CoP members are open opportunity based on volunteer interest, their participation should be reified with formation of unprecedented solutions or practices. Participation is not solely based on the number of visits to the online forum or postings to the message threads, but also on the context of the involvement. For example, if the participation is on text-based forum, the text analysis on the discourse shall reveal the context of subject and persistency of the individual in dwelling the subject. The collective effort of all the members in the community will form the mesh network with certain depth and density. The reification is produced from the collective effort of members as a result of repetitive participation. The reification exhibits the gradual development of the learning object during the community learning process. Figure 15.6 shows the process of repetitive participation of the community in building the learning objects (left side) and the reification as the final outcome (right side). The learning objects are evolutionary where the contents grow after the iterative participatory cycles. For example, meeting minutes or software applications are progressive learning objects where the progression can be recorded and traced. For example, the process of shaping and formulating a departmental policy is done in a progressive manner through a series of meetings and brainstorming. The Learning Object Analyzer examines the progression based on the chronological information, the members' participation threads, and the context of the subjects.

The learning object analyzer performs the following tasks:

(a) Content analysis – learning objects are in computer-accessible format where the content can be recognized or annotated using artificial intelligence and information-retrieval techniques.

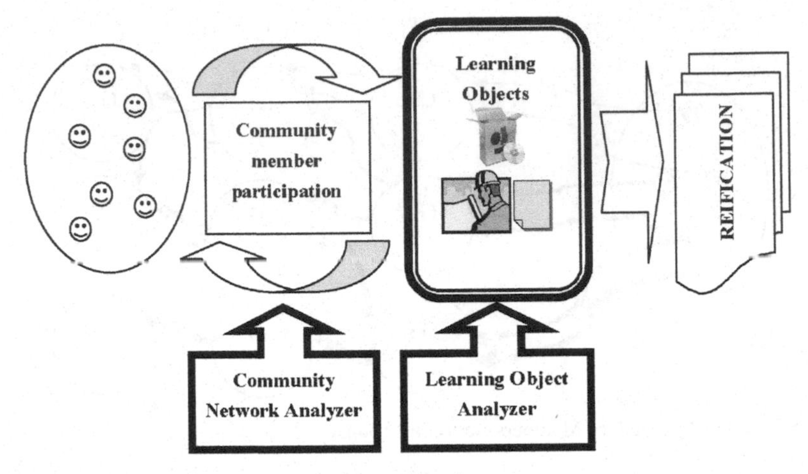

Fig. 15.6 Reification process of community participation

(b) Chronicle analysis – the reification substantiates that the learning objects have undergone the process of development, which involves modification, alteration, deletion, insertion, and refinement through community participation. The process captures the chronological information alongside the content development details.
(c) Unanimity analysis – the ultimate product of the learning object is to reach a consensus and unanimity among the community. The ultimate product development is reached once the Learning Object Analyzer detects the cessation of learning object development and community involvement for a period of time.

The community network analyzer performs the following tasks:

(a) Contribution density – the contribution of each member is measured against the maximum number of contributions of any member in the same group, denoted as γ. The contribution density is the number of members whose number of participation exceeds the minimal threshold of the required number of participation. The threshold is commonly set as half the γ value.
(b) Response density – the contribution of each member is measure against the maximum of responses of any member in the same group, denoted as β. The response density is the number of members whose number of received responses exceeds the minimal threshold of the required number of received responses. The threshold is commonly set as half of the β value.
(c) Member density – the density of a member, A, is measured by the number of members that a member A is connected to against the number of members available in the network. The connectivity is described as a member who performs certain participatory activity with another member such as replies to a posted message, collaborates on the similar learning object, or makes a reference to the other member.

15.4.3 Negotiation of Meaning

Negotiation of meaning describes the process of learning among the community members through textual or verbal discourse, textual comments on specific learning object such as policies, rules, political statement or news, or collaborative design and development on tangible products of a team. The process of learning involves unraveling the confusion, clarifying ambiguity, answering query, developing new thought, proposing alternative, opposing idea, contradicting belief, unleashing tension, agreeing to a solution, resolving conflict and supporting argument. In short, the negotiation of meaning is not only about the evolutionary of community artifacts in the content due to repetitive visit but also the evolutionary of community cognitive process over a subject matter.

The negotiation of meaning in the Web-based environment takes place in the control environment where a user's profile is known, his or her action can be traced, and his or her communication can be captured. The basic component that is essential for identifying the negotiation of meaning that has taken place is shown in Figure 15.7. In Figure 15.7a, the negotiated object is an object that is chosen among the learning

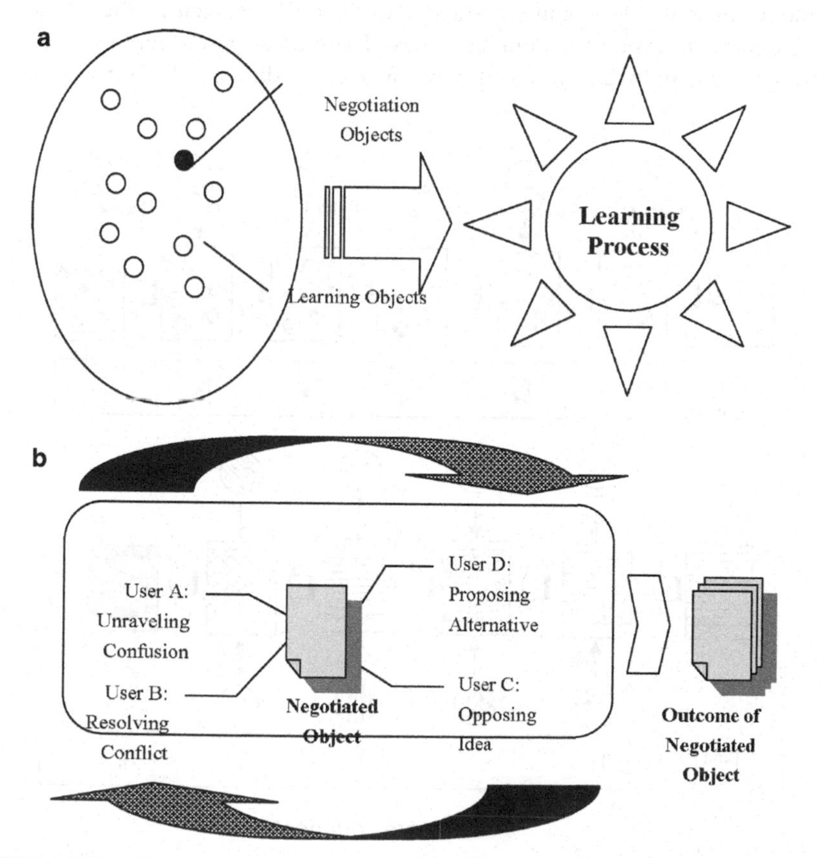

Fig. 15.7 **(a)** Negotiation of meaning process. **(b)** Negotiated object learning cycle

object. For example, if designing new academic regulation is the negotiated object, then the learning objects will be the university academic practices, university council meeting minutes, and comments from the community members. In Figure 15.7b, the learning object is the central object of negotiation among the communities where discussions, clarifications, objections, suggestions, appreciations, and other cognitive manipulation take place repetitively on the negotiated object. The users are members who are identified through their user profiles, their actions are captured in the log files, and their communications are transcribed in the textual format. The outcome of the negotiation of meaning is a formation of an agreed and unprecedented learning object.

15.4.4 Learning as Temporal

The learning takes place over a period of time where the progress development of learning objects can be annotated with the time series. The content of the learning objects are analyzed and differentiated for every timescale. The learning objects are the members of the community and also the artifacts such as the documents, blogs, forums, or even programming codes. Therefore, the learning evidences are measured on the individuals as well as the artifacts as shown in Figure 15.8.

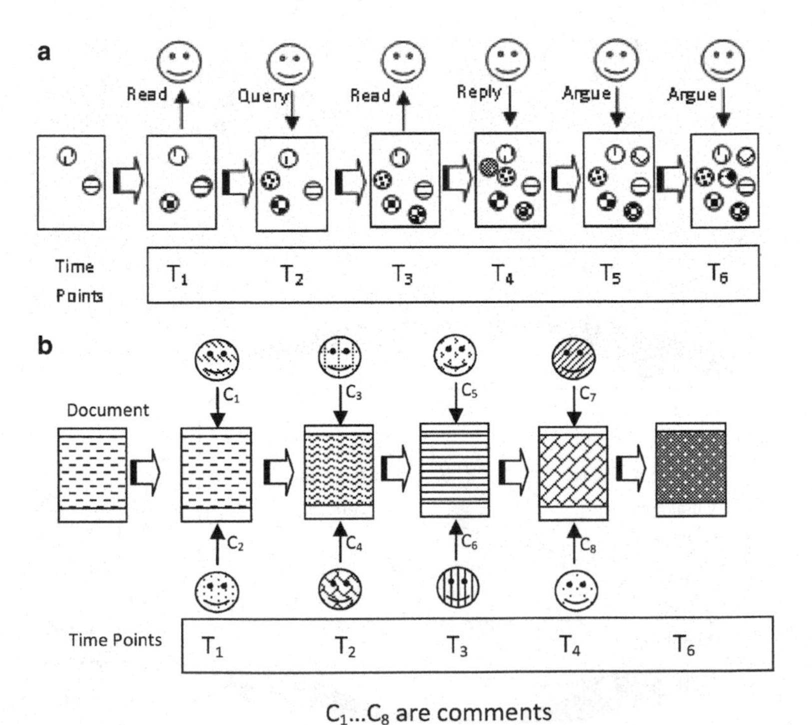

Fig. 15.8 Temporal learning of **(a)** a member in a community and **(b)** artifact of a learning object

Figure 15.8a shows the learning progress of an individual who begins and ends in different modes of learning (i.e., reading query ... arguing) at different time points. The learning objects grow over the time as they are contributed by different members. The purported member begins with reading and enhances the understanding about the domain subject before leveling up the status to be able to argue. The progression of the person in learning is captured against the timescale. Figure 15.8b shows the development of a document where various community members contribute comments (C1 ⋯ C8) in shaping up the document. The document undergoes a progressive change in the content that is captured periodically by the system. The example given in Figure 15.8b is an example for a document and the same concept could be applied to other learning objects and resource repositories. The growth in resource can be measured over the time period in order to detect the learning process.

15.4.5 Boundary Objects and Boundary Encounters

The boundary objects are those that are referenced by two or more distinctive group with different interpretations within the CoP. The subsequent effect to that is the issuance of new learning objects from the group encounters so-called boundary encounters. In the Web-based environment, the great challenge is to gauge the physical encounters of the members, physical sharing of resources, and the newborn learning objects that are not in computational form. Other forms of boundary objects could be the vocabulary in the language, the documents and social media interfacing the two disparate CoP groups, and programmable and editable resources such as architect design plan or programming source code. In the Web-based environment, these boundary objects are traceable items as they can be computationally processed by sophisticated algorithms and techniques.

Figure 15.9 shows two sets of community members who possess independent learning object repositories. The object matching engine searches from both repositories for the common learning objects based on the similar file names, media format, and content. The matching learning objects have associations with the community forum where the communities frequently refer to in the forum (i.e., text-based discussion and hyperlinked to the learning objects). The discussion in the forum reflects how the community perceived, argued, clarified, and explained a learning object, which leads to different types of interpretation. The boundary objects are found when the interpretations (shown in Figure 15.9 – interpretation A and B) of the common learning objects are different made by the two communities. The boundary object identification engine compares the interpretation and determines the difference. For text-based forum, the text-processing techniques can be used for this purpose.

The indication of boundary encounters is the issuance of a newborn learning object as a result of two distinct groups encounter. In the Web-based environment, the boundary encounters can occur when the community channels such as forums

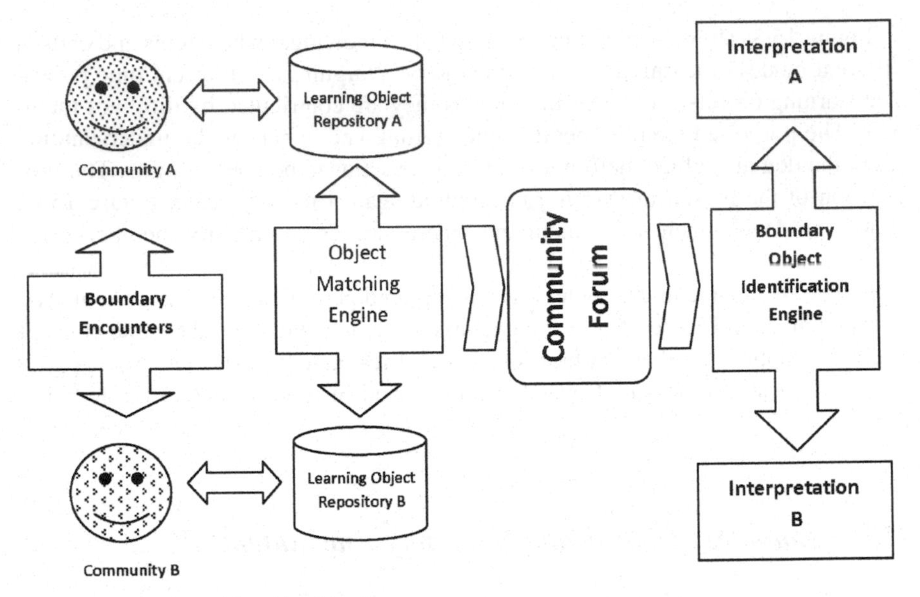

Fig. 15.9 Boundary object identification process

or blogs of distinct CoP groups subscribe to the same learning object as shown in Figure 15.10. This encounters result in multiple interpretations as they are from different community groups.

15.4.6 Mutual Engagement, Joint Enterprise, and Shared Repertoire

The CoP participatory is highly relevant with the structure of the community (cf. community structure) as discussed earlier. The mutual engagement that requires the Web-based system to support agreed community activity can be done through the *volatility of the membership*, mutual significance in contribution through *temporal domination in the community participation hierarchy*, mutual interest of every members through *existence of common interest*, and same level of participation through *fluid movement between groups* (refer to the previous section on the discussion detail). The joint enterprise is a community's perception and consensus about one's undertaking in the social responsibility such as sense-making, coordination, cooperative, responsibility, ownership, interactivity, and relationship. In a Web-based environment, these can be gauged through the community's comments and public polling. Figure 15.11 shows how a member participates in a joint enterprise.

The community channel is a platform for communication among CoP members, where expressions can be exhibited and articulated by labeling their postings. In the

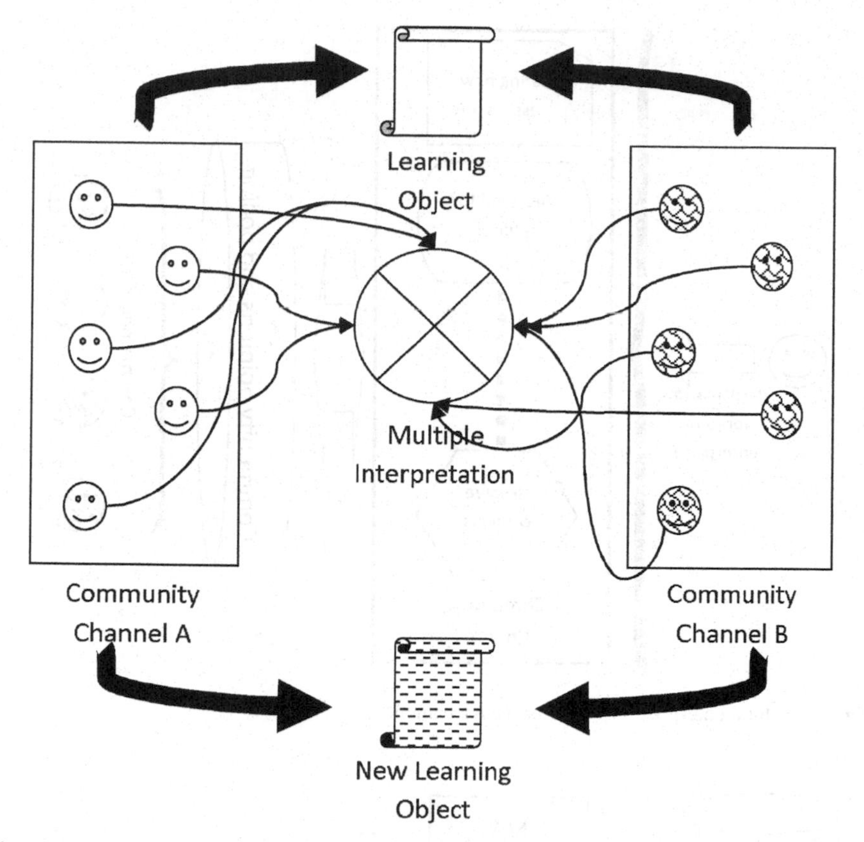

Fig. 15.10 Boundary encounters in issuing new learning object

Web-based environment, the individual member demonstrates sense-making through certain responses in the community channel such as clarifying ambiguities, resolving conflicts, or explaining queries. Community channels can be designed to track the actions taken by the members on the Web. Certain tasks can be offered online openly to members and the takers can be articulated as being cooperative to the community. The takers then coordinate with other members toward achieving the goals of the assignment. This process involves discussing, planning, and negotiating that reflects the coordination activities. The completion of the assigned task to the assignee can be marked to implicate that the responsibility and ownership has taken place. The community's polling and opinions regarding an individual are measures of strong relationship and degree of interactivity among the community members.

Shared repertoire of the CoP members' are shared in their communication platform. The attributes of the properties are that they are common to the specific group and rare in the others.

Figure 15.12 demonstrates two blocks of community repositories where key features that describe the most common words, jokes, clichs, documents, and references

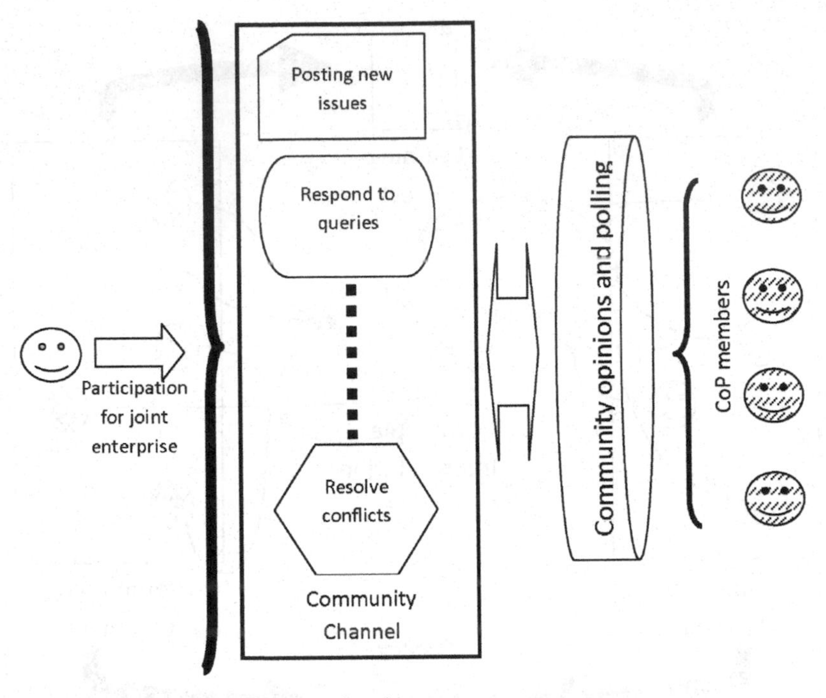

Fig. 15.11 Joint enterprise in Web-based environment

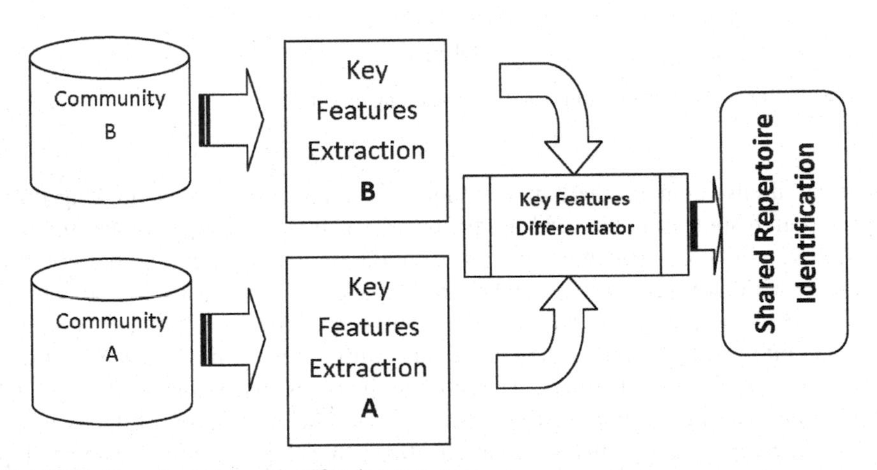

Fig. 15.12 Shared repertoire identification process

to the community are extracted. The two sets of extracted key features from the respective repositories are compared to determine the differentiation. The key features that are unique are identified as the shared repertoire for the specific community.

15.4.7 Identity

Each of the CoP members carries a community-recognizable identity by the practices exhibited to the community. The identity is self-disposition where the individual is unanimously categorized by the others given a long time duration of engagement in the community activities. The informal setting in CoP environment does not limit the categorization of an individual to the professions but also personal attributes. In a profession categorization, an individual can be known as expert in a certain job description or technical field while the personal attributes describe one's sense of humor, temperament, and other social behavior. The categorization for both types of identity of an individual is derived from the status quo of the community. The reason is that in CoP orientation, there is no formal acknowledgment of an individual's identity that is determined by an organization in comparison to the formal organizational setting. The community's perception and utilization are the basis in categorizing the identity of the individual. In the Web-based environment, the identification for the two categories is shown in Figure 15.13.

The community channels such as the blogs, forums, online discussions, or collaborative design tools are the communication platform that allows the community members to initiate dialogue or to provide feedback on various topics or domains. The various types of responses and interactions are the source of information for categorizing the identity of the individual in the community. The domains are areas where the member can choose and partake in such that the interests and expertise on those domains can be drawn from analysis. The community contribution analysis measures one's familiarity on the area based on the responses and interactions that are labeled at various degrees. Posting a query for clarification or a problem for a

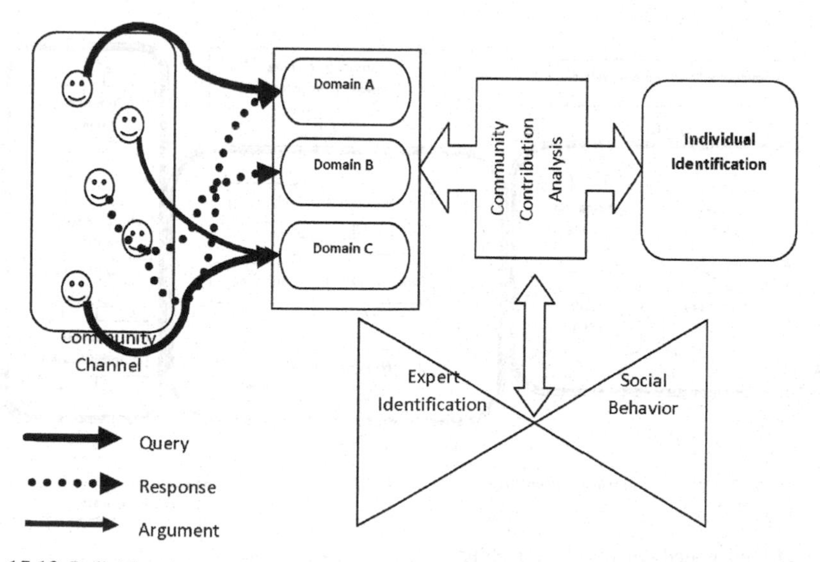

Fig. 15.13 Individual identification from community contribution

solution ascribes ones to be the lowest degree of expert compares to one's postings in providing solution or arguments. The analysis can be extended to factor in the community's judges on the solutions provided by the members as reliable and valid where the quantification is used for expert identification. That means a member who has a high polling rate in having a reliable solution is favored by the community to be an expert in the particular domain. The communication platform also allows the members to exhibit social expressions for arguments such as disagree, agree, support, hatred, uncertainty and others through iconic or labeled buttons that can be used to annotate the posted messages. The social expressions are reactions to the posted messages of other members and the mesh of responses forms the social behavior network of an individual and the community. A member can be characterized as active or passive in group participation, aggressive in quelling arguments, vigorous in creating new issues, or versatile in skills and knowledge among the members in the community.

15.5 Integrated Schema of the Entire System

The previous discussions on various parts of the CoP sensing framework can be viewed in an integrated schema as shown in Figure 15.14. The Web-based CoP sensing system is coordinated by the four types of management systems. The Community Management System builds the user profiling repository and records the community activities in the repository. It traces the movements and actions of each individual that are essential information feeds to the other management systems. Community Learning Management System focuses on the learning aspects of the

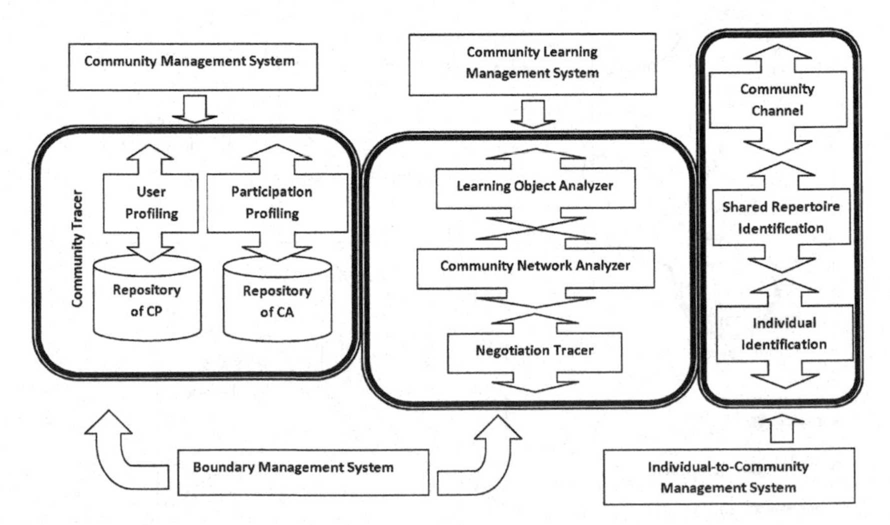

Fig. 15.14 Integrated schema of the Web-based CoP sensing system

community in managing the learning objects, the social network among the learning community members and analyzes the negotiation of meaning among the members. Individual-to-Community Management System manages the individualities of the members by identifying the commonness among them through shared repertoire checking, community channel involvement, and individual identification in informal organizational setting. Finally, the Boundary Management System computes the differences and similarities among the distinctive groups in order to determine the encountered boundaries among the groups. The Boundary Management System accesses the data and information from both the individual/community profile and community learning process.

15.6 Conclusion

The CoP has received serious attention and extensively adopted by multinational organizations. The geographical separation of their worldwide branches has made the implementation of CoP a challenging task where the social learning has to take place in computer-assisted environment. Traditionally, the CoP experts can be physically present to observe and analyze the social learning practices in an organization that exists in single location. The cost of CoP implementation can be exorbitant by the fast growing number of physical offices and branches that an organization has designated for. Considering the needs to monitor that the social learning activities do take place continuously in an organization, the CoP experts have to be contracted periodically for this task that is not affordable for any organization. The Community of Practice Sensing Framework is a computer-supported CoP that is designed to capture automatically the CoP values using the Web-based platform. The CoP characteristics are identified and the values are defined such that they can be sensed and measured by computational model. The discussion in this chapter reviews each component of the framework and highlights the state of the art of the Web-based technologies that can be used to support the social learning sensing framework.

References

1. Alani, H., Dasmahapatra, S., O'Hara, K., and Shadbolt, N. (2003). Identifying Communities of Practice through Ontology Network Analysis. *IEEE Intelligent Systems*, 18 (2), 18–25.
2. Connery, A., and Hasan, H. (2005). Social and Commercial Sustainability of Regional Web-based Communities. *Journal of Web-based Communities,* 1 (3), 246–261.
3. Da Lio, E., Fraboni, L., and Leo, T. (2005). TWiki-based facilitation in a newly formed academic community of practice. *WikiSym '05* (pp. 85–111). San Diego, USA: ACM.
4. Das, S., and Krishnan, M. (Nov 10–13, 1999). Effective Use Of Web-Based Communication Tools In A Team-Oriented, Project-Based, Multi-Disciplinary Course. *29th ASEE/IEEE Frontiers in Education Conference.* San Juan, Puerto Rico.

5. Davies, J., Duke, A., and Sure, Y. (2003). OntoShare: A Knowledge Management Environment for Virtual Communities of Practice. *K-CAP 03* (pp. 20–27). Sanibel Island, Florida, USA: Oct 23-25.
6. Donath, J., Karahalios, K., and Viegas, F. (1999). Visualizing Conversation. *Proceedings of the 32nd Annual Hawaii International Conference on System Sciences,* 4(4), p. 9. Maui, Hawaii.
7. Duvall, J. B., Powell, M., Hod, E., and Ellis, M. (2007). Text Messaging to Improve Social Presence in Online Learning. *EDUCAUSE Quarterly: The IT Practitioner's Journal,* 30 (3), 24–28.
8. Hildreth, P., Kimble, C., and Wright, P. (2000). Communities of Practice in Distributed International Environment. *Journal of Knowledge Management,* 27–38.
9. Hoadley, C., and Kilner, P. (2005). Using technology to transform communities of practice into knowledge-building communities. *ACM SIGGROUP Bulletin Special issue on online learning communities,* 25(1) January 2005, pp. 31–40.
10. Kafai, Y., and Resnick, M. (1996). *Constructionism in practice: Designing, thinking, and learning in a digital world.* Mahwah, NJ: Lawrence Erlbaum Associates.
11. Karel, K., Kirschner, P., Jochems, W., and Hans van, B. (2004). Determining Sociability, Social Space, and Social Presence in (A)synchronous Collaborative Groups. *CyberPsychology & Behavior,* 7 (2).
12. Kimble, C., and Ribeiro, R. (2008). Identifying 'Hidden' Communities of Practice within Electronic Networks: Some Preliminary Premises. *Proceedings of 13th UKAIS Conference.* Bournemouth.
13. Lave, J. (1991). *Situated Learning: Legitimate Peripheral Participation.* Cambridge University Press.
14. Lee, K. M., and Nass, C. (2003). Designing social presence of social actors in human computer interaction. *Conference on Human Factors in Computing Systems archive,* (pp. 289–296). Ft. Lauderdale, Florida, USA.
15. Llinares, S. (2003). Participation and Reification in Learning to Teach: The Role of Knowledge and Beliefs. In E. P. GILAH C. LEDER (Ed.), *Beliefs: A Hidden Variable in Mathematics Education?* (Vol. 31, pp. 195–209). Springer Netherlands.
16. Mao, Q., Zhan, Y., Wang, J., Xie, Z., and Song, S. (2005). The Shared Knowledge Space Model in Web-based Cooperative Learning Coalition. *The 9th International Conference on Computer Supported Cooperative Work in Design Proceedings,* (pp. 1152–1157). UK.
17. May, M., George, S., and Prévôt, P. (2006). A Web-based System for Observing and Analyzing Computer Mediated Communication. *IEEE/WIC/ACM International Conference on Web Intelligence, 2006 (WI 2006),* (pp. 983-986). 18–22 Dec, Hong Kong.
18. Mojta, D. (2002). Building a Community of Practice at the Help Desk. *SIGUCCS 02,* November 20–23, 2002, (pp. 204–205). Providence, Rhode Island, USA.
19. Ormrod, J. (1999). *Human learning.* Upper Saddle River, NJ: Prentice-Hall.
20. Paal, S., Brcker, L., and Borowski, M. (2006). Supporting On Demand Collaboration in Web-Based Communities. *Proceedings of the 17th International Conference on Database and Expert Systems Applications (DEXA'06),* (pp. 293–298).
21. Rohde, M., Klamma, R., and Wulf, V. (2005). Establishing communities of practice among students and start-up companies. *Proceedings of th 2005 conference on Computer support for collaborative learning: learning 2005: the next 10 years!* (pp. 514–519). Taipei, Taiwan: International Society of the Learning Sciences .
22. Shekary, M., and Tahririan, M. (2006). Negotiation of Meaning and Noticing in Text-Based Online Chat. *The Modern Language Journal,* 90 (4), 557–573.
23. Star, S., and Griesemer, J. (1989). Institutional Ecology,Translations and Boundary Objects: Amateurs and Professionals in Berkeley's Museum of Vertebrate Zoology. *Social Studies of Science,* 19 (4), 387–420.
24. Ubon, A., and Kimble, C. (2004). Exploring Social Presence in Asynchronous Text-Based Online Learning Communities (OLCS). *Proceedings of the 5th International Conference on Information Communication Technologies in Education.* Samos, Greece.
25. Wenger, E. (1998). *Communities of Practice: Learning, Meaning and Identity.* Cambridge: Cambridge University Press.

26. Yao, J. (2008). An Introduction to Web-based Support Systems. *Journal of Intelligent Systems,* 17 (1–13), 267–281.
27. Yao, J. (2005). Design of Web-based Support Systems. *8th International Conference on Computer Science and Informatics (CSI),* July 21–26, (pp. 349–352). Salt Lake City, USA.
28. Yao, J. (2005). On Web-based Support Systems. *Proceedings of the 2nd Indian International Conference on Artificial Intelligence,* (pp. 2589–2600). Pune, India.
29. Yao, J., Kim, D., and Herbert, J. (2007). Supporting Online Learning with Games. In B. V. Dasarathy (Ed.), *Proceedings of SPIE Vol. 6570, Data Mining, Intrusion Detection, Information Assurance, and Data Networks Security 2007,* (pp. 1–11). Orlando, Florida, USA.
30. Yao, J., Liu, W., Fan, L., Yao, Y., and Yang, X. (2006). Supporting Sustainable Communities with Web-Based Information Systems. *Journal of Environmental Informatics,* 7 (2), 84–94.
31. Zhou, W., Wang, H., Wu, H., and Ji, X. (2007). Research on Web-based Community and Real-world Community in IT Education for Public. *Proceedings of the Second International Conference on Innovative Computing, Information and Control* (pp. 29–32). Washington DC: IEEE Computer Society.

Chapter 16
Designing a Successful Bidding Strategy Using Fuzzy Sets and Agent Attitudes

Jun Ma and Madhu Lata Goyal

Abstract To be successful in a multi-attribute auction, agents must be capable of adapting to continuously changing bidding price. This chapter presents a novel fuzzy attitude-based bidding strategy (FA-Bid), which employs dual assessment technique, i.e., assessment of multiple attributes of the goods as well as assessment of agents' attitude (eagerness) to procure an item in automated auction. The assessment of attributes adapts the fuzzy sets technique to handle uncertainty of the bidding process as well use heuristic rules to determine the attitude of bidding agents in simulated auctions to procure goods. The overall assessment is used to determine a price range based on current bid, which finally selects the best one as the new bid.

16.1 Introduction

The emergence of electronic marketplaces has dramatically increased the opportunities for online auctions (e.g., eBay and Amazon). Intelligent agent technology [1, 3, 9, 10] provides a powerful mechanism to address the complex problems of dynamic pricing in automated auctions. The agents can use different auction mechanisms (English, Dutch, Vickery, etc.) for procurement of goods or reaching an agreement between agents. The agent makes decisions on behalf of the consumer and endeavors to guarantee the delivery of items according to the buyer's preferences. In these auctions buyers are faced with the difficult task of deciding the amount to bid in order to get the desired item matching their preferences. For this reason, the formalization of bidding mechanism has received a great deal of attention from the agent community for the past decade. These software agents should be smart enough to bargain a favorable deal for the user. In order to be called an

J. Ma and M.L. Goyal
Faculty of Engineering and Information Technology University of Technology, Sydney
PO Box 123 Broadway NSW 2007 Australia
e-mail: Junm@it.uts.edu.au; madhu@it.uts.edu.au

J.T. Yao (ed.), *Web-Based Support Systems*, Advanced Information and Knowledge Processing, DOI 10.1007/978-1-84882-628-1_16,

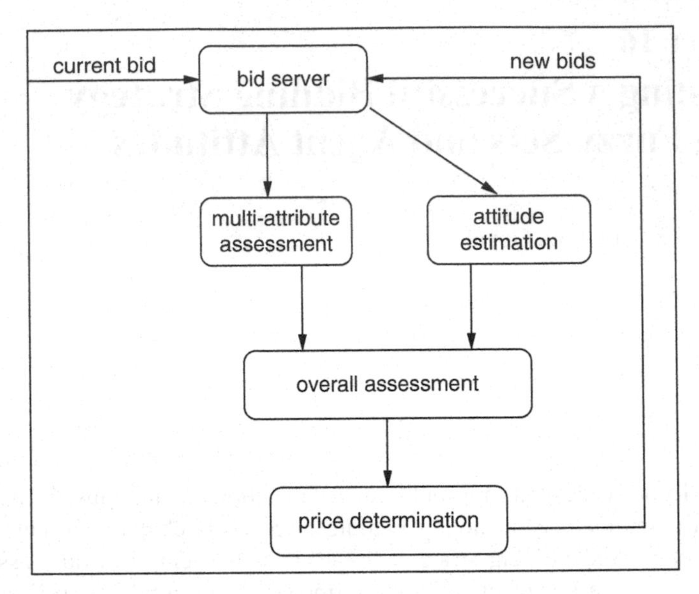

Fig. 16.1 A Fuzzy Bidding Strategy (FA-Bid) model

"intelligent agent," the software must satisfy several criteria like autonomy, temporal continuity, communication, and cooperation. To this end, a number of researchers [14, 15, 17, 19, 21, 26] have reported different frameworks that help an autonomous agent to tackle the problem of bidding in auctions. Currently, no single implementation satisfies all the criteria, but there are several promising results for bargaining intelligent agents.

In this chapter, a fuzzy bidding strategy (FA-Bid) is designed in an automated auction based on the dual assessment of multiple attributes of items as well as agents' attitude on bidding item. To quantify attitudes and to deal with uncertainty of attribute assessment fuzzy sets technique is applied in the presented strategy. The basic procedure of the strategy is shown in Figure 16.1. The remainder of the chapter is organized as below. First, related works are reviewed in Section 16.2. Then, the detail of the presented strategy is illustrated in Section 16.3. Finally, conclusions are discussed in Section 16.4.

16.2 Related Works

There have been several approaches from computer scientists for developing intelligent software methods and protocols for automated negotiation. In particular, the first trading agent competition (TAC) was held in Boston in July 2000 [12]. TAC agents acted as simulated travel agents and had to procure goods for their clients in different types of auctions, bidding against autonomous agents. Priest et al. [20]

proposed an algorithm design for agents who participate in multiple simultaneous English auctions. The algorithm proposes a coordination mechanism to be used in an environment where all the auctions terminate simultaneously, and a learning method to tackle auctions that terminate at different times. Byde et al. [3] presented another decision theoretic framework that an autonomous agent can use to bid effectively across multiple auctions with various protocols. The framework uses an approximation function that provides an estimate of the expected utility of participating in the set of future auctions and it can be employed to purchase single or multiple items. Anthony et al. [1] proposed an approach for agents to bid for a single item in English, Dutch, and Vickrey auctions. The agent decides what to bid based on four parameters: (i) the remaining time; (ii) the number of remaining auctions; (iii) the desire for bargain; and (iv) the desperateness of the agent. The overall strategy is to combine these four tactics using a set of relative weights provided by the user. In an extension to this model [19], a genetic algorithm is used to search the effective strategies so that an agent can behave appropriately according to the assessment of its prevailing circumstances. The machine-learning techniques [21] are also used to obtain a model of the price dynamics based on the past data (e.g., the data in the seeding round) to predict the closing prices of the hotels in the future. It also uses mixed-integer linear programming to find the optimal allocation of the goods.

Considering the uncertainty that exists in auctions, the fuzzy techniques are used to manage an agent's interactions. Faratin et al. [7] used fuzzy similarity to compute trade-offs among multiple attributes during bilateral negotiations in order to find a win–win solution for both parties. Kowalcyzk and Bui [13, 14] modeled the multi-issue negotiation process as a fuzzy constraint satisfaction problem. Their approach performs negotiation on individual solutions one at a time. During negotiation, an agent evaluates the offers, relaxes the preferences and constraints, and makes counteroffers to find an agreement for both parties. Luo et al. [15, 16] developed a fuzzy constraint-based framework for bilateral multi-issue negotiations in semi-competitive trading environments. The framework is expressed via two knowledge models: one for the seller agent and the other for the buyer agent. The seller agent's domain of knowledge consists of its multidimensional representation of products or services it offers. The buyer agent's domain of knowledge consists of the buyer's requirement/preference model (a prioritized fuzzy constraint problem) and the buyer's profile model (fuzzy truth propositions). The buyer and seller agents exchange offers and counteroffers, with additional constraints revealed or existing constraints being relaxed. Finally, a solution is found if there is one.

Different researchers have also provided alternatives to fuzzy reasoning for coping with the uncertainties in bidding. For example, the possibility-based approach [18] has been used to perform multi-agent reasoning under uncertainty for bilateral negotiations where uncertainties due to the lack of knowledge about other agents, behaviors are modeled by possibility distributions. The Bayesian learning method has also been used to model multi-issue negotiation in a sequential decision-making model. In [27], a Bayesian network is used to update the knowledge and belief each agent has about the environment and other agents, and offers and counteroffers between agents during bilateral negotiations are generated based on Bayesian probabilities.

Attitude is a learned predisposition to respond in a consistently favorable or unfavorable manner with respect to a given object [8]. In other words, the attitude is a preparation in advance of the actual response, constituting an important determinant of the ensuing behavior. In AI, the fundamental notions to generate the desirable behaviors of the agents often include goals, beliefs, intentions, and commitments. Goal is a subset of states, and belief is a proposition that is held as true by an agent. Bratman [2] addresses the problem of defining the nature of intentions. Crucial to his argument is the subtle distinction between doing something intentionally and intending to do something. Cohen and Levesque [4], on the other hand, developed a logic in which intention is defined as a commitment to act in a certain mental state of believing throughout what one is doing. Thus, to provide a definition of attitude that is concrete enough for computational purposes, we model attitudes using goals, beliefs, intentions, and commitments. From the Fishbeins [8] definition it is clear that when an attitude is adopted, an agent has to exhibit an appropriate behavior (predisposition means to behave in a particular way). The exhibited behavior is based on a number of factors that depend on the nature of the dynamic world. Once an agent chooses to adopt an attitude, he or she strives to maintain this attitude, until a situation arrives where the agent may choose to drop his or her current attitude toward the object and adopt a new attitude toward the same object. Thus, an agent's attitude toward an object refers to persistent degree of commitment toward achieving one or several goals associated with the object, which gives rise to an overall favorable or unfavorable behavior with regard to that object.

16.3 A Fuzzy Bidding Strategy (FA-Bid)

16.3.1 Basic Scenario

In this section, we will present an automated bidding strategy using fuzzy sets and attitudes. Our strategy is discussed based on the following scenario.

1) Suppose a travel agent wants to book tickets for some clients. These clients have different preferences and requirements of the possible tickets. Assume that six factors are concerned in this situation, i.e., ticket price (c_0, e.g., from $800 to 2000), departure time (c_1, e.g., 18:00 PM, Wednesday), arrival time (c_2, e.g., 10:00 AM, Friday), number of stops (c_3, e.g., at most three), seat positions (c_4, e.g., near window, aisle, etc.), and travel season (c_5, e.g., off-peak season).
2) Suppose the identified perspective of an agent is summarized as below:
   - The agent prefers a cheaper ticket and agrees that the cheaper it is, the better.
   - The agent prefers to travel at the weekend rather than on a working day.
   - The agent prefers a no-stop travel.
   - The agent prefers an aisle seat rather than a window seat.
   - The agent prefers to travel during off-peak season rather than peak season.

- The agent thinks the flight price is the most important factor, secondly the travel season, and other factors are of same importance.

3) Based on the client's perspective, the agent evaluates a flight ticket using several terms (such as "very bad," "bad," "slightly bad," "acceptable," "fairly good," "good," "very good," etc.).

Using this scenario, an agent is required to bid for a flight ticket based on his or her attitude.

16.3.2 FA-Bid Overview

In an automated auction, an agent's bidding activity is influenced mainly by two aspects: (1) the attributes of goods and (2) the agent's attitude. Any agent prefers to make a bid for quality goods by adopting an appropriate bidding strategy. Considering the existence of uncertainty in a real auction situation, this chapter focuses on how to make a bid by using the agent's personal perspective.

To make a bid for a unit of goods, the agent should balance between his or her assessment of the goods and his or her attitude to win an auction. Generally speaking, an agent has stronger attitude to make a bid for a high-quality goods rather than a lower one. The attitude is mainly based on the assessment of the goods. Moreover, an agent's attitude is also influenced by the bids because price is the unique factor through which agents negotiate till they make a deal. To win an auction, an agent must balance the price (bid), the assessment of the goods, and the attitude to win a bid.

Roughly speaking, the bidding procedure runs as follows:

- Firstly, evaluation on each related attribute is determined.
- Then these evaluations are aggregated to form an overall assessment of the goods.
- Next, the attitude of the agent is determined.
- Overall assessment is conducted.
- Finally, a new bid is determined.

Since in a real situation uncertainty exists ubiquitously in expressing assessments, attitudes as well as their relationships with price, this chapter uses the fuzzy-set-based method to process uncertainty in assessment and attitude. First of all, a satisfactory degree of measure is used as the common universe of assessment, i.e., an assessment is treated as a fuzzy set on the satisfactory degree. Second, an attitude is expressed as a fuzzy set on the set of assessments, i.e., the assessment set is the universe of attitude.

In the following sections, details of the strategy are illustrated.

16.3.3 Attribute Evaluation

Attribute evaluation includes two kinds of processes. The first one is individual attribute assessment, and the second one is assessment aggregation. To implement attribute evaluation, three issues are concerned, i.e., attribute weights (relative importance) determination, assessment expression, and assessment aggregation.

16.3.3.1 Weights Determination

Weights indicate different preferences of an agent on each identified factor. In this chapter, we use the analytic hierarchy process (AHP) method [22] to determine the weight for each factor because this method has been validated in practice, although it may induce inner inconsistency. Suppose the obtained initial weight vector is W.

16.3.3.2 Assessment Expression

Since uncertain expressions are often used in a real situation, linguistic terms are used to express assessments [11,23]. These linguistic terms are illustrated by a fuzzy set. Moreover, the universe of these fuzzy sets is unified to a real interval [0,1], which means the satisfactory degree of the agent to a particular attribute. Therefore, all fuzzy sets have the same universe, which is convenient for aggregating assessments.

Suppose g_k $(k = 0, 1, \ldots, K)$ is the satisfactory degree of measure for attribute c_k. Then an agent's opinion of the goods in terms of attribute c_k is denoted by $g_k(u)$, where $u(\in U_k)$ is the real attribute value of attribute c_k, and U_k is the real universe for attribute c_k. For instance, departure time is an attribute for a flight ticket. The possible departing time in a day is from 0:00 to 23:59. For any time slot u, an agent may present a satisfactory degree, such as departing at 7:30 is with satisfactory degree 0.9 and departing at 3:00 is with 0.3.

In the following, let $A = \{a_1, \ldots, a_n\}$ be the set of used assessment terms which are fuzzy sets on satisfactory degree [0,1]. Then a numeric satisfactory degree is transformed to a linguistic term. Continuing the above example, suppose the assessment set is as shown in Figure 16.2. Notice that a_7 is the biggest membership degree for 0.9, the assessment for departing at 7:30 is a_6 by the maximum membership degree principle. Similarly, the assessment for 0.3 is a_2.

16.3.3.3 Assessments Aggregation

An aggregated assessment is the agent's overall opinion/preference of the goods in terms of multiple attributes. The change of an attribute value may lead to the alteration of an assessment. Instinctive natures of different attributes increase the difficulty and uncertainty for obtaining an overall assessment. Notice that an agent's

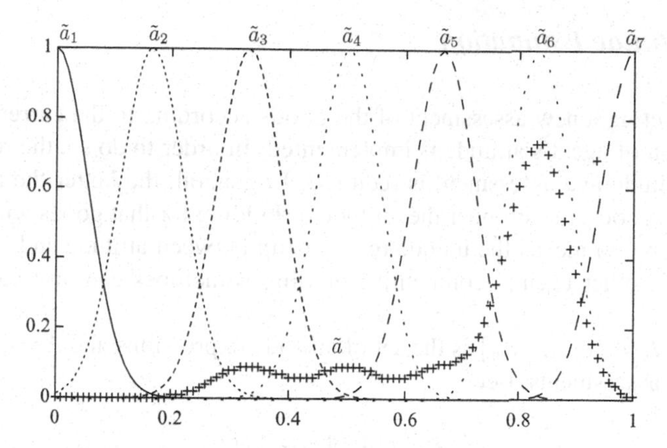

Fig. 16.2 Assessment aggregation

preference of an individual attribute can be expressed through the agent's satisfactory degree of that attribute. This chapter uses a satisfactory degree of measure as the common universe of assessment.

Based on the assessment on each individual attribute, an overall assessment can be obtained as follows. Suppose the individual assessments of all attributes are $v_0, v_1, \ldots, v_K$ and the weights of them are $w_0, w_1, \ldots, w_k$ respectively. Then an overall assessment is obtained by

$$a = \text{Agg}\{(v_0, w_0), (v_1, w_1), \ldots, (v_K, w_K)\} \tag{16.1}$$

where Agg is a selected aggregation method and $v_k \in A(k = 0, 1, \ldots, K)$ is the linguistic assessment on attribute c_k.

To get an overall assessment in terms of a set of criteria, an aggregation method Agg is applied. Some existing methods can be used here, such as OWA operator [24, 25], 2-tuples linguistic aggregation [5, 6, 11], and weighted-sum. For convenience, we use the weighted-sum-based method to obtain an overall assessment as follows.

First, we construct a fuzzy set $\tilde{a}$ on [0,1] through

$$\tilde{a}(u) = \sum_{k=0}^{K} w_k \cdot v_k(u), \qquad u \in [0, 1], \tag{16.2}$$

where $v_k(u)$ is the membership degree of u in v_k.

Next, we calculate the distance between $\tilde{a}$ and $a_i \in A$ by

$$d(\tilde{a}, a_i) = \int_0^1 |\tilde{a}_\lambda - a_{i\lambda}| \, d\lambda. \tag{16.3}$$

Finally, we select the nearest term(s) a to $\tilde{a}$ as the overall assessment.

For example, A has seven terms, namely, $a_1, a_2, \ldots, a_7$, as shown in Figure 16.2. Suppose $\tilde{a}$ is the obtained fuzzy set. By comparing the distances between $\tilde{a}$ and each element in A, we know that a_6 is the nearest item to $\tilde{a}$. Hence, a_6 will be taken as the overall assessment.

16.3.4 Attitude Estimation

After conducting a new assessment of the goods according to the current price p_c, an estimation of agent's attitude is implemented. In order to do so, the relationship between attitude and assessment is required. In general, the better the assessment on the given goods, the stronger the attitude of bidding for that goods will be. However, this is by no means the unique relationship between attitude and assessment. For instance, other agents' competitive bidding sometimes can also cause strong willingness.

Suppose $E = \{e_1, \ldots, e_m\}$ is the set of attitude expressions and $A = \{a_1, \ldots, a_n\}$ is the set of assessments. Let

$$r: \quad (a_i \Rightarrow e_j, \alpha_{ij}) \tag{16.4}$$

be a given rule from an agent, where $a_i \in A$, $e_j \in E$, and α_{ij} is the reliability degree of the rule. Such a rule depicts the approximate degree of an agent's attitude e_j to which the agent can win the bid under the assumption that the overall assessment is a_i. Furthermore, these rules can be treated as a set of fuzzy sets on A such that the membership degree in a fuzzy set f_j corresponding to eagerness e_j is α_{ij}. Obviously, f_j is an integration of rules $(a_i \Rightarrow e_j, \alpha_{ij})(i = 1, \ldots, n)$, which may be treated as an alias of e_j. Hence, the fuzzy set f_j is also called attitude in the following without other specification.

Based on the rules in R, an agent can estimate the possible attitude of the agent when it learns the current overall assessment. A set of fuzzy sets is obtained in the following way: suppose the overall assessment is a_c, then the attitude at the moment is determined by the maximum membership degree principle

$$e_c \in E(a_c) = \{e_j \in E | f_j(a_c) \geq f_i(a_c) \text{ if } i \neq j\}. \tag{16.5}$$

Notice that such determined e_c may not necessarily be unique. In the following, we call $E(a_c)$ the candidate attitude set under a_c.

Once the current attitude of the agent is determined, requirements for searching new bids can then be determined. The main requirements include identifying the required overall assessment and finding the candidate prices.

16.3.5 Overall Assessment

The prerequisite of overall assessment is the basic requirement of the goods such that the agent has the highest possibility to win a bid under the current attitude. To find the prerequisite of overall assessment, an order is firstly defined in E according to the strength of attitude. Without loss of generality, suppose $e_i < e_j$ if $i < j$. Therefore, it is possible to select the strongest element from $E(a_c)$. Then the strongest element in $E(a_c)$ is chosen as the first candidate attitude to determine the prerequisite of overall assessment.

Suppose a set of rules $\bar{R}$ is determined such that any $\bar{r} \in \bar{R}$ is of the form

$$\bar{r}: \quad (e_j \Rightarrow a_i, \bar{\alpha}_{ij}), \tag{16.6}$$

where $e_j \in E$, $a_i \in A$, and $\bar{\alpha}_{ij}$ is the reliability degree. These rules indicate to what extent an assessment a_i is obtained given an attitude e_j.

Based on the maximum membership degree principle, a set of candidate assessment is determined such that

$$A(e_c) = \{a_i \in A \,|\, \bar{f}_i(e_c) \geq \bar{f}_j(e_c) \text{ if } i \neq j\}, \tag{16.7}$$

where $\bar{f}_i$ is the counterpart to f_i. Each element a in $A(e_c)$ is called a candidate assessment under eagerness e_c.

16.3.6 Agent Price Determination

An agent's assessment demonstrates some expectation of the quality of the goods. As other criteria except the price are seldom changeable in an auction, this is regarded in terms of price.

Suppose $U_0 = [p_l, p_u]$ is the real range of price. A price range $U(a)$ corresponding to a candidate assessment a is a subset of U_0 such that for any $u \in U(a)$, the assessment based on u and W is a. Notice that an assessment is a fuzzy set on the satisfactory degree [0,1], which is the bridge between assessment and price; a price range is determined by the following steps.

Step 1: We divide the satisfactory degree [0,1] into n subsets $D_1, D_2, \ldots, D_n$ such that

$$a_i(d) \geq a_j(d) \tag{16.8}$$

for any $d \in D_i$ and $j \neq i$, i.e., element in D_i with biggest membership degree in a_i.

Step 2: For D_a corresponding to a candidate assessment a, we select a price in U_0 such that $g_a(u) \in D_a$. U_a is called a candidate bid set. Considering that the satisfactory degree is continuously changing with the price, we assume that U_a is an interval in U_0. Hence, let p_{la} and p_{ua} be the left and right boundary of U_a.

Thus, a candidate price range for assessment a is determined. For instance, Figure 16.3 indicates that the price range corresponding to assessment a_6 is \$900–1,000.

Suppose for any element in $A(e_c)$ we have obtained a corresponding candidate price range. Because a new bid should be higher than the present price p_c, a candidate price set for $A(e_c)$ is determined by

$$U_A = \{p_{li} \,|\, p_{li} > p_c, a_i \in A(e_c)\} \\ \cup \{p_{ui} \,|\, p_{ui} > p_c, a_i \in A(e_c)\}.$$

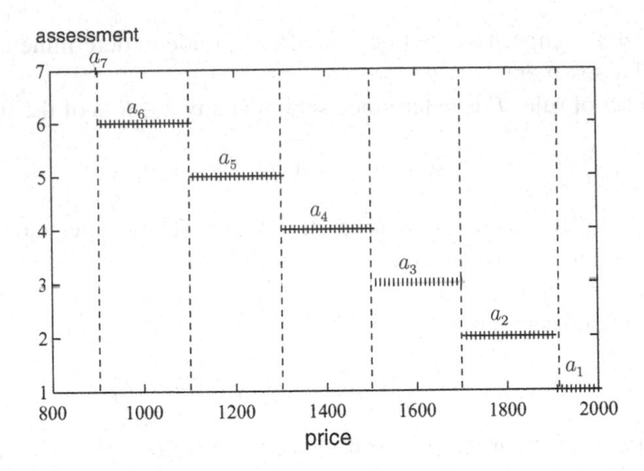

Fig. 16.3 Candidate price range

As it can be seen that the candidate price range may not exist under some assessments, in these cases, a weaker attitude is selected to repeat the candidate price determination process until a range is found or the attitude is weaker than an acceptable level.

Suppose U_A is a found price range, there must be a smallest element b in it. Then b is selected as the new bid. For example, in Figure 16.3, suppose the current flight price is $900, the prerequisite assessment is a_6, and the least increment is $50, then the new should be $950.

16.4 Conclusions

In this chapter, a novel fuzzy attitude-based bidding strategy (FA-Bid) is discussed. It was noticed that agents who adopt attitudes behave more flexibly and efficiently than agents without attitude, and adapt more easily to dynamic situations. Another unique idea presented is that to deal quantitatively the imprecision or uncertainty of multiple attributes of items that can be acquired in auctions, a fuzzy set technique is used. The fuzzy logic provides attitude-based agents resources in the decision-making process of the bidding agent. The bidding strategy also allows for flexible heuristics both for the overall gain and for individual attribute evaluations. It also explores the relationships between evaluations of different attributes using the analytic hierarchy process method [22].

There are a number of areas of further investigation. In the future we would further like to explore the development of strategies for multiple auctions. We would also like to compare our bidding techniques with other decision theoretic approaches to determine the relative strengths and weaknesses of these methods. Different strategies may perform well in some environments but may perform poorly in others. The numbers of strategies that can be employed are endless and the search space is huge. To address this issue, we intend to use learning techniques to obtain a model of the price dynamics based on the past data and to search for most successful strategies in predefined environments in an offline fashion.

References

1. Anthony, P., N.R.Jennings: Evolving bidding strategies for multiple auctions. In: Proceedings of 15th European Conference on Artificial Intelligence, pp. 187–192. Netherlands (2002)
2. Bratman, M.E.: Intentions, Plans and Practical Reason. Harvard University Press, Cambridge, MA (1987)
3. Byde, A., Priest, C., N.R.Jennings: Decision procedures for multiple auctions. In: Proceedings of the First International Joint Conference on Autonomous agents and Multiagent Systems, pp. 613–620. Bologana, Italy (2002)
4. Cohen, P.R., Levesque, H.J.: Teamwork. Cognitive Science and Artificial Intelligence **25**(4) (1991)
5. Delgado, M., Herrera, F., Herrera-Viedma, E.: Combining linguistic information in a distributed intelligent agent model for information gathering on the internet. In: P. Wang (ed.) Computing with words, pp. 251–276. John Wiley and Sons, Inc. (2001)
6. Delgado, M., Herrera, F., Herrera-Viedma, E., Verdegay, J.L., Vila, M.A.: Aggregation of linguistic information based on a symbolic approach. In: L. Zadeh, J. Kacpryzk (eds.) Computing with Words in Information/Intelligent Systems I. Foundations, pp. 428–440. Physica–Verlag (1999)
7. Faratin, P., Sierra, C., Jennings, N.: Negotiation decision functions for autonomous agents. International Journal of Robotics and Autonomous Systems **24**(3–4), 159–185 (1998)
8. Fishbein, M., Ajzen, I.: Belief, Attitude, Intention and Behaviour: An Introduction to theory and research. Addison-Wesley, Reading, MA, USA (1975)
9. Greenwald, A., Stone, P.: Autonomous bidding agents in the trading agent competition. IEEE Internet Computing pp. 52–60 (2001)
10. He, M., Leung, H.F., Jennings, N.R.: A fuzzy-logic based bidding strategy for autonomous agents in continuous double auctions. IEEE Transactions on Knowledge and data Engineering **15**(6), 1345–1363 (2003)
11. Herrera, F., Herrera-Viedma, E., Chiclana, F.: Multiperson decision–making based on multiplicative preference relations. European J. Operational Research **129**, 372–385 (2001)
12. Jennings, N., Faratin, P., Lomuscio, A., Parsons, S., Sierra, C., Wooldrige, M.: Automated negotiation: prospects, methods and challenges. Group Decision and Negotiation **10**(2), 199–215 (2001)
13. Kowalcyzk, R.: On negotiation as a distributed fuzzy constraint satisfaction problem. In: Proceedings DEXA e-Negotiation Workshop, pp. 631–637 (2000)
14. Kowalcyzk, R., Bui, V.: On fuzzy e-negotiation agents: Autonomous negotiation with incomplete and imprecise information. In: Proceedings Dexa e-Negotiation Workshop (2000)
15. Luo, X., Jennings, N., Shadbolt, N., Leung, H., Lee, J.: A fuzzy constraint based model for bilateral, multi-issue negotiation in semi-competitive environments. Artificial Intelligence **148**(1–2), 53–102 (2003)
16. Luo, X., Zhang, C., Jennings, N.: A hybrid model of sharing between fuzzy, uncertain and default reasoning models in multi-agent systems. International Journal of Uncertainty, Fuzziness Knowledge Based Systems **10**(4), 401–450 (2002)
17. Ma, H., Leung, H.F.: An adaptive attitude bidding strategy for agents in continuous double auctions. Electronic Commerce Research and Applications **6**, 383–398 (2007)
18. Matos, N., Sierra, C.: Evolutionary computing and negotiating agents. In: Agent Mediated Electronic Commerce, *Lecture Notes in Artificial Intelligence*, vol. 1571, pp. 126–150. Springer-Verlag, New York (1998)

19. P. Anthony, N.R. Jennings: Developing a bidding agent for multiple heterogeneous auctions. ACM transactions on Internet Technology **3**(3), 185–217 (2003)
20. Priest, C., Bartolini, C., I. Philips: Algorithm design for agents which participate in multiple simultaneous auctions. In: In Agent Mediated Electronic Commerce III (LNAI), pp. 139–154. Springer, Berlin, German (2001)
21. P. Stone, Littman, M., S. Singh, M. Kearns: Attac-2000: An adaptive autonomous bidding agent. Journal of Artificial Intelligence Research **15**, 189–206 (2001)
22. Saaty, T.: The Analytic Hierarchy Process. McGraw Hill, NY (1980)
23. Yager, R.R.: On ordered weighted averaging aggregation operators in multi-criteria decision making. IEEE Transactions on Systems, Man, and Cybernetics **18**(1), 183–190 (1988)
24. Yager, R.R.: Families of OWA operators. Fuzzy Sets and Systems **59**, 125–148 (1993)
25. Yager, R.R.: OWA aggregation over a continuous interval argument with applications to decision making. IEEE Transactions on Systems, Man, and Cybernetics–Part B: Cybernetics **34**(5), 1952–1963 (2004)
26. Yao, J.: An introduction to web-based support systems. Journal of Intelligent Systems **17**(1–3), 267–281 (2008)
27. Zeng, D., Sycara, K.: Bayesian learning in negotiation. International Journal Human Computer Studies **48**, 125–141 (1998)

Chapter 17
Design Scenarios for Web-Based Management of Online Information

Daryl H. Hepting and Timothy Maciag

Abstract The Internet enables access to more information, from a greater variety of perspectives and with greater immediacy, than ever before. A person may be interested in information to become more informed or to coordinate his or her local activities and place them into a larger, more global context. The challenge, as has been noted by many, is to sift through all the information to find what is relevant without becoming overwhelmed. Furthermore, the selected information must be put into an actionable form. The diversity of the Web has important consequences for the variety of ideas that are now available. While people once relied on newspaper editors to shape their view of the world, today's technology creates room for a more democratic approach. Today it is easy to pull news feeds from a variety of sources and aggregate them. It is less easy to push that information to a variety of channels. At a higher level, we might have the goal of collecting all the available information about a certain topic, on a daily basis. There are many new technologies available under the umbrella of Web 2.0, but it can be difficult to use them together for the management of online information. Web-based support for online communication management is the most appropriate choice to address the deficiencies apparent with current technologies. We consider the requirements and potential designs for such information management support, by following an example related to local food.

D.H. Hepting and T. Maciag
Department of Computer Science, University of Regina, 3737 Wascana Parkway, Regina, Saskatchewan, Canada
e-mail: dhh@cs.uregina.ca; maciagt@cs.uregina.ca

J.T. Yao (ed.), *Web-Based Support Systems*, Advanced Information and Knowledge Processing, DOI 10.1007/978-1-84882-628-1_17,

17.1 Introduction

Over the last several years there has been a shift in the way we communicate, as we have adopted and accepted the Web as a preferred communication medium and information source [1,3,25]. Online modes of communication are quite diverse [9,17], ranging from simple e-mail lists, where communication is conducted via personal e-mail clients; to online content management systems (CMSs), where communication is conducted through forums, blogs, newsgroups, wikis, and calendars; to social networking sites (e.g., Facebook (http://www.facebook.com) and MySpace (http://www.myspace.com)), where communication is conducted using available networking tools (status updates, events, private messages, public comments, and notifications of others' activities). Keeping track of what is going on within our communities has never been more difficult. Often we are required to shift our communication preferences and adapt to unfamiliar, potentially unsatisfactory modes of communication. New tools, and variations on tools, emerge with surprising rapidity, but many are still searching for problems to solve. From a human–computer interaction perspective, this approach is far from ideal [25]. Rather, as Norman [23] suggests, the focus should be on designing for activities and harmonizing that design to incorporate the inherent strengths of technology and end users.

Many of us, either by preference or necessity, have become accustomed to specific modes of communication. These modes of communication, which include e-mail, instant messaging (IM), and short message service (SMS), all have strengths and weaknesses, which means that it is ultimately preferable, if not required, for the user to be fluent in more than one mode. To be truly informed, we are often required to track several sources, with different modes of interaction or with different membership requirements. As such, *Death by Blogging* [26] can be seen as a real cause for concern. As well, the potential to increase the likelihood of information overload [14, 29] is increased as each source requires its user to access and then filter its content, some of which is duplicate and some of which may be considered spam [27]. Many barriers [13] can remain between these modes, including lack of standards and data portability. Some progress has been made with respect to system infrastructure standards, as seen in the vocabularies such as SIOC [5] (Semantically Interlinked Online Communities) and FOAF [21] (Friend of a Friend) with which it is possible to annotate our Webpages.

Managing our expanding online presences is becoming more of an issue since good end user support is not readily available. Although there presently exist Web-based support technologies that provide support in varying degrees, none of these technologies provides complete end user support and satisfaction. Consider the FriendFeed (http://friendfeed.com/) plugin for Facebook, Twitter (http://twitter.com/), and iGoogle (http://www.igoogle.ca) that provide end users with the capability to share updates to the content that they create across multiple platforms. Content authors may have difficultly in configuring the service for their own situation, because it requires the following process: register for the FriendFeed service and list which content should be shared; install the application on the particular platform where the content is to be shared; configure the platform application. Also, FriendFeed only allows content to be pushed to (and not pulled from) Facebook, for example. Even within these individual platforms, end users must still adapt to the interfaces provided by the platform applications that have not yet been standardized in any meaningful way. Given the lack of complete and satisfying support solutions, more focus on the end users and understanding their needs and preferences is required. For this goal, we describe the foundation of a Web-based support system (WSS) that would enable such user support.

The general aim of WSS is to assist end users in conducting activities on the Web [30]. In its most general sense, WSS could include support for many types of Web-based activities, including those common in Web-based business applications (e.g., advertising, online form submission), Web-based information dissemination and retrieval (e.g., searching, blogging, browsing), Web-based consumer decision-making support (e.g., purchasing, selling, browsing), and Web-based communication (e.g., IM, social networking), among others [30, 31]. For this chapter, we focus on WSS for Web-based communication. Specifically, the vision that guides our research is the development of a WSS that would allow people to effectively manage their incoming and outgoing online communication in a satisfying way that is not overwhelming.

For this goal, we describe a framework for WSS that is designed to support end users by enabling them to use their preferred mode of communication for all of their Web-based communicative activities. For example, a recommender system such as News@Hand [6] is a WSS that recommends news items based on user preferences. This system helps users gather information that they may wish to share with others. Yet, this sharing requires a lot of work, including manual editing and posting to other platforms. By allowing end users the freedom to choose their preferred mode of communication through this WSS, they would improve their level of satisfaction.

This chapter will discuss a scenario-based design process, and results thereof, used to examine how online communication management might be supported by a Web-based system. The rest of this chapter is organized as follows. Scenario-based design is introduced in Section 17.2. Section 17.3 applies scenario-based methods to examine design opportunities within the users' current workflows. Section 17.4 describes different current communications technologies and how they may help to address the identified opportunities. Section 17.5 describes designs through which information management support might be achieved. Finally, Section 17.6 provides a discussion of future directions for this work.

17.2 Scenario-Based Development

In 1994, Poltrock and Grudin [24] observed obstacles to interface design. Most designers reported that they gave some consideration to users, but few involved users directly. Poltrock and Grudin found that the principles of interactive system design, as proposed by Gould and Lewis [12], were found to be both obvious to and largely unheeded by system developers. Their four principles are:

(a) Early focus on users
(b) Early – and continual – user testing
(c) Iterative design
(d) Integrated design

Rosson and Carroll [28] popularized the idea of scenario-based design (SBD) as an approach to support better design of interactive systems. SBD is a user-centered process that involves the creation of stories, called scenarios, to illustrate and communicate about issues with current or future software. These stories relate the experiences of different stakeholders. Understanding and respecting those different perspectives is important to designing satisfying experiences. Each stakeholder group is embodied by an actor, which is an abstraction between a real user, or an amalgamation of several real users, and the audience discussing the system. It is important to understand the background, expectations, and preferences of the stakeholder groups and communicate it. We need to create real, common situations for actors in order to test the software system and explore any problems or opportunities for design. Likewise, an actor must be believable while performing his or her given tasks. There are more formal methods, such as the rational unified process (RUP) [18], but the advantage of SBD is that it communicates information in plain language that is therefore accessible to users.

Consider the following example: six blind men encounter an elephant. Each of them touches a different part of the elephant and expresses what the elephant is. Although they are touching the same elephant, each man's description is completely different from that of the others [11]. Here, each man's description presents a different scenario, all of which must be integrated to understand the elephant.

The SBD process begins by conducting a requirements analysis, which comprises a root concept, detailing the project vision, rationale, assumptions, and stakeholders, field studies to gather end user data, summaries that detail stakeholders' backgrounds, expectations, and preferences, and finally, problem scenarios which describe the end user's experiences. Each phase in the SBD process includes claims analysis, which highlights the positives and negatives of different features brought out in the scenarios. A useful lens for this analysis of scenario features is provided by Raskin's [25] two laws of interface design:

- The computer shall do no harm to the user's work; nor, through inaction, allow harm to come to the user's work.
- The user shall do no more work than is absolutely necessary.

The need for a strong focus on usability is reinforced in a recent study of the technology acceptance model (TAM), which relates the perceived usefulness and the perceived ease of use to the intention to adopt the technology. Chan and Teo [7] found empirical evidence that usability helps with acceptance, in many cases.

Based on the analysis of the existing practice, new designs and innovations are described through the writing of activity scenarios that describe various aspects of system functionality, to information and interaction scenarios that describe the detailed interactions between the technology and the end user. Low- and high-fidelity prototype designs are developed based on results found, which are then evaluated with end users. Based on user feedback, further insights are gained and areas requiring re-design are identified. In this chapter, we are focused on the analysis and design phases of the SBD process.

17.3 Understanding Design Opportunities

As presented in Section 17.1, the vision for this research is to develop a Web-based support system that will allow people to effectively manage their incoming and outgoing online communication. We begin the design process for the new WSS by seeking to understand how online communication is managed presently. Local food (or, more broadly, SOLE: sustainable organic local ethical food) is the subject area that grounds our discussion. It is an increasingly popular area of concern for many, and the *New Oxford American Dictionary* made the term locavore (defined as someone who seeks to eat only foods grown locally) its word of the year in 2007.

The current context is explored through the use of problem scenarios that describe real tasks done by real people. Those real people are represented by actors, who are representative of real stakeholders in the issue of online communication management.

- Activist users who are proactive in seeking out, creating, and disseminating content: represented by the actor Joe (introduced in Table 17.1).
- Largely passive consumers of information and content: represented by the actor Emily (introduced in Table 17.2).

For purposes of illustration, we consider the two stakeholder groups described below. Beyond this chapter, it may be appropriate to consider other stakeholder groups that would be handled in the same way as those we discuss here.

Given the different stakeholder groups that they represent, it is not surprising that Joe and Emily have different roles in their various social networks. Emily can connect the local food communities in the two towns in which she lives, but she is not an active participant in either. Joe, however, is at the center of his local food community and is responsible for keeping it active and informed. Because Joe is recognized, he has been invited to join some lists and he has joined others to keep himself better informed. In terms of social network analysis, Emily is a receiver who has the potential to act as a carrier or boundary spanner and Joe is a transmitter or central connector [20].

Social networks have become an important tool when considering personal knowledge networks, for example, so the relative position of our actors in their networks is important information. Information overload is a common complaint today, but part of this overload is caused by the need to filter duplicate information and to separate the "wheat from the chaff." Spam filters are now quite capable of removing

Table 17.1 Problem scenario that introduces Joe, the actor who represents activist users and central connectors in their social networks

Joe is an active local citizen who is technologically literate. He is committed to the cause of local food because of what he sees as its economic, environmental, health, and societal benefits. He is willing to stand up to be counted and take a leadership role on this issue, interest in which has been growing rapidly over the past few years. To promote his interest, Joe created an e-mail list of people interested in local food. To offer more features to his local food community, he decided to start a Web site devoted to local food. The Web site was more difficult to configure than he imagined, not all of his readers were comfortable moving away from e-mail, and he also got several comments about the site being boring and difficult to use. It was always his intention to fix the Web site, but since it seemed to require a lot of work, little progress has been made. In the meantime, as Facebook became more popular, he decided to create and maintain a group there. To keep information flowing to all these channels requires a lot of effort on his part, sometimes too much of it. Recently, he accepted an invitation to join a ning (http://www.ning.com) network for a different topic. He liked the features that he found there, but decided that it would be too much effort, with likely too much attrition within his group, to move all of his own network to such a service.

Situation feature	Possible pros (+) and cons (−)
E-mail list	+ Easy to set up and maintain – May be restricted to plain text – May be difficult to advertise Web site
Web site	+ Provides the opportunity for more features – Provides more targets for criticism + More easily discoverable, from a Web search – Difficult to set up and maintain – Unfamiliar technology for some users to adopt
Facebook	+ Easy to set up and maintain – Separate interface that requires more effort
Information on several outlets	+ More chance for people to see it – Much more work to keep everything current

junk mail from consideration, but there is another class of communication that is more difficult to manage, something that is now called bacn – not spam, but still something you don't want to see right now. This word was a runner-up to locavore for word of the year in 2007. These communications also raise ethical issues [27].

Armed with this background knowledge, we construct a number of problem scenarios involving our actors. These are found, along with corresponding claims analyses, in Tables 17.1–17.5.

To summarize the results from our problem scenario analyses, we see while e-mail is good as far as it goes, there may be much better communication technology available now. As Joe moves to embrace the best available technology for him to create and to share content, he must be aware that some members of his network will not want to leave their e-mail. He would like to welcome anyone to his network, regardless of technology, but the reality is that adding more communication modes adds a great deal of work. This situation is illustrated in Figure 17.1. Emily is able to do what she needs with her e-mail client, yet the work is becoming more burdensome. From both their perspectives, there is an understandable tendency to

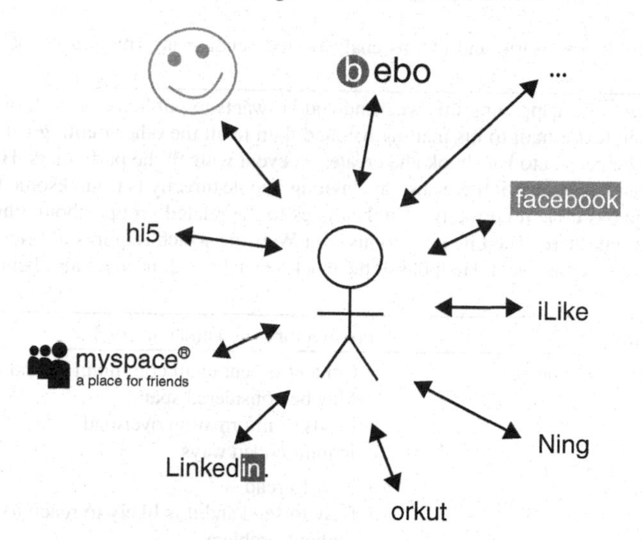

Fig. 17.1 An illustration of the problem scenarios (Tables 17.1–17.5): the user must interact separately with each information provider or service

Table 17.2 Problem scenario that introduces Emily, the actor who represents passive consumers of content and receivers (and possibly carriers) in their social networks

Emily is an undergraduate student of environmental sciences. She attends college in another part of the country but for the summer months she returns to her hometown, which is also where Joe lives. She has known of Joe for a while because she was an early subscriber to Joe's local food mailing list. One of her class projects involved local food and so she was able to make connections within her university town and she continues to be a subscriber to a few mailing lists there that are dedicated to local food. She likes getting information to keep herself informed about big issues and local events, both in her hometown and her university town. She is a very busy student so she prefers to remain on the periphery of these various groups, at least for now. Her e-mail client is configured to place her local food messages in a separate folder, so they do not clutter her inbox. She reads the messages when she has free time, or needs a change of pace. She is not concerned about any other, more advanced, communication options to get news about local food. In fact, she wonders whether the e-mail is too much. Her local food folder often fills up faster than she can read it and she gets bothered when she gets the same announcement from a few different sources. She wonders if she should drop her subscriptions to all but one of her lists, even though she realizes that she'd be missing out on some things.

Situation feature	Possible pros (+) and cons (–)
E-mail subscriber	+ Comfortable with e-mail technology – May be uncomfortable with other modes
Prefers periphery	+ Stays informed – Hard to get her involved in activities
E-mail client configuration	+ Able to manage messages, not overwhelmed – Easy to locate messages – Rules can sometimes be complex and work in unintended ways
Same items from different sources	– Hard to manage multiple subscriptions + Extra reminders about important items – May tune-out very important items – May be encouraged to unsubsribe

Table 17.3 Problem scenario and claims analysis: Joe sending information about an upcoming event

Joe learns of an event happening this weekend and he wants to publicize it to all of his contacts. He sends a plain text e-mail to his mailing list, and then to all the other mailing lists to which he belongs. Then he goes onto Facebook and creates an event with all the particulars. His group isn't sponsoring it but he attaches it there anyway. Inviting people directly is troublesome for him (he is never sure if he has done it correctly). He then goes to the related groups about which he knows and posts the event there. Then he goes to his own Web site which requires a different format of information to create the event. He looks at the clock when he is done and can't believe that it has taken so long.

Situation feature	Possible pros (+) and cons (−)
Publicize to all of his contacts	+ Content is sent to all communities and networks − May be considered spam − Leads to information overload in unintended ways
Plain text e-mail sent	+ Easy to read + Easy to send and it is likely to reach its destination without problem − Not easy to transfer plain text into calendar format, without retyping − May be encouraged to unsubsribe
Creating event on facebook	+ Facebook users have content in their preferred form − Must learn to use Facebook interfaces, which don't always match his own model
Sending invitations	+ Explicit notification may be appreciated − May be considered as spam
Posting to other websites	+ Increases visibility of event − Time-consuming − Must learn to use other Web sites' interfaces, which don't always match his own model

consolidate information in as few places as possible. Yet, this may not always be feasible, as illustrated in the United Kingdom after their recent food crises [8].

It was extremely difficult to get producer groups to form any consensus that allowed a common message to be delivered to consumers. Rather, each found it in their best interests to protect their niche market at the expense of others promoting similar products. This phenomenon can also be seen in Facebook with its proliferation of groups devoted to niche interests. It can be very difficult to find a common message to unite these groups because those who administer the groups seem to cherish the autonomy of their own group and the platform that it provides. Rather than trying to reduce the number of sources of information, no matter how noisy they may be, it may be a better approach to support the end user in managing this information. Fischer [10] considered this type of situation when describing the utility of weak input filters with strong output filters.

The two actors have some real issues with the technology that they use, which need to be addressed in future designs. Section 17.4 examines technological solutions that are presently available, in the context of our actors' needs.

Table 17.4 Problem scenario and claims analysis: Joe sending new content out to his social network

Joe is always interested in finding new, relevant information for his network. However, he doesn't always have time to search the Internet for new items and news sources pertaining to local food. He decides to create a Google news query for "local food." Instead of setting up an alert, he sees that he can read the RSS feed for this page. He has always preferred e-mail but he's embraced RSS because he is now able to read those within his e-mail client. He sees that a lot of interesting results have been returned and he wishes that he had more time to read them. He picks an interesting-looking story and it seems applicable so he sends across the URL along with some text from the story to his network.

Situation feature	Possible pros (+) and cons (−)
Using Google news	+ Aggregates a great deal of relevant content − Must place some trust in google regarding results
Reading RSS feed from e-mail client	+ Brings Web information to him inside his favoured e-mail client − RSS feed may still miss important results [?]
Sends content to network	+ Network benefits from content filtered by Joe − Network may rely too much on Joe for new content − Extra messages may cause some to unsubscribe − Time-consuming to reformat content before sending

Table 17.5 Problem scenario and claims analysis: Emily tunes her e-mail client rules

Because Emily spends summers at home and is away at school from fall to spring, she wants to be less involved in her other town while she is away from it. Unless it comes from Robert or Joe, the organizers of the local food networks in her university and home town, respectively. She would still like e-mail and other notifications, but not so intrusively. She does this with some more involved rules for her email client – which must be changed every fall and spring as she changes towns. She'd really like to have a way to have better control over the local food e-mail she gets, but resigns herself to her current setup for some time.

Situation feature	Possible pros (+) and cons (−)
Desires complex filtering	+ May lead to "perfect" environment for her − Knowledge and overhead to maintain − If unrealized, may unsubscribe completely

17.4 Current Technologies

In the larger context of Web 2.0, there are many interesting trends. Aggregators of different types are becoming more popular. There are increasing efforts toward interoperability and data portability. Each of the communication media types in use has different strengths and weaknesses. And within each type, there are a variety of options. Audiences are fragmented by delivery channels and also by the perspective of the authors. The expectation, or the hope, that people will use only a single social networking site, for example, is becoming outdated.

The capabilities of our online communication technologies continue to grow. For example, e-mail clients now readily support retrieval of e-mail from a variety of different addresses and service providers. Apple Mail version 3.5, for example, also incorporates RSS (rich site summary, or really simple syndication, or RDF site summary) feed reading. Many of the technologies fulfill some part of our requirements. Here, we consider them broadly according to the following general headings.

17.4.1 Input

What are the new items that should be presented to the network? User-generated content has become an important source of information. As Fischer describes, Web 2.0 has moved us from the expert editor to crowd-sourced information. Visiting digg.com (`http://www.digg.com`), for example, one gets an idea of what others deem important as the crowd exercises editorial control. In blogs, the authors exercise their own editorial control. If readers agree, they give those bloggers more authority in a variety of ways, including linking to the original posts. Blogs have infiltrated all areas of society: the *New York Times* operates several blogs, including dot earth (`http://www.dotearth.blogs.nytimes.com`). All these blogs contribute their own voices to the discussion, which allows for more democratic participation. With the increased participation, there is a growing need for filters [10], however they are applied. No matter what is done, even searching the Web with Google, for example, represents a choice of filter and what we see is shaped by those filters [22].

The wiki, a tool which allows the editing of Web pages on the Web, has become an important part of the online culture. One of the first major applications was the wikipedia (`http://www.wikipedia.org`), but it has been plagued by questions about the expertise of its possibly anonymous contributors. Citizendium (`http://www.citizendium.org`) is a response to wikipedia which requires all of its contributors to be known. In both wikipedia and citizendium, there is but one official version of an entry. Google's knol (`http://knol.google.com`), currently under development, is said to have a model that allows for multiple versions, or perspectives, on the same topic. Squidoo (`http://www.squidoo.com`) is a Web site that offers this basic model in a relaxed way. Users can create lenses about their favourite topics and the Web site provides different tools to allow inclusion of various interactive elements, like polls. Klein [16] has proposed a moderated wiki, as part of MIT's climate change collaboratorium. In that model, the last post does not have special favor. Groups work with moderators to define the structure of arguments so that once a point is located, it can be elaborated. The removal of the temporal component helps to balance the discussion. Yet all of these things are static. Rollyo (`http://rollyo.com`) might represent a better chance at a dynamic quality, since it is more of a live search. The others collect and/or represent the work you might do to research a particular topic. Citizendium clearly has experts involved: either they have credentials or they are pro-am [19] (expert amateur). With Rollyo, it is possible to list up to 25 URLs to search for a particular topic. However, when searching, the user receives a list with many duplicates. Therefore, while the concept is very nice, much effort would be expended to identify the unique results in such a list. Furthermore, repeating such a search daily would not always come up with new results within the top results. Bernstein and Zobel [2] do, however, describe a method for filtering duplicates. Tagging, and the creation of folksonomies [15], is a different way to handle this issue.

17.4.2 Output

There is a lack of integration between different media. It is easy to pull with RSS, and the FriendFeed (`http://friendfeed.com`) site is one example of this approach. At present, it is not easy to push content to a variety of sites, but SIOC may enable the easy repurposing of content, as described by Bojars et al. [4]. However, not all the Web sites may have this annotation. Furthermore, allowing transmission of complete content archives may adversely impact business models that enable current services to be offered in their current forms.

17.4.3 Portability

With respect to calendar events, the Internet Engineering Task Force (IETF) has created the iCalendar standard (published first as request for comment (RFC) 2445 – `http://tools.ietf.org/html/rfc2445`) to allow for the interchange of event information. Exactly how this will be realized in practice does not yet seem clear. Microformats (`http://microformats.org/`) may fill this need, but at present there are tools to output the format and none that accepts it as input. There is an event container in the SIOC vocabulary, but few details about its use. The Socializr site (`http://www.socializr.com`) is well regarded, and it provides the means to share or promote an event to some social network sites. The sharing is limited; for example, with Facebook one can only post to one's profile and not to groups. Sharing with a blog requires copying and pasting. There are pieces of interesting technology that can improve interoperability. For example, Blogmailr (`http://www.blogmailr.com`) provides a general interface to enable blog postings via e-mail. However, because it is a Web service, the asynchronous advantages of e-mail are lost to a degree: we cannot compose a blog post offline in an e-mail message to be sent when an Internet connection becomes available.

As Web-based communities evolve, the need to more adequately support the diversity of end users is becoming more apparent. Modes of communication technology are plentiful, but they are not always satisfactory. With new tools emerging at a rapid rate, to be on the cutting edge we must be flexible in adapting to new technology that places more strain on our cognitive abilities to communicate. Ideally, we should have more freedom of choice based on our individual needs. Web-based support structures could assist in such regards, enabling greater accessibility and higher portability of information sources.

17.5 Towards New Designs

The actors discussed in this examination are not people who seek detachment from the world around them but just better tools to help their interactions be as effective as possible. Some of the technologies described could enable an automatic solution to the problems we identified earlier, but this is too much. The augmentation of peoples' abilities and interests is what we seek.

In the process of transforming our problem scenarios into design scenarios, we have begun to examine the issues in terms of how to innovate system functionality, how to more effectively organize the information displayed to end users, and how to better develop the physical interactions between the end users and the technology. In this abbreviated SBD process, we examine potential design approaches at a macro level in order to keep a wide focus on the issues.

We envision a Web site where people log in to manage their online communications, which allows information to be both gathered and distributed. Based on the variety of problem scenarios created, we present two design scenarios (one for each actor) in Tables 17.6 and 17.7. Figure 17.2 provides a conceptual view of the image of this WSS.

17.6 Discussion

Scenario-based design is an effective process for determining design directions based on a thorough understanding of the end user. Such an understanding comes not only through the hard work of the designers but also because end users are involved in the entire process, and the plain language of the scenarios constructed in Sections 17.3 and 17.5 gives them full access.

The semantic Web, or the gigantic global graph as Berners-Lee described it in a recent blog post (`http://dig.csail.mit.edu/breadcrumbs/node/215`), will likely bring great benefits to the citizens of the web. However, at what point will those benefits be of sufficient perceived usefulness and perceived ease of use to encourage widespread adoption [7]?

At the moment, this brave new world is the domain of the fearless early adopters. Even when the semantic Web has been widely adopted, it is interesting to consider whether everyone will be interested in partaking of these schemes to make their information more public. Interactions with privacy legislation in various jurisdictions must also be considered.

Table 17.6 Design scenario and claims analysis for Joe

Joe logs into the topic manager Web service to see what has been going on in the world with respect to local food. He is greeted by a summary page that tells him about the activity on this topic. First he clicks on the link for his own subscriptions, where he can see what has arrived from his e-mail and from his RSS feeds. He likes the idea that all of that traffic is directed here instead of his e-mail client on his laptop, which seems to be getting overwhelmed by all the e-mail he receives. When he signed up for this service, he was given a new e-mail address, which he then subscribed to his e-mail lists in place of his personal address. The content filtering on the site has found several duplicates in the day's messages and has put them aside, while recording the multiple sources for each document. He finds a notification of a blog posting that looks interesting and visits the blog to read more. He decides that he would like to share this with his community so he clicks on the "share" button in interface. He is then given some tools to compose an e-mail message that will go out to his community. Joe is able to select which destinations will receive the message, and he is also shown from which of these he has received the message. Sometimes he would like to post a comment on a blog posting, and this allows him to tailor the message for each destination. He also finds an event that he wants to share with his community. Rather than go to each site and create a specially formatted event for the various calendars within his community, he adds the event to his own local food calendar on this site and then sends a message out with the details, and a link to the feed for his calendar. In addition to the calendar feed, Joe likes that he has a number of options to publish what he has shared in an RSS feed. People who like Joe's treatment of the local food topic can also sign up for other notifications from the topic manager, depending on their needs. He likes the fact that people have found his local food community by first finding his feed on the topic manager site. Another link on his homepage is for new content found on the Web. He has configured the search engine on the site to look through a number of other sites for stories about local food. He checks digg.com, del.icio.us, and many others. The results from here are filtered for duplicates and also checked against the content from his subscriptions. When he finds a new site with interesting content, he adds it to his subscriptions. When interesting one-off stories appear from this search, he is able to share them with his network just like any other item. Joe is happy that he can handle everything to do with local food from one place. He logs off from the site, glad that he has one less thing to worry about.

Situation feature	Possible pros (+) and cons (−)
E-mail lists going to Web service	+ Reduces traffic on e-mail client on his laptop − Must trust service to handle email correctly duplicate messages removed + Able to concentrate more fully on remaining items − Must trust that important details are not removed from topic-related content on Web + Less clutter in his e-mail client − Does not allow for offline work as accustomed
Calendar feed	+ Reduces complexity of event maintenance − Some people in network may balk at new process
Subscribers can follow topical RSS directly	+ Lower barrier to entry for new people − Without contact, people may become disengaged
Search engine	+ Configuration gives user more control − May give user too many options + Duplicate removal potentially very valuable

Table 17.7 Design scenario and claims analysis for Emily

Emily has heard about the topic manager Web service and so she decides to give it a try. Once she signs up and defines the local food topic that she would like to manage, she is given an e-mail address along with instructions about how to subscribe that address to all of her current mailing lists. While doing that, she also unsubscribes her personal address from those lists. She can feel her inbox get lighter almost immediately. She is then asked if there are any RSS feeds she would like to track and she enters a few that she has heard about. She is also asked about her presence on a few social networking sites, and any groups she may have joined there. Then she is asked if she would like to add any topic streams from the topic manager service. Because she has identified local food as her topic, she is presented with a list of topic streams that she might like to follow. She is not surprised to find a stream set up by Joe, but she also finds a local food topic stream run by Robert, who has Joe's role in her university town. She subscribes to both of their streams. This will let her stay informed without having to wade through any duplicates, she thinks, and smiles to herself. She realizes that she also has some storage associated with her account and she can keep some messages here without having to read through everything about her home town when she is at university and vice versa. She's also read that this service uses SIOC, but she doesn't want to deal with that now.

Situation feature	Possible pros (+) and cons (−)
Configuration	+ User has a lot of control − User may not understand all the technical issues
Storage of messages on server	+ Reduced burden for her e-mail client − Stronger dependence on active Internet connection
Support for semantic Web	+ Permits more features + Does not impose requirements on incoming sources − User may not understand technological implications

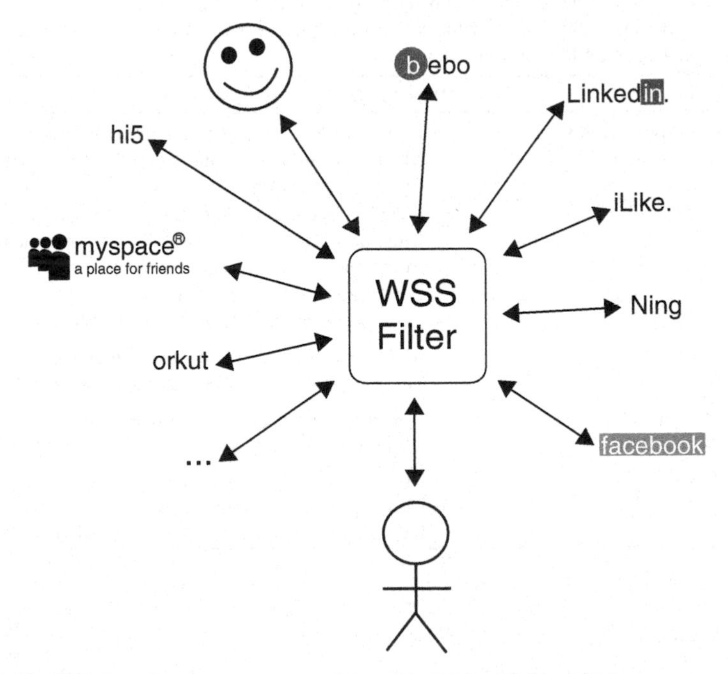

Fig. 17.2 The Web-based support system envisioned in Tables 17.6 and 17.7

Acknowledgments The authors would like to acknowledge the support of the Natural Sciences and Engineering Research Council (NSERC) of Canada. Emma Findlater made several helpful suggestions to improve the chapter. Daryl Hepting acknowledges the inspiration gained from his visit to Gerhard Fischer's Center for Lifelong Learning and Design (L3D) at the University of Colorado Boulder.

References

1. Backstrom, L., Kumar, R., Marlow, C., Novak, J., Tomkins, A.: Preferential behavior in online groups. In Proc. Web Search and Web Data Mining pp. 117–128 (2008)
2. Bernstein, Y., Zobel, J.: Redundant documents and search effectiveness. Proceedings of the 14th ACM international conference on information and knowledge management pp. 736–743 (2005)
3. Bishop, J.: Increasing participation in online communities: A framework for human–computer interaction. Computers in Human Behavior **23**(4), 1881–1893 (2007)
4. Bojars, U., Passant, A., Breslin, J., Decker, S.: Social network and data portability using semantic web technologies. Science Foundation Ireland pp. 5–19 (2007)
5. Breslin, J., Harth, A., Bojars, U., Decker, S.: Towards semantically-interlinked online communities. In Proc. European Semantic Web Conference pp. 500–514 (2005)
6. Cantador, I., Bellogín, A., Castells, P.: News@hand: A semantic web approach to recommending news. In: Proceedings of the 5th international conference on Adaptive Hypermedia, pp. 279–283. Springer (2008)
7. Chan, H., Teo, H.H.: Evaluating the boundary conditions of the technology acceptance model: An exploratory investigation. Transactions on Computer-Human Interaction (TOCHI) **14**(2) (2007)
8. Duffy, R., Fearne, A., Healing, V.: Reconnection in the UK food chain. British Food Journal **107**(1), 17–33 (2005)
9. Dwyer, C., Hiltz, S., Widmeyer, G.: Understanding development and usage of social networking sites: The social software performance model. In Proc. Hawaii International Conference on System Sciences (2008)
10. Fischer, G.: Distances and diversity: Sources for social creativity. Proceedings of the 5th conference on Creativity & cognition pp. 128–136 (2005)
11. Go, K., Carroll, J.: The blind men and the elephant: views of scenario-based system design. interactions **11**(6), 44–53 (2004)
12. Gould, J., Lewis, C.: Designing for usability: key principles and what designers think. Communications of the ACM **28**(3) (1985)
13. Heyman, K.: The move to make social data portable. Industry Trends pp. 13–15 (2008)
14. Himma, K.E.: The concept of information overload: A preliminary step in understanding the nature of a harmful information-related condition. Ethics and Information Technology **9**(4), 259–272 (2007)
15. Hunter, J., Khan, I., Gerber, A.: Harvana: harvesting community tags to enrich collection metadata. JCDL '08: Proceedings of the 8th ACM/IEEE-CS joint conference on Digital libraries pp. 147–156 (2008)
16. Klein, M., Malone, T., Sterman, J., Quadir, I.: The climate collaboratorium: Harnessing collective intelligence to address climate change issues pp. 1–23 (2006)
17. Kolbitsch, J., Maurer, H.: The transformation of the web: How emerging communities shape the information we consume. Journal of Universal Computer Science **12**(2), 187–213 (2006)
18. Kruchten, P.: The Rational Unified Process: An Introduction, 3rd edn. Addison Wesley (2003)
19. Leadbeater, C., Miller, P.: The Pro-Am revolution: how enthusiasts are changing our economy and society. Demos, London (2004)
20. Liebowitz, J.: Social Networking: the Essence of Innovation. Scarecrow Press (2007)

21. Mika, P.: Flink: Semantic web technology for the extraction and analysis of social networks. Web Semantics: Science, Services, and Agents on the World Wide Web **3**(2-3), 211–223 (2005)
22. Mowshowitz, A., Kawaguchi, A.: Bias on the web. Communications of the ACM **45**(9), 56–60 (2002)
23. Norman, D.: The design of future things. Basic Books (2007)
24. Poltrock, S., Grudin, J.: Organizational obstacles to interface design and development: two participant-observer studies. Transactions on Computer-Human Interaction (TOCHI **1**(1) (1994)
25. Raskin, J.: The Humane Interface: New Directions for Designing Interactive Systems. Addison Wesley (2000)
26. Richtel, M.: In web world of 24/7 stress, writers blog till they drop. New York Times (2008)
27. Rooksby, E.: The ethical status of non-commercial spam. Ethics and Information Technology **9**(2), 141–152 (2007)
28. Rosson, M., Carroll, J.: Usability engineering: Scenario-based development of human-computer interaction. Morgan Kaufmann (2002)
29. Suzuki, D.: Selective information overload. URL: [http://www.davidsuzuki.org/about_us/Dr_David_Suzuki/Article_Archives/weekly03210801.asp] (Accessed May 2008) (2008)
30. Yao, J.: Design of Web-Based Support Systems. In Proc. International Conference on Computer Science and Informatics pp. 349–352 (2005)
31. Yao, J.: An Introduction to Web-Based Support Systems. Journal of Intelligent Information Systems **17**(1-3), 267–281 (2008)

Chapter 18
Data Mining for Web-Based Support Systems: A Case Study in e-Custom Systems

Liana Razmerita and Kathrin Kirchner

Abstract This chapter provides an example of a Web-based support system (WSS) used to streamline trade procedures, prevent potential security threats, and reduce tax-related fraud in cross-border trade. The architecture is based on a service-oriented architecture that includes smart seals and Web services. We discuss the implications and suggest further enhancements to demonstrate how such systems can move toward a Web-based decision support system with the support of data mining methods. We provide a concrete example of how data mining can help to analyze the vast amount of data collected while monitoring the container movements along its supply chain.

18.1 Introduction

In a fast developing world, the design and implementation of advanced information systems is a continuous challenge for modern, dynamic organizations. Information systems featuring intelligence have been recently implemented as complex applications with modular architecture relying on Web services, Web technology, integrating data mining approaches, and/or smart seals technology. Within this endeavor, Web-based support systems (WSS) are an emerging multidisciplinary research area that studies the support of human activities on the Web as platform, medium, and interface. WSS are a natural evolution of the studies on various computerized support systems such as decision support systems (DSS), computer-aided design, and computer-aided software engineering [1]. Web technology provides a number of benefits for support systems in various application domains [2]:

L. Razmerita
Copenhagen Business School, Dalgas Have 15, Frederiksberg 2000, Denmark
e-mail: liana.razmerita@cbs.dk

K. Kirchner
Department of Business Information Systems, Faculty of Business and Economics, Friedrich-Schiller-University of Jena, 07743, Jena, Germany
e-mail: kathrin.kirchner@uni-jena.de

J.T. Yao (ed.), *Web-Based Support Systems*, Advanced Information and Knowledge Processing, DOI 10.1007/978-1-84882-628-1_18,

- The Web provides a distributed infrastructure for information processing.
- The Web enables timely, secure, ubiquitous access to information.

A very important area of WSS is DSS. According to [3], decision support implies the use of computers to assist managers in their decision processes in semi-structured tasks. Core components of DSS are the database, the model, the method, and the user interface [4]. Five types of DSS can be distinguished [5]: data-driven, model-driven, communication-driven, knowledge-driven, and document-driven. While data-driven DSS focus on accessing and manipulating data, e.g., time-series, model-driven DSS use statistical, simulation, or optimization models to aid the decision-making process. Communication-driven DSS support communication, coordination, and information sharing between people and can support group decision making. Knowledge-driven DSS provide recommendations for user actions according to hidden patterns in large amounts of data. Document-driven DSS manage, retrieve, and manipulate unstructured information like oral, video, or written information. The Web has now become the platform of choice for building DSS [6]. These Web-based and Web-enabled DSS became feasible in the mid of 1990's [7]. Web-enabled DSS can be defined as computerized systems that deliver decision support information or decision support tools to a manager or business analyst using a "thin-client" Web browser [5]. Web-based DSS means that the entire application is based on Web technologies like Web server and Web client, HTML, CGI, and databases.

The Web has expanded the possibilities of traditional DSS. Communication-driven DSS rely on electronic communication technologies to link decision makers with relevant information and tools. Knowledge-driven and document-driven DSS are based on large amounts of data or documents to support decisions. Web technologies help to reach a broader user community and to select more documents. For a Web-based DSS, data, models and methods can be accessed from a single Web server, whereas the user interface is a Web site. But data sources and methods could be distributed over several Web servers as well [8].

Bharatia and Chaudhury [6] argue that Web technology brought three major changes in DSS – the changes in the user community, problem areas, and underlying architecture. A Web-enabled DSS can include a broader user community and does not only involve managers making business-related decisions, but also book readers (with, e.g., Amazon) or children deciding about the look of their teddy bear (shop.vermontteddybear.com).

Web-based DSS have emerged as an integral part of e-commerce [9]. They can be used for supporting purchase decisions, helping to find relevant information or products for a user (e.g., Amazon.com), customizing products (e.g., gateway.com) or helping with post-purchase troubles [10].

Web-support systems for governmental decisions are complicated by factors such as multiple stakeholders (often with conflicting objectives), diverse information needs, and a variety of decision-making places [11].

This chapter introduces WSS as a case study for international trade. The objective of this chapter is to present innovative Web-support interorganizational systems that will dramatically reduce administrative costs, while at the same time reduce fraud in international trade. Within this chapter we argue that emergent, novel technologies are the main driver for introducing innovation within e-Governmental solutions and we exemplify our argument through an analysis of a concrete case study.

The chapter focuses on the redesign of custom systems toward more secure and paperless solutions using WSS. The focus is placed on the advantages of use of the novel technologies for the design of advanced e-custom systems. In particular we highlight the use of service-oriented architecture (SOA), Web services, and TREC (tamper-resistant embedded controller) device in an integrated framework and we analyze the benefits of using such a system. EPCIS (electronic product codes information system) is implemented based on a SOA using an advanced track and trace technology. The system ensures a more secure and advanced solution for monitoring containers through the supply chain and it raises alerts in case of problems (exceptions) for cross-border trade. The chapter analyzes the advantages and perspectives of future e-custom systems using WSS such as a SOA, Web services, data mining, and a TREC device. The TREC is an intelligent wireless monitoring device mounted on a container communicating with a (GPS) global positioning system in order to position the container precisely. Apart the physical location, the TREC device collects information related to the state of the container (e.g., temperature, humidity, ambient light, acceleration, and door status).

The remainder of the chapter is organized as follows: Section 18.2 provides an overview of data-mining technologies and their potential role for WSS in e-business and e-customs. Section 18.3 discusses the novel technologies applied for the redesign of the e-custom/e-business system toward advanced WSS. In Section 18.4 we introduce the EPCIS architecture. Section 18.5 discusses the evaluation of the system, advantages of its use as a web-based business support, and outlines future possible developments using data mining techniques. Section 18.6 summarizes our conclusions.

18.2 Data Mining as a Part of the Decision Making Process

The Web grows tremendously every single day. Using the Web any information can be distributed worldwide very fast. Nowadays the Web is the largest knowledge base in the world and everybody can use it. However, some problems occur for both users and Web site owners [12]:

- The difficulty to find and filter out relevant information.
- The creation of new knowledge out of Web-enabled data is not easy.
- The information on the Web should be personalized, so people can get information and suggestions according to their interests and needs.

Data mining can address these problems based on data analysis methods that allow to discover novel patterns and trends in a huge amount of data. It is a part of the knowledge discovery process along with preprocessing, subsampling, transformation of data, and the evaluation of the discovered pattern [13]. Data-mining techniques can be classified as

- Descriptive data mining that characterizes the general properties of data.
- Predictive data mining that performs inferences on the current data in order to make predictions.

Different types of algorithms enable the mining of different types of data like relational, text, spatial, spatio-termporal, Web content or multimedia [14]. Classical data-mining algorithms are commonly classified in four categories:

- Classification algorithms typically find a function that maps records into one of several discrete, predefined classes [15]. This can be used to understand preexisting data as well as to predict a class association for new data sets. A typical representative is decision tree induction, where a flowchart-like structure is learned from class-labeled training data. Each internal node represents a test of the data attribute and each branch is an outcome of such a test. The top node is the root, and each leaf holds a class label. For a new data tuple, where the class label is unknown, its attributes are tested against the tree to predict the class membership [16].
- Association rules algorithms [17] find patterns, associations, and correlations among sets of items or objects. The a priori algorithm [18], for example, finds frequent item sets that often occur together in a given data set with a frequency above a given threshold.
- Clustering algorithms [14] typically identify groups within data sets that are similar within each other but different from the remaining data. For example partitioning algorithms group [20] data sets-based on their attributes into *k* partitions, whereas the number *k* is given by the user.

Data and Web mining is an integral part of the whole decision-making process as represented in Figure 18.1. After a data preparation step, data-mining algorithms try to find patterns in data in order to make predictions on new data and/or optimize business processes. This process provides a basis for efficient decision making. During the decision-making step new knowledge is created that helps to optimize business processes or improve the quality of predictions. Data mining extracts and

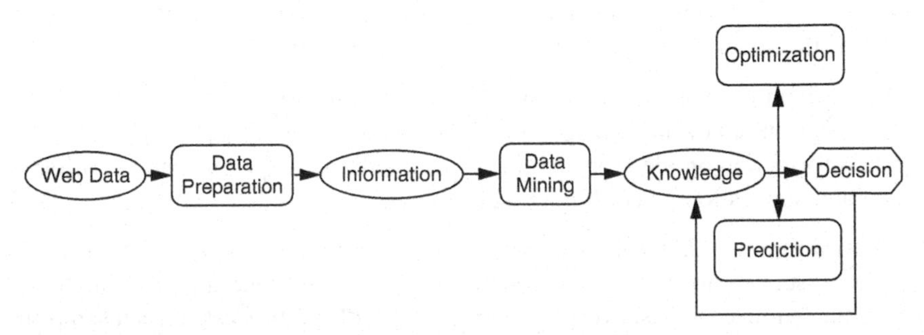

Fig. 18.1 Data mining as a part of the decision making process [21]

analyzes useful information from huge amounts of data stored in different data sources to provide additional information and knowledge that are crucial for the decision-making process [19].

In the following, we will present several examples of Web decision making using data mining. In e-commerce, retailers often need decision support for pricing their products. Pricing depends on the range of prices of other e-retailers for similar products, the reachability of the Web site, and customer's perceptions of service quality [22]. For preparing the decision, the first step is to collect data from the Web. Via BizRate.com online customers rank products, delivery time, or support. Alexa.com shows a number of visitors of different online stores. Information about other competitors and products can be found on the Internet. Focused Web crawler start from Web pages with interesting information for the retailer and recursively explores all linked pages. In Web-content mining, clustering algorithms can be used to group similar information. It is also possible to find out rules when a shop is especially successful (according number of customers). These rules – and the grouped Web site information – can help the retailer to make decisions in his or her own shop, especially about product prices.

Another example is the online auction platform. If a company or a private person wants to sell products by auction, a lot of decisions have to be made. Of course the person wants to find the best auction platform for his or her product [22]. Finding interesting bidding patterns and strategies is also important [23]. This can also be done via data-mining. Data about lots of auctions can be collected automatically. Data from back-end databases or application servers could be included too. A variety of data mining techniques such as clustering, association rule mining, or sequential pattern discovery [24] can be applied to answer questions such as: When is the best time to start and finish auctions? What is a good starting price? Which product descriptions are especially useful to get high prices?

The characteristics of those data mining applications in e-business contexts are the processing of vast distributed data volumes, computational cost of algorithms, and the diversity of users. SOA have the flexibility to add large data volumes from several sources and expand the computing capacity simply and transparently by advertising new services to the selected interface [25].

Heterogeneous structured Web data that are stored in databases but can also occur as video, sound, and images can lead to problems [26]. Semi-structured web data hinder algorithms from working efficiently; dynamic Web pages and inflexible data-mining systems do not always lead to good mining results. To solve these problems, Li and Le [26] propose the Web service technology and build a new Web mining system that can share and manage semi-structured data from different kinds of platforms, strengthen the specialty of mining algorithms, improve the mining efficiency, and eliminate the closed character of mining results.

As we will exemplify further in the chapter, data mining algorithms can be used for different types of decisions including optimization of traffic and routes, trajectory for trucks or containers, optimal display of products on the shelves, etc.

18.3 Building Blocks for New Web-Based Support Systems: Web Services, SOA, Smart Seals

Security threats, tax-related fraud, interoperability, and simplified, paperless trade transactions constitute important challenges to be addressed by modern custom/ business WSS. E-infrastructures as WSS can be defined as networks and Webs that enable locally controlled and maintained systems to interoperate more or less seamlessly. Three technologies can contribute to support a novel type of infrastructures: Web services, SOA and smart seals.

The focus is placed on the advantages of advanced WSS. In particular we highlight the use of SOA, Web services and TREC devices in an integrated framework. The TREC is an intelligent wireless monitoring device mounted on a container communicating with a GPS in order to position the container precisely. Apart from monitoring the physical location, the TREC device collects information related to the state of the container (e.g., temperature, humidity, ambient light, acceleration, and door status).

The EPCIS architecture [27] is implemented based on an SOA using an advance track and trace technology. The EPCIS system ensures a more secure and advanced solution for monitoring containers through the excise supply chain. It raises alerts in case of problems (exceptions) for cross-border trade. EPCIS system is designed to be compliant with EPCGlobal standards (see: www.epcglobalinc.org) and it is integrated in the EPCNetwork.

18.3.1 Web Services

Web services are software applications based on XML and standard Web protocols that can be discovered, described, and accessed over intranets, extranets, and the Internet [28].

Initially the Web service efforts focused on interoperability, standards, and protocols for performing B2B transactions. Various types of advanced information systems can apply Web services technology. Web services enable a new way to establish more flexible, low-cost connections across applications. Web services viewed as a next generation of Web-based technology communicate using messages. Web services are emerging to provide a systematic and extensible framework for application-to-application interaction built on top of existing Web protocols and based on open XML standards [29]. Typically, messages exchanges are using SOAP (Simple Object Access Protocol) and their description is made with WSDL (Web Service Description Language).

The next challenge for Web services is the use of semantic annotations for the description of services. This added semantics would enable to orchestrate them to achieve enhanced query mechanisms, advanced service choreographies, and complex functionality, and thus to achieve service intelligence.

Web services are built on a foundation set of standards and protocols, including Internet protocols (TCP/IP, HTTP). These standards enable the establishment of automated connections between applications and databases. The design and implementation of Web services is associated with three basic aspects: communication protocols, service descriptions, and service discovery. Industry specifications build on top of XML and SOAP standards. Basic technologies associated with the implementation of Web Services are the following protocols:

- **SOAP** (Simple Object Access Protocol) is a XML-based protocol for exchanging information in distributed environments. It relies on XML as an encoding scheme and HTTP as a transport protocol.

  > SOAP is a lightweight protocol for exchange of information in a decentralized, distributed environment. (W3C: www.w3.org/TR/soap/)

  SOAP is a basic building block that serves as a basic foundation for many Web Service standards, especially WSDL.
- **WSDL** (Web Service Description Language) is a metadata language used to design specifications for Web services.

  > WSDL is an XML format for describing network services as a set of endpoints operating on messages containing either document-oriented or procedure-oriented information. (W3C: www.w3.org/TR/wsdl)

  WSDL describes where the service is located, what the service does, and how to interact with the service. The description language uses XML format. It has been developed by combined efforts of IBM, Microsoft, and Ariba.
- **UDDI** (Universal Description Discovery & Integration) is a specification for Web-based information registries of Web services. Through UDDI specifications, services can be located dynamically.

18.3.2 Service-Oriented Architecture (SOA)

The SOA is an architectural style that has emerged and become popular in recent years [29–31]. Service oriented computing is a flexible computational model that can simplify software development [30]. SOA relies on the concept of service. A service is a piece of software that implements a well-defined functionality that can be consumed by clients or other services regardless of the application or business model. Services communicate with each other by means of message exchanges.

A basic SOA is represented in Figure 18.2. This architecture involves three main parts: a provider, a consumer, and a registry. A service provider publishes the service profile for a Web service in a registry. A service requester, which can be a human agent, a software agent, or a business application, finds the service based on the service description provided in the service registry and can execute the service. A business application subsequently sends a request for a service at a given URL using SOAP over HTTP.

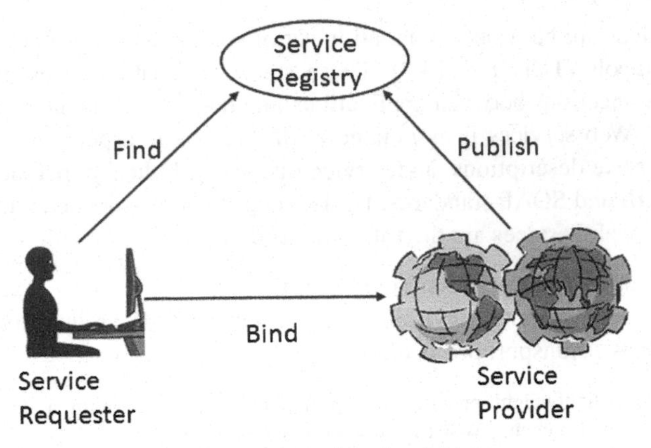

Fig. 18.2 Distributed computing using service-oriented architectures

SOA provides a number of features that make them very advantageous for employment in different business applications. The main characteristics of a SOA are interoperability, loose coupling, isolation, composability, and machine processability:

- **Interoperability** is a very important feature of software applications providing seamless communications among different applications. Programs written in different programming languages deployed over different platforms and using different protocols can communicate mediated by services.
- **Loose coupling** or decoupling refers to the fact that services are described and defined independently of each other, reducing the mutual dependencies.
- **Isolation** deals with the possibility to modify services and their definition without an impact on other services with which they interact. Service consumers are decoupled of details of service implementation and location of service providers.
- **Composability** or service composition refers to the possibility to combine several simple services in order to achieve more complex or new functionality.
- **Machine processability** deals with the ability to process the data attached to the service description automatically. So services can interact with other services without human intervention.

18.3.3 Smart Seals: TREC or RFID Technology

The TREC is an intelligent wireless monitoring device that can be mounted on containers. It has been developed by IBM Zurich research labs. The TREC enables the creation of an audit trail of container movements and events from the point of origin to the destination. The TREC device collects and transmits information via a satellite network, a cellular system (GSM/GPRS), or a Wireless Personal Area Network (WPAN) based on ZigBee/IEEE 802.15.4 radio [32]. The information that can be collected and transmitted includes the location of the container and other parameters of the container (e.g., temperature, humidity, door status, velocity). Information related to the container's precise location can be sent to the different supply chain partners including custom officers. Handheld devices can also communicate with the TREC device. The TREC can process information, analyze events, and control certain actions or, predefined events (e.g., to open the door based on input from an authorized person in certain geographic areas, or to trigger alerts in case of deviations or abnormal physical conditions of the container).

RFID (Radio Frequency Identification) is an electronic tagging technology for capturing source data. RFID tags can act as wireless bar-codes and can be used on various types of goods. By tagging items and containers companies would be able to monitor them wirelessly and thus streamline their distribution and delivery systems.

RFID has been hailed as a breakthrough technology that would revolutionize logistics. This technology has received a lot of attention in the last few years mainly due to its applications in supply chain management [33, 34] as well as its potential to improve efficiency.

18.4 Web-Based Support Systems for e-Business and e-Custom

EU governments are under pressure to increase the security of international trade and to reduce its administrative burden [35]. Single-window (SW) and authorized economic operators (AEO) are crucial to solve these two important objectives. SW concept involves providing online SW access points where businesses and public administrations can exchange the data required by legislation for the EU cross-border movements of goods. By January 2013 SW is scheduled to replace the current silo solutions [36]. The AEO status involves a special partnership between the private sector and customs administration. Certified AEO will enjoy simplified trade procedures and fast custom clearance. The transformation of paper trade documents to electronic ones along with the redesign of customs procedures and the implementation of advanced e-custom systems is necessary to accomplish the above-mentioned objectives.

The European Commission has defined three strategic goals for e-Custom development in Europe: achieving pan-European interoperability, establishing SW access points, and granting Authorized Economic Operator (AEO) status to trading partners. These are defined as follows:

- Pan-European interoperability involves different European governmental systems and will be able to communicate and exchange data independent of their heterogeneity and the different implemented standards and proprietary solutions.
- The SW initiative involves providing online SW access points where businesses and public administrations can exchange the data required by legislation for the EU cross-border movements of goods. By January 2013, it is planned that the SW will replace the current "silo solutions" [36].

- AEO status may be obtained by businesses that fulfill certain business process requirements. This involves a special partnership between the private sector and custom administration. Certified AEO will enjoy simplified trade procedures and fast custom clearance.

The EU envisages that it will achieve these objectives through the transformation of paper trade documents to electronic ones along with the redesign of customs procedures and the implementation of new e-custom systems.

This section introduces the technologies described above in an integrated architecture named EPCIS prototyped within one of the four living labs of the ITAIDE project. This solution has been designed, implemented, and tested within the beer living lab [37].

A concrete scenario of use in trade procedures is presented in Figure 18.3. In this scenario, businesses with AEO status will no longer have to submit the many export declarations to customs authorities. As mentioned earlier in the paper, AEO status is granted to businesses that meet certain requirements regarding their security, quality, transparency, and audibility of their business processes. The advantages to the business are simplified customs and tax procedures.

The international trade scenario involves three stakeholders: a business that exports goods, a carrier responsible for the shipment of goods, and the customs administration. Currently the trade procedures based on silo solutions require businesses to submit various declarations containing commercial data. These various declarations often require a large, redundant amount of data from businesses and represent a heavy administrative burden [38].

As represented in Figure 18.3, BusinessCo makes its commercial data available from its enterprise resource planning (ERP) system via an EPCIS database. Using specialized Web services, customs administrations can access the commercial data at any time from the EPCIS database. In addition, customs administration can audit

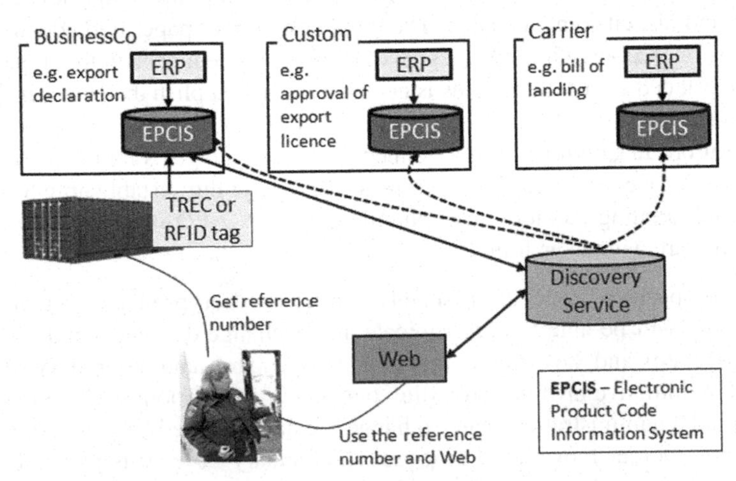

Fig. 18.3 EPCIS distributed architecture, modified from [38]

BusinessCo periodically in order to make sure that the data provided are valid and updated. The architecture relies on an SOA that enables the deployment of the solution in a geographically dispersed environment. The solution makes use of EPC-global (Electronic Product Code Global) [39] standard for capturing shipment data.

EPCglobal is the globally accepted standard that ensures universal applicability and optimal functionality across the globe for all industry sectors. This standard is intended to facilitate the development of interoperable systems for industries, commercial sectors, and/or geographic regions as well as to facilitate collaboration across industries independent of physical location and geography.

The EPCIS standard [27] provides the foundation necessary for the capture, communication, and discovery of EPC event data. In order to enforce security policies, the EPCIS portals are located behind firewalls that can be only passed by users from restricted IP addresses. EPCIS databases capture events related to the export declarations and other predefined events related to shipment data. The events are stored in EPCIS repositories of the parties that own the data and are made available to authorized parties.

Web services enable the tracking of goods as they move through the supply chain. The TREC is mounted on a container and transmits data about the precise location of a container along its supply chain. The TREC is configured to signal when the container enters or leaves a predefined location and can detect the loading or unloading of the container onto or from a vessel. In the pilot scenario the TREC was set to monitor the containers' inside temperature and generate alert whenever the container deviates from its intended route. Specialized Web services capture various events related to the shipment of goods, their location, and/or container condition. Notifications and alerts can be triggered when containers deviate from their predefined route or other abnormal conditions occur, thus enhancing security of shipment of goods.

The query interface is designed for obtaining and sharing data about unique objects in the supply chain within and across organizations based on an authorized access [40]. Compliant with the world custom organization recommendations, each container has assigned a Unique Consignment Reference (UCR) number and an identifier. This number is used by discovery services to search for data associated with the container within EPCIS database. Thus, specialized Web services enable authorized parties in the supply chain to retrieve timely data related to the status of the container. The stakeholders can also be notified by the system when the container arrives at its final destination. This kind of notification will enable the elimination of the current paper-based export evidence procedure.

18.5 Evaluation and Discussion

In the following, outline evaluation results, analyze the main benefits of using SOA, and discuss further possible developments of Web-based support solutions using data mining techniques.

The EPCIS system has been tested in relation with the export of beer from Heineken in the context of ITAIDE project. EPCIS has been implemented by IBM between July and November 2006. Subsequently, between November 2006 and January 2007, 14 containers have been equipped with TREC devices and EPCIS pilot has been tested in two phases. The first pilot test has been shipped from the Netherlands to the United States (nine containers) while the second one has been sent to the United Kingdom (five containers). During the transportation phase the TREC could help as a "witness device" for import–export procedures. When an export declaration is published, an alert event is triggered to the custom authorities, who can then calculate the risk and decide on whether a physical inspection is warranted.

Even though the overall system worked according to its overall specifications, some glitches have been identified in the technical solution and further developments are in progress. For instance, the TREC functionality and its battery lifetime have to be improved, the register and query interfaces associated with the discovery service are only in beta version, and the discovery service is still to be finalized within the pilot. In the current version of EPCIS system, the portal supports only simple queries which are events related to a specific container with a specific UCR. In the future, more complex queries are planned to be implemented (e.g., which containers are addressed to a particular consignee? Do the quantities and types of goods, declared for export match those declared for import?).

EPCIS enables not only the publishing of custom declarations but also the tracking of products through the whole supply chain. EPCIS is a massively distributed integrated solution based on SOA and Web services. It allows authorized parties (custom administration) to track products through the whole supply chain (each party in the supply chain can track products). The TREC device also enables ubiquitous access to information about the location of the container and it can trigger events/alerts in case of deviations from the predefined routes or unauthorized container access, thus improving security and control. Web services enable ubiquitous, timely access to data about the container and support customs-related queries. Web services integrated in SOA enable the implementation of the Single Window access concept.

Finally, SOA and the use of EPCglobal standard ensure the interoperability of the proposed solution. The EPCglobal network, also called the *Internet of things*, is the ideal backbone for tracking goods moving along a supply chain [41]. It enables to create an open-standard, SOA-based data-sharing mechanism between the trading partners. SOA facilitates integration with other backend systems used by businesses and customs administration.

The technological solution is just a first step toward the diffusion and adoption of such an advanced e-custom system. The development of new e-custom solutions is a very complex issue that involves not only technical aspects outlined in this chapter, but also organizational and political decisions. A complementary picture of the whole process associated with innovation and transformation of e-customs focusing on political, organizational aspects, and control procedures is further discussed in publications from the ITAIDE project [38, 42, 43].

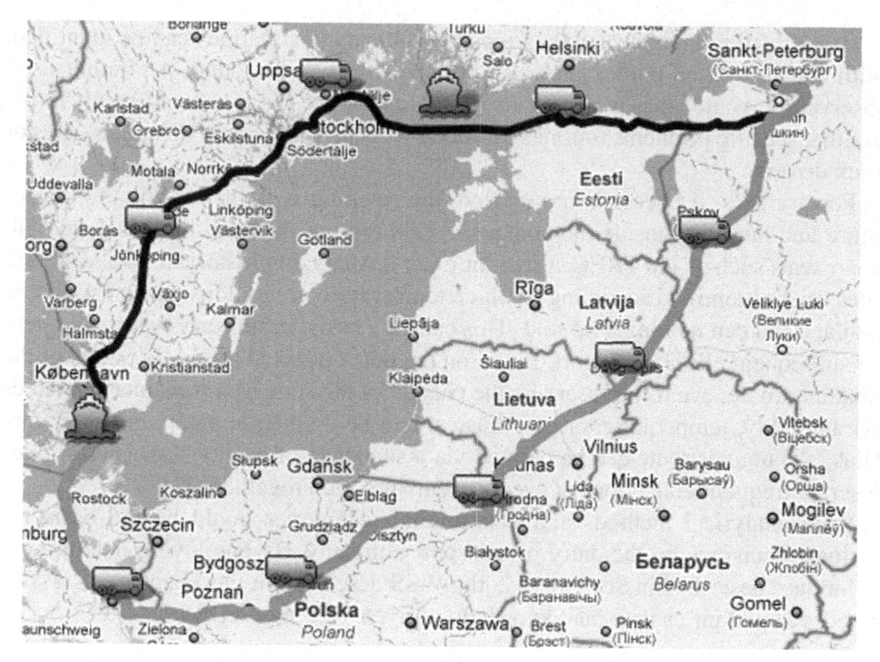

Fig. 18.4 Container route example (Google maps)

In the following paragraph, we discuss how WSS and data-mining techniques could be applied in the ITAIDE food living lab. The ITAIDE food living lab is centered on the dairy industry and has as a partner one of the largest dairy companies worldwide [44].

The smart seals employing either a TREC or a RFID-enabled device enable the collection of large amounts of data about the container as well as its characteristics at certain time stamps and certain places. Underlying spatial and spatial–temporal data-mining methods can be used to find patterns and optimize different parameters [45]. Figure 18.4 represents two hypothetical cheese transport routes from Copenhagen to Saint Petersburg that could be monitored along its supply chain using smart seals. The data could be further analyzed with data mining algorithms to find typical patterns and optimize the transport routes.

The optimization can include the improvement of the container's routes according to drive time analysis. Drive time analysis could include typical times for traffic jams, streets quality, or the necessity to pass high mountains or border crossings at certain locations. Different parameters can help to redefine the truck route. Optimal routes reduce the costs and time, which is critical for perishable food transport. An overview of time–space methods for activity-travel patterns can be found in [46].

Optimal routes also could be planned according the location of customers and the maximum transportation time for perishable foods. This also includes the number of containers that are needed at different time periods and in different places.

By analyzing the collected data typical problematic situations can be identified. With distance-based outlier analysis, unsual routes can be found. An outlier is an observation (route) that is distant from the rest of the data. It could result in the fact that specific problems (outliers) especially occur with certain container types or truck drivers.

For example, if cheese is transported, the ripening is influenced by the temperature and humidity inside the container. This type of data can be measured with smart seals such as the TREC system or others. Analyzing historical data, different rules can be identified according to which temperature or humidity can cause cheese spoilage and can no longer be sold. This can be a basis for an early-warning system for subsequent cheese transport. Based on the data collected at certain points of the container route, events representing the change of the state of a container (variables like humidity, temperature, or coordinates) can be defined at a certain timestamp. Thus, a container route can be defined via a sequence of events. Association rules describe frequent collections of events occurring often together [47].

Such analytical methods and data-mining techniques could be the basis for business decisions by the dairy or transport company. By employing data-mining techniques described in Section 18.2, the WSS described in this chapter can be further developed into a data- and knowledge-driven Web-based DSS.

18.6 Conclusions

This chapter is primarily concerned with the description of a Web-based software infrastructure to support interorganizational business processes for international trade. The chapter has presented a WSS operating as a service-oriented infrastructure. It has highlighted the advantages of such a system in report with increasing security, facilitating the cross-border trade, reducing transaction cost. The system is designed to conform to strategic objectives of the World Trade Organization and the customs authorities in the United States and European Union.

The chapter discusses how WSS can be further enhanced by analyzing the data collected from smart seals using data-mining techniques. These techniques would enable to integrate DSS within future WSS for advanced e-custom or e-business systems.

Acknowledgements This research is part of the integrated project ITAIDE (nr. 027829), which is funded by the 6th Framework IST programme of the European Commission (see www.itaide.org). We are greatly indebted to all participants of the project for their valuable contributions to the work presented here.

References

1. Yao, J., Yao, Y.: Web-based support systems. In: Proceedings of the Workshop on Applications, Products and Services of Web-based Support Systems (WSS'03), pp. 1–5. (2003)
2. Yao, J.: An Introduction to Web-based Support Systems. Journal Of Intelligent Systems 17(1/3), 267. (2008)
3. Keen, P.G.W., Scott Morton M.S.: Decision Support Systems: An Organizational Perspective. Addison-Wesley, Inc., Reading, Mass. (1978)
4. Turban, E., Aronson, J.E.: Decision support systems and intelligent systems. Prentice Hall, Upper Saddle River, NJ. (2001)
5. Power, D.J.:Decision support systems. Westport, Conn., Quorum Books. (2002)
6. Bharatia, P., Chaudhury, A.: An empirical investigation of decision-making satisfaction in web-based decision support systems. Decision Support Systems, 37:187–197. (2004)
7. Bhargava, H.K., Power, D.J., Sun, D.: Progress in Web-based decision support technologies. Decision Support Systems, 43:1083–1095. (2007)
8. Zhang, S., Goddard, S.: A software architecture and framework for web-based distributed decision support systems. Decision Support Systems, 43:1133–1150. (2007)
9. Holsapple, C., Joshi, K., Singh, M.: Decision support applications in electronic commerce. In: Shaw, M., Blanning, R., Strader, T. (Eds.): Handbook on Electronic Commerce. Berlin: Springer. (2000)
10. Zashedi, M.F., Song, J., Jarupathirun, S.: Web-based Decision Support, In: Burstein, F., Holsapple, C.W. (Eds.): Handbook on decision support systems. Basis Themes. Berlin, Springer, pp. 315–338. (2008)
11. Ault, J.T., Gleason, J.M.: U.S. Government decision makers' expectations and patterns of use of emerging and existing information technologies. Government Information Quarterly, 20:63–76. (2003)
12. Kosala, R., Blockeel, H.: Web mining research: a survey. ACM SIGKDD Explorations Newsletter, vol. 2, no. 1, pp. 1–15. (2000)
13. Fayyad, U.M., Piatetsky-Shapiro, G., Smyth, P.: From Data Mining to Knowledge Discovery: An Overview. In: Usama, M. Fayyad, U.M., Piatetsky-Shapiro, G., Smyth, P., Uthurusamy, R. (Eds.): Advances in knowledge discovery and data mining. Menlo Park, Calif.: AAAI Press, pp. 1–34. (1996)
14. Han, J., Kamber, M.: Data mining. Concepts and Techniques. Elsevier/Morgan Kaufmann, Amsterdam. (2006)
15. Tan, P.-N., Steinbach, M., Kumar, V.: Introduction to Data Mining. Addison-Wesley. (2006)
16. Quinlan, J.R.: C4.5: Programs for Machine Learning. Morgan Kaufmann Publishers. (1993)
17. Agrawal, R., Imielinski, T., Swami, A.: Mining association rules between sets of items in large databases. In: Proc. 1993 ACM-SIGMOD Int. Comf. Management of Data (SIGMOD'93), pp. 207–216. (1993)
18. Agrawal, R., Srikant, R.: Fast algorithms for mining association rules in large databases. In: Research Report RJ 9839, IBM Almaden Research Center, San Jose. (1994)
19. Lee, J.H., Park, S.C.: Agent and data mining based decision support system and its adaptation to a new customer-centric electronic commerce. In: Expert Systems with Applications, Volume 25, Issue 4, November 2003, pp. 619–635. (2003)
20. Kaufman, L., Rousseeuw, P.J.: Finding Groups in Data: An Introduction to Cluster Analysis. John Wiley and Sons. (1990)
21. Michalewicz, Z., Schmidt, M., Michalewicz, M., Chiriac, C.: Adaptive business intelligence. Springer, Berlin. (2007)
22. Marsden, J.R.: The Internet and DSS: massive, real-time data availability is changing the DSS landscape. Information Systems and E-Business Management, vol. 6, no. 2, pp. 193–203.(2008)
23. Bapna, R., Goes, P., Gupta, A.: Replicating Online Yankee Auctions to Analyze Auctioneers' and Bidders' Strategies. Information Systems Research, vol. 14, no. 3, pp. 244–268.(2003)

24. Mobasher, B.: Data Mining for Personalization. In: Brusilovsky, P., Kobsa, A., Nejdl, W. (Eds.): The Adaptive Web. Methods and Strategies of Web Personalization, LNCS Vol. 4321, pp. 90–135.(2007)
25. Guedes, D., Meira, W., Ferreira, R.: Anteater: A Service-Oriented Architecture for High-Performance Data Mining. Internet Computing, IEEE, Volume 10, Issue 4, pp. 36–43. (2006)
26. Li, B.B., Le. J.J.: Applications of web services in web mining. In: Hang, J.Z., He, J.-H., Fu, Y. (Eds.): Proceedings of First International Symposium, CIS 2004, Shanghai, China, December 16–18, 2004, Lecture Notes in computer Sciences. vol. 3314, pp. 989–994. (2005)
27. EPCIS: EPC Information Service (EPCIS) version 1.0 specification, EPCGlobal. (2006)
28. Daconta, M., Obrst, L., Smith, K.: The Semantic Web: A guide to the future of XML, Web Services and Knowledge Management. Wiley Publishing Inc., Indiana. (2003)
29. Curbera, F.D., Khalaf, R., Nagy, W., Mukhi, N., Weerawarana, S.: Unraveling the Web Services Web: an Introduction to SOAP, WSDL, and UDDI. Internet Computing, IEEE, vol. 6, pp. 86–93, March-April. (2002)
30. Huhns, M.N., Singh, M.P.: Service-Oriented Computing: Key Concepts and Principles. IEEE Internet Computing, vol. 9, pp. 75–81.(2005)
31. Chung, J.-Y., Chao, K.-M.: A view on service-oriented architecture. Service Oriented Computing and Applications, vol. 1, pp. 93–95.(2007)
32. Dolivo, F.: The IBM Secure Trade Lane Solution. ERCIM (European Research Consortium for Informatics and Mathematics) News, vol. 68, pp. 45–47. (2007)
33. Niederman, F., Mathieu, R.G., Morley, R., Kwon, R.: Examining RFID Applications In Supply Chain Management. Communications of the ACM, vol. 50, no. 7, pp. 93–101. (2007)
34. Songini, M.: Wal-Mart offers RFID updates. In: Computerworld (2006) www.computerworld.com/mobiletopics/story/0,10801,109418,00.html.
35. Bjørn-Andersen, N., Razmerita, L., Zinner-Henriksen, H.: The streamlining of Cross-Border Taxation Using IT: The Danish eExport Solution. In: Macolm, G.O.J. (Ed.): E-Taxation: State & Perspectives – E-Government in the field of Taxation: Scientific Basis, Implementation Strategies, Good Practice Examples. Linz, Trauner Verlag, pp. 195–206. (2007)
36. E.C. Taxation and Customs Union: Electronic Customs Multi-Annual Strategic Plan (MASP Rev 7), Working document TAXUD/477/2004 – Rev. 7 – EN. (2006)
37. Rukanova, B., van Stijn, E., Zinner-Henriksen, H., Baida, Z., Tan, Y.-H.: Multi-Level Analysis of Complex IS Change: A Case Study of eCustoms. Paper presented at the Electronic Commerce Conference, Bled, Slovenia. (2008)
38. Baida, Z., Rukanova, B., Liu, J., Tan, Y.-H.: Rethinking EU Trade Procedures – The Beer Living Lab. In: Bled eCommerce Conference, Slovenia. (2007)
39. EPCglobalinc.(2007) vol. 2007
40. Koldijk, F.: eTaxation and eCustom demonstrator, Deliverable 5.1.4. ITAIDE consortium. (2006)
41. Schaefer, S.: Secure trade lane: a sensor network solution for more predictable and more secure container shipments. OOPSLA Companion, pp. 839–845. (2006)
42. Tan, Y.-H., Klein, S., Rukanova, B., Higgins, A., Baida, Z.: e-Customs Innovation and Transformation: A Research Approach. In: Bled eCommerce Conference, Slovenia. (2006)
43. Liu, J., Baida, Z., Tan, Y.-H.: e-Customs control procedure redesign methodology: model-based application. In: Proceedings of the 15th European Conference on Information Systems. St Gallen, Switzerland. (2007)
44. Bjørn-Andersen, N., Razmerita, L.: Advanced e-Governmental Solutions for Global Trade. 3rd International Conference on ICT: from Theory to Applications (ICTTA), Damascus, Syria, April. (2008)
45. Roddick, J.F., Lees, B.G.: Paradigms for spatial and spatio-temporal data mining. In: Miller, H., Han, J. (Eds.): Geographic Data Mining and Knowledge Discovery. Taylor and Francis. Research Monographs in Geographic Information Systems. (2001)
46. Timmermans, H., Arentze, T., Joh, C.: Analysing space-time behaviour: new approaches to old problems. Progress in Human Geography, 26:175–19. (2002)
47. Manila, H.,Toivonen,H., Verkamo, A.I.: Discovery of frequent episodes in event sequences. In: Data Mining and Knowledge Discovery. Vol. 1, pp. 259–289. (1997)

Chapter 19
Service-Oriented Architecture (SOA) as a Technical Framework for Web-Based Support Systems (WSS)

Vishav Vir Singh

Abstract Software-oriented architecture (SOA) is a very powerful Internet-based framework which can be targeted towards Web-based Support Systems (WSS). SOA equips these systems to supplement their intelligence, decision-making, and processing potential with the universality and ubiquity of the Internet. It makes the WSS 'agile', implying that these support systems can adapt their intelligence very quickly due to the swift access of data available in the form of a 'service' through SOA. No matter how distributed and specialized a data source is, and regardless of whether it is domain-independent or domain-dependent, these systems can just 'plug' into any technology they want by acting as consumers of a 'service'. This truly amplifies the performance and system response of WSS. This chapter follows this vision of defining the architectural technique of SOA and how its nuances interplay with the concept of WSS. A complete case study outlining the marriage of SOA and WSS has been provided to further illustrate the merits of such a combined system.

19.1 Introduction

The metamorphosis of monolithic, self-sufficient support systems (SS) into a collaborative, adaptive, and well-networked system is inevitable. The motion of all of artificial intelligence along the pathways of the Internet is imperative to overcome the basic processing limitation of the human brain.

Service-oriented architecture (SOA) is a very powerful Internet-based framework which can be targeted towards Web-based Support Systems (WSS). SOA equips these systems to supplement their intelligence, decision-making, and processing potential with the universality and ubiquity of the Internet. It makes the WSS 'agile',

V.V. Singh
Intersil Corporation, 1001 Murphy Ranch Road, Milpitas, CA 95035, USA
e-mail: vsingh@intersil.com

J.T. Yao (ed.), *Web-Based Support Systems*, Advanced Information and Knowledge Processing, DOI 10.1007/978-1-84882-628-1_19,

implying that these support systems can adapt their intelligence very quickly due to the swift access of data available in the form of a 'service' through SOA. Efficiency is another reason that SOA fits perfectly with the ethos of WSS. No matter how distributed and specialized a data source is, and regardless of whether it is domain-independent or domain-dependent, these systems can just 'plug' into any technology they want by acting as consumers of a 'service'. This truly amplifies the performance and system response of WSS.

Since SOA works with XML, the language of the Internet, WSS witness and receive 'better' data due to XML, which renders structure to all kinds of data. This has been made possible through the rise of the concept of a 'tag' that in all its capabilities transformed the Internet forever.

This chapter delves into a study of all the above-mentioned concepts and concerns and follows a very close track to the premise of ensuring that the recent advancements of the Web technologies are explained within the purview of execution of WSS.

19.2 Support Systems: A Historical Perspective

Now let us trace some of the historical notions and concepts that fed the vertical building of the term 'support systems'.

The first term of significance is 'datum'. *Datum* is a Latin word and is the neuter past participle of 'dare' and 'to give', hence 'something given' [25]. The past participle of 'to give' has been in use since times immemorial with Euclid unleashing a pioneering publishing work titled *Dedomena*, which in Latin means *Data*, in circa 300 BC. Hence, a *datum* implies the profoundly indivisible atom of the 'the given' yielded to us by nature. This 'given' piece of datum can act as a message, sensory input [25], state, and static occurrence of something existent in nature, or just a reference from which we define the beginnings of annals of data.

The word 'data' is the plural of Latin 'datum'. Some authorities and institutions deem 'data' as the absolute plural of 'datum', while some carry the view of obsoletion of the word 'datum' and hence use 'data' as both the plural and singular forms.

However, philosophically data is not just an English plural of 'datum'. Data may be defined as a 'processable' collection of 'datums'. The term 'processable' has the following markers:

- Atomicity: Datums are 'atomic', implying they are the most indivisible microelements at the finest level of granularity.
- Factness: A 'datum' is always a fact, implying that it is a known and constant reference.

Now let us delve into the term 'decision theory'. Decision theory is 'an area of study of discrete mathematics that models human decision-making in science, engineering and indeed all human social activities' [20]. The underlying foundation of decision theory is a branch of discrete mathematics called the information theory.

Information theory is a science that involves the discretization or qualification of data. Discretization is basically the fragmentation of data into smaller, machine representable constructs that symbolize a uniform and universal method of representation on a much larger system scale across any technology platform. Underlining a finer and more pertinent nuance of information theory as propounded by Shannon [22], information theory is a set of tools exposing the science of mensuration as it extends to measuring, storing, and communicating the structures needed to discretize data. The popular model of bit and byte representation of data is a perfect example of a model that employs information theory to quantify processable collection of 'datums', that being data [5].

The term 'support system' or analogously 'decision support systems' has been used in the history of computing in multifarious contexts. This diversity of schools of thoughts essentially derives itself from the various doctrines that apply to the process of decision-making in general. Different perspectives on the process of decision-making and hence support systems generically include the learning, engineering, psychological, anticipated utility, and deterministic perspectives etc [10,27].

A support system, in substance, is a mechanized system capable of storing decision-making data, and then applying a very wide range of nimble processing capabilities on this data set that are required in order to reach a decision. In other words, it is a digitized and automated assembly that consumes decision theories and data and effects decisions as output. The processing capabilities that were alluded to earlier could span across a host of conceptual patterns. For example, the following capabilities can be leveraged in decision-making: mathematical and quantitative models, communicative or collaborative data processing that fosters shared decision-making, data representation where data are stored in a myriad of formats to achieve intelligence and pattern-based decision-making, or knowledge-based models spread across all sciences that are pertinent to decision-making in any given context [11]. An example of this would be Boolean logic theories, chemical science reactions, mathematical and statistical tools, etc. that could well be amalgamated to achieve an integrated process of decision-making.

19.3 Service as a Medium of Information Exchange for Web-Based Support Systems

The notion of databases is a very primitive one where the chief labour of theory is on 'reservoiring' of data with a score of processing capabilities applicable to the data. Little would one have thought that with the onset of distributed computing, the data would set in an incessant wave of flux that will change the face of the world forever. This manifestation of data flux came to be known as what we today call the Internet.

One of the most fundamental purposes of the Internet and one that has rooted the genesis of location transparency and reachability is the 'communication' of data, or in other words, the 'exchange' of data. Data exchange has been a familiar concept to mankind ever since the prehistoric ages when the modes of communication were disparate, yet the purpose has been the same ever since [6].

The Internet forms the crucial backbone of all support systems today as data continues to be produced, synthesized, coagulated, processed, and most importantly specialized. The name 'Web-based support systems' implies support systems based on the World Wide Web (WWW) abbreviated to just the 'Web', which is an information-sharing model with the Internet as its operative medium [26].

According to the Data Management Association, data management is defined as 'the development and execution of architectures, policies, practices and procedures that properly manage the full data lifecycle needs of an enterprise' [23]. Data management is fairly inclusive of data creation, modification, updating, modelling, storage, processing, etc. Some of the formal topics in data management that are linked to any Web-based decision support system (DSS) are [25]:

- Data analysis
- Data flux
- Data mining
- Data warehousing
- Data modeling

Data management would essentially be pertinent to the entire data clock in a support system context from the locus of input of data into the system, its modeling, processing, and then its locus of output in the new form of a 'decision'.

Similarly, the notion of model-management infuses a feature of arranging, simulating, and engineering a decision-making chain of steps with certain model parameters that are steered over the data, acting as the 'food' of the model to achieve learning and systemic functioning [16]. Any type of support system, including WSS, can have copious amounts of models functioning together as a cohesive unit to achieve a well-grounded decision. An example of a model management framework is that of neural network, which is a network of intelligent agents called 'neurons' which constantly exchange information between them and produce an output as a unit.

19.3.1 Genesis of a 'Service' as Data/Input Access Characteristic

A very important basis of a support system has always been the 'characteristics' of its data source component. A 'characteristic' is simply a property or an aspect of an entity. A characteristic of a data source could be an aspect or a property of its data domain (e.g. scientific, historical, or religion) or an aspect or a property of its data dissemination mechanism (e.g. aggregational or pluralistic), where data streams are clustered as they head out of the source or singular where each unit of data is treated individually.

Data domain characteristic is pertinent to systems that rely heavily on aggregational or pluralistic information dissemination framework [17], for example, a library system that consumes massively aggregated data about all kinds of books to provide the user with the choice that is nearest to the book sought for. Fine granularity is not sanctioned. However, in a support system, which by nature tends to be highly specialized, the data dissemination characteristics are much more prescribed than the data domain characteristics, although domain characteristics do decide the data field in which a support system is operating.

An important point to be noted in this context is that a data source implies not only a generic database but any source of input that a support system of decision-making assemblies consumes to thrust the workflow of decision-making. Any component regardless of whether it is external or internal to the support system can act as a source of data or input.

19.3.2 'Service': A Short Primer

In the everyday world around us, we have been quite familiar with the word 'service' in various contexts. We understand that a 'service' is a capability or a function that an entity can perform. For example, a cook has cooking as its service, a driver has driving as its service, and a teacher has teaching as its service. So we can conclude that a service is an official capability or a specialized function that an agent is known to possess [4].

In technology, service has an analogous meaning along the same lines. A 'service' in technology is a function or an execution capability that a technical agent (e.g. software, server, line of code, device) is expected in a technical landscape to possess. This is usually achieved through a mandatory segment of any service agent which is called the solution or business logic. A solution or business logic is the constituent technical code or constructs that are executable and actually perform some action which could be called monolithically the capability of an agent and hence ultimately a service or this could be a part of a bigger 'service' offering.

Figure 19.1 illustrates the notion of business logic.

19.3.3 Service-Oriented Inputs as Digestible Units for a Support System

A service agent is always atomic in its pattern of service. What this essentially means is that for its goals to be met, an agent offering a 'service' has to make it indivisible at the most elementary level.

For example, a pizza delivery person offers his service of pizza delivery. However, for his service to meet its purpose, which initially is to get hired somewhere, he has to offer the service of driving, company representation, and delivery. These three services are 'digestible' units in that these underlying intrinsic 'services' undergo composition to form a 'compound service'. A service has no context if it is devoid of unit-based consumption or consideration.

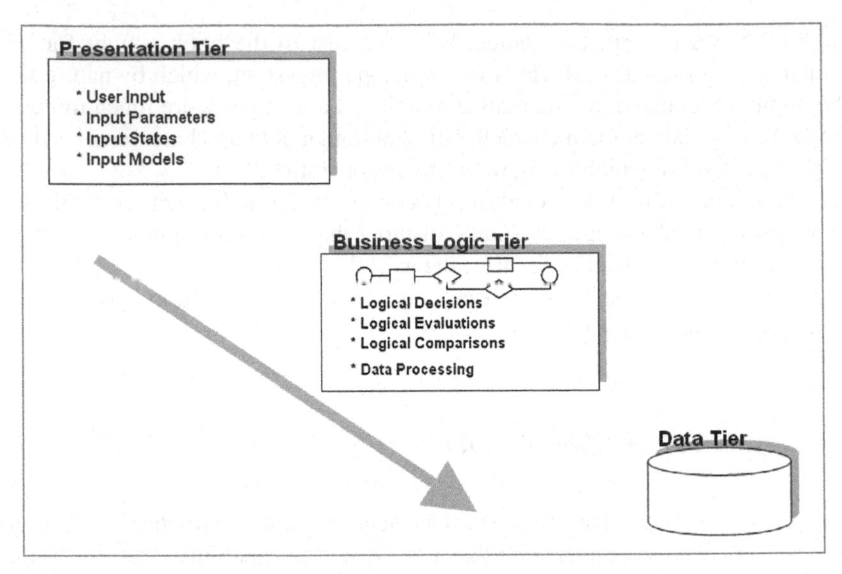

Fig. 19.1 The notion of business logic

The concept of 'digestibility' applies to a service in the support system framework only too absolutely. What this notion semantically means is that any component or a system collaborating in a support system framework will always provide service which is indivisible in nature, and all the functions of the SS can be represented by a finite combination of these individual indivisible services. There cannot be any function that cannot be expressed as an aggregate of these atomic services.

The reason behind the atomicity of a service provider is that service functions are always built bottom-up and never top-down because a top-level service will always consist of smaller and concentrated services and not vice versa. Composite services are ubiquitous in an SS in that it is built up as an aggregate sum of individual atomic services from multiple sources like other internal components, external components and systems, and external or internal decision-making engines.

19.3.4 Service-Oriented Information as Fine-Grained Output Decision Stream Services

A service has a segment called the business logic, as was delineated earlier. The business logic is the machine analogue of cognitive thinking in humans. Business logic fulfils a request or executes a task based upon the predicate framework that steers a host of programmatic actions contained in the business logic body. The output of business logic is what is called the 'service' of a service. This exposed resultant wrapping which is the result of business logic execution is called the 'output stream info' of a service. When this output stream information is rested upon a decision-oriented model that is operational in the server upon a request by the client, the stream is called 'output decision stream service' [12].

A direct consequence of output decision stream service is that within one single support system itself, one component can yield service offerings to other components within the same support system as can components outside a support system to components inside a support system in a conventional setting.

This notion of an output decision stream service is illustrated in Figure 19.4.

19.4 Service-Oriented Architecture (SOA): An Architectural Evolution for Data Access

By nature, any system that is modestly large scale is heterogeneous. This heterogeneity is introduced and attributed to a multitude of factors. A system could differ significantly from its peers in its mission statement, time of conception, design patterns, operational differences, etc. The variation in the technological sketches is due to the continual affixing of varied hardware and software constructs and agents that are a direct result of the evolution of programming languages, as well as middleware and hardware improvements.

An important system of such nature is the database management system (DBMS) that is the backbone of every Web-based and non-Web-based support system. A salient feature of any DBMS is that the endurance longevity and shelf-life of any stored data construct tends to be quite large and hence the dissemination mechanism and data relevance may lose its state-of-the-artness. In other words, as DBMS are loaded with ever-increasing data, they become clunky and their mammoth loads need to be supervised much more effectively [13].

19.4.1 Service-Oriented Architecture (SOA)

Thomas Erl in [8] defines SOA as 'an architectural model that aims to enhance the efficiency, agility and productivity of an enterprise by positioning services as the primary means through which solution logic is represented in support of the realization of strategic goals associated with service-oriented computing'.

To better illustrate what SOA implies, we explore the term 'Full-blown Architecture'. We ask the following question to understand this notion:

- How do we structure the collaboration between systems, human resources, information, and business goals in a way that no matter what technological platform or system we are on, we get results with agility and speciality?
  A question related to this is
- How do we design a technological architecture in such a way that it boosts the features of realignment at any point of technology lifecycle at minimal cost,

interoperability between any combinations of technological architectures, and alterable focus in terms of the business goals that are served?

The answer to the first question is full-blown architecture. Full-blown architecture is the concert of systems, human resources, information, and business goals. It is a form of architecture that ties together the above-mentioned agents so that this entire combined supersystem is composed of services coming out of one subsystem and entering another. The services are technologically neutral streams of bytes that circulate within the full blown architecture. Figure 19.2 shows a full-blown architecture.

Now, let us analyze the answer to the second question because this is where we achieve the formalization of the concept of SOA. Nicolai Josuttis [14] describes three technical concepts of SOA that provide the answer to the second challenge question posed previously:

1. Service: We defined a service earlier in this chapter. To revisit, a service is just a technological capability of a business which is executable. Josutti notably comments that a service is an abstract construct whereby the external and exposed interfaces are represented very intuitively and hide all the embedded technical details from the user, thereby providing the user a very light, efficient, and clear usage experience with a technology.
2. Interoperability: The term 'interoperability' refers to the notion of plug-and-play by principles of which the underlying technical architectures are neutralized and treated transparently by the operating services in a heterogeneous environment. It is impossible to characterize the underlying technical framework just by looking at a service because they are interoperable services which can function through any level of dissimilarity of technical framework.

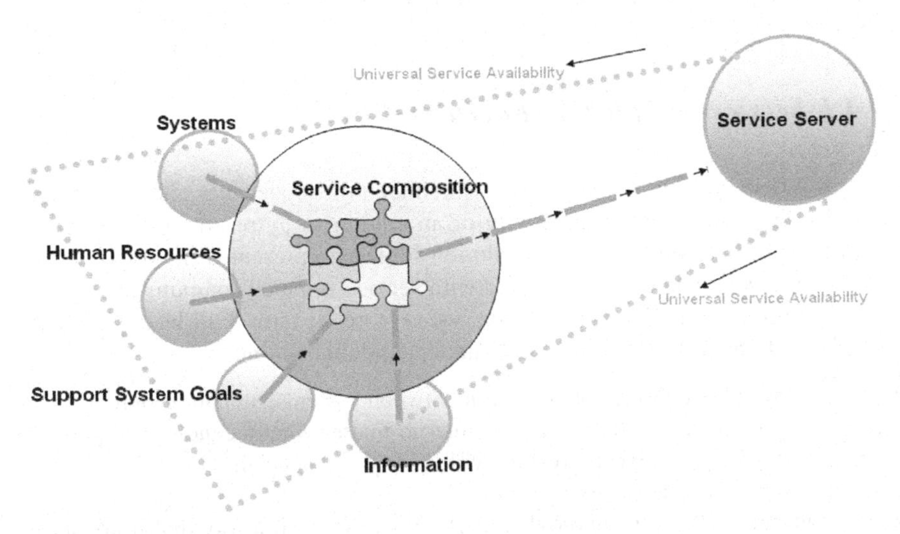

Fig. 19.2 Full-blown architecture

3. Loose coupling: Loose coupling is all about boosting independence and hence curbing dependencies and intertwinings. A salient feature of loose coupling is that it is exercisable at different levels of interaction in a system – physical, data-based, model-based, etc. The core concepts still remain – decouple various agents to form autonomous assemblies that can function without dependencies. The data sources of a WSS need to be loosely coupled so fault-tolerance is achieved through transfer of load to alternate systems in case of a disaster.

19.4.2 Rise of a Web Service: Software as a Service (SaaS) Over the Internet

A marked departure in the modus operandi of how software is deployed or made operational for a client has been the advent of the notion of software as a service (SaaS). In this software application delivery model, the commodity being sold to the client is not the software package but its execution power in the form of a service. So the focus is not on owning the software but rather on its usage.

With the technology portability and interoperability effected by the Web, SaaS enables a software vendor to develop software for the Web, host it natively on servers owned by him, and then sell 'usage lines' of the same piece of software to multiple customers over the Web. So in a manner of looking, the software executes 'inside the Internet shell'.

In the context of WSS, SaaS has a great deal of pertinence owing to the previously expensive access to proprietary databases owned by vendors that only sold once solid piece of coalesced databases, whereas SaaS really induced a high degree of cost-effectiveness access due to the ability of WSS vendor to buy selective access to a databases domain as dictated by the realm of knowledge in which the support system functions [1].

The commoditization of computing itself has revolutionized the data access domain, something that WSS relies heavily on. It has led to the following merits:

1. Standardization: For SaaS to be put in place, the coalitions and standardization committees came up with standards that facilitated SaaS delivery model. A great example of this has been the rise of XML standard. The XML standard is so powerful that it has become the de facto standard for every data communication scheme existing behind the SaaS delivery. Similarly other standards are coming into being incessantly as most of the computing moves over the Web and the major vendors scurry for a monopoly in this strategic domain.
2. Fast access: With exceptional Internet bandwidths and ever-improving Internet Service Provider (ISP) offerings, data access is just a click away and is all instantaneous owing to the Web. Moreover, with the ever-increasing competition, the speed of access is the foremost feature that a vendor targets as it is one of the benchmarks that lends a software its alacrity and a visible edge over all its peers. The lightweight orientation of a SaaS is another reason that speed improvement becomes a natural resultant.

3. IT infrastructure flexibility: Organizations no longer need to invest billions of dollars in platform-switching exercises that used to be a nightmare to the IT teams. With SaaS, all you do is switch services which is all done over the Web. Various kinds of adapters are available all across the Internal domain that helps in this exercise. For organizations the magic component is an Enterprise Service Bus, which is illustrated in the next section. This Bus can transform any input in any possible technological form to the standard that any organization follows and works on. Therefore, technology translation is a conventional and managed operation in the SaaS scenario [9].

19.5 SOA: The Information Gateway for Support Systems

SOA has enabled the support systems to work in collaboration with any existing software machinery at any scale. The fact that SOA has fragmented the execution power of any system into divided services that can be considered as separate operating entities and can be decoupled has really opened up unprecedented avenues to tap for support systems in terms of reaching out and extending their knowledge base exponentially [15]. This results in not only a better and wider basis for a support system to make decisions but also injected a lot of broadness in the decision models that any one support system can utilize.

19.5.1 Enterprise Service Bus: SOA's Elixir for Data Access for Support Systems

The Enterprise Service Bus (ESB) is the communication headquarters for a service in SOA. It is the medium over which services travel. They provide universal connectivity to different types of components implemented or operated over a large number of technical platforms [2]. The responsibilities of an ESB are as follows:

- Impart connectivity
- Smart routing of services
- Service brokership
- Infrastructure services such as security, quality of service, etc.
- Record of participating services and channels

An ESB is usually a very bulky element of SOA architecture. This is partly because over time, a deluge of functions and capabilities were added to the ESB so they are robust and can cover any technological context that may arise due to different working combinations of all the technologies existing out there. Another primary reason for this was that SOA parties wanted to place their trust and effort in one intermediary that could really plug into any type of middleware and other components and could completely take over the duty of connectivity, intelligent routing, and other transport functions of services.

19.5.1.1 Inside the Enterprise Service Bus

ESB is the 'miraculous' component that provides the necessary connection of any service onto a common channel that is serviceable by a WSS. Labelling an ESB as the most intelligent and responsible component of SOA will be no exaggeration. Let us take a look inside the ESB and what makes this intelligence possible.

An ESB is composed of smaller subcomponents that carry out individual functions that add up together to create a technology translation and service plugging effect. These subcomponents are

1. Translation component: This component transforms one technology to another to make sure that the incoming service complies with the technology being used by the service consumer. This may be composed of the more commonplace assemblers, compilers, and translators.
2. Interface Service component: This component essentially matches the interface method signature exposed in the incoming service to what was called initially by the service consumer. An interface method signature is basically a declaration of a programmatic method or a function without any specification of its body. Method signatures simply tell what a function accepts as arguments, what the name of the function or a method is, and what type of value it returns after the execution of its business logic. This is a configuration task that is important for pluggability.
3. Routing component: This component keeps track of the location of the business logic module inside the service consumer, which could be a WSS, which originally called up the service. After this is identified, the Bus automatically routes the service to this module using a variety of routing algorithms.
4. Security component: This component supervises the scan of security over all the operations that the Bus executes. It does encrypt or decrypt operations, virus checks, firewall scans, etc.

These constituent parts of an ESB are shown in Fig. 19.3:

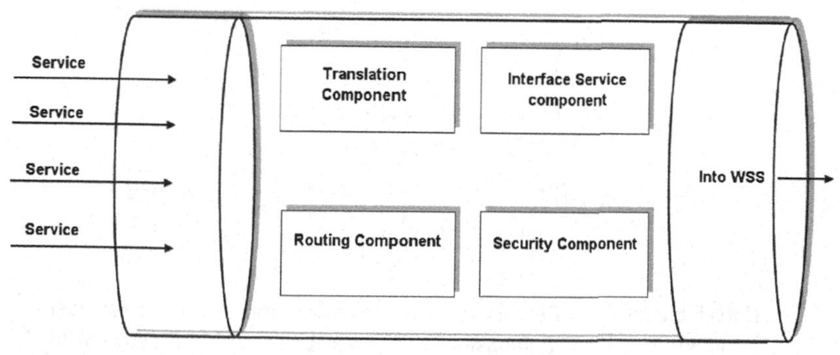

Fig. 19.3 An Enterprise Service Bus (ESB) from the inside

19.5.2 *AirMod-X: A Support System Example*

Let us consider a WSS called AirMod-X which is a hypothetical airline reservation support system. Now, this system has the capability to manage and rebook flight itineraries in case of any changes from the airlines. AirMod-X is a composite WSS that has the following characteristics:

- It works over the Web.
- It is connected directly to heterogeneous sources of data namely other airline companies, food providers, hotel industry players, rental car companies, cargo management service providers, etc. This is shown in Figure 19.4.
- Business logic of AirMod-X always uses parametric data from direct data sources it is connected to for its operation. In other words, it is tightly coupled with each data source and there is a one to one correspondence.
- The nature of middleware in picture in the communication between a business logic module and a data source is heterogeneous. Some interactions are COBOL based, some may be JAVA Remote Method Invocation (RMI) based and some may be C++ based.

19.5.2.1 Challenge Scenario – 1: The Traditional Way

From the description of point 4 above, it is fair to assess that each module of business logic in AirMod-X is a 'jack-of-one-trade' implying that each module runs an

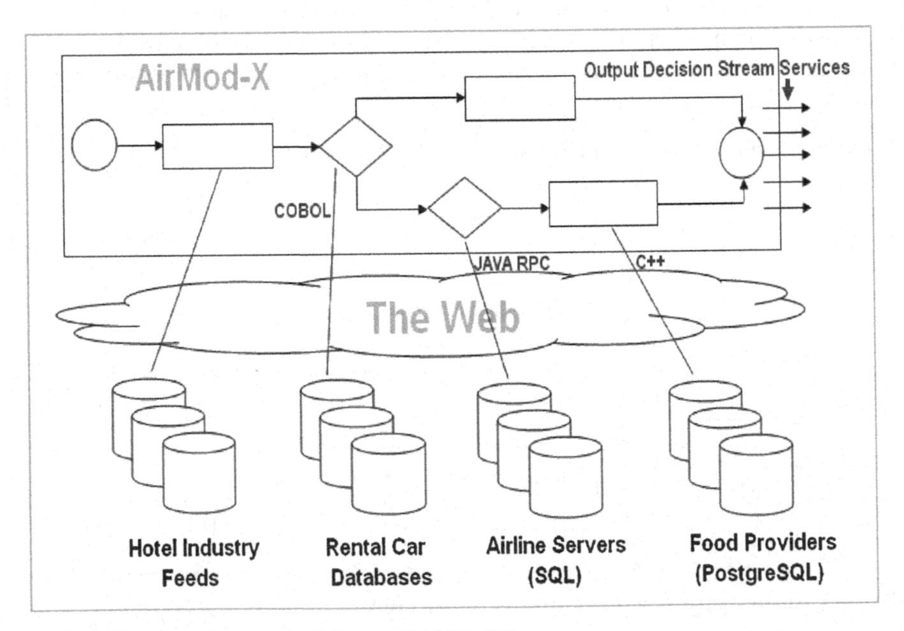

Fig. 19.4 Heterogeneous source linkages of AirMod-X

interpreter/compiler that allows it to understand the language that the data source talks to it in. The compilers/interpreters have been in place for quite a few years and they do a reliable job of translation.

Let us figure this:

- What if a new airline has just begun its operation and it shares data with systems in any new generation language format?
- What if a food provider, previously implementing its data services in C++ now switches over to JAVA without a prior notice?
- What if a car rental company shuts down and no longer has data service share?

A common problem that will be presented to AirMod-X in all scenarios is that of interoperability or flexibility. Either AirMod-X has to significantly deploy new compilers/interpreters to keep in sync with the changing technical landscape of data shares or has to be in a dilemma whether or not to unwind a compiler/interpreter just because a service talking in that language has shut down.

19.5.2.2 Challenge Scenario – 2: The SOA Way

In this scenario, we introduce a mediator or a gateway called an Enterprise Service Bus that provides a N $\rightarrow$ N Mapping in terms of languages/platforms translated from and to. This is achieved by conversion of any existing language to a Web service as shown in Figure 19.5.

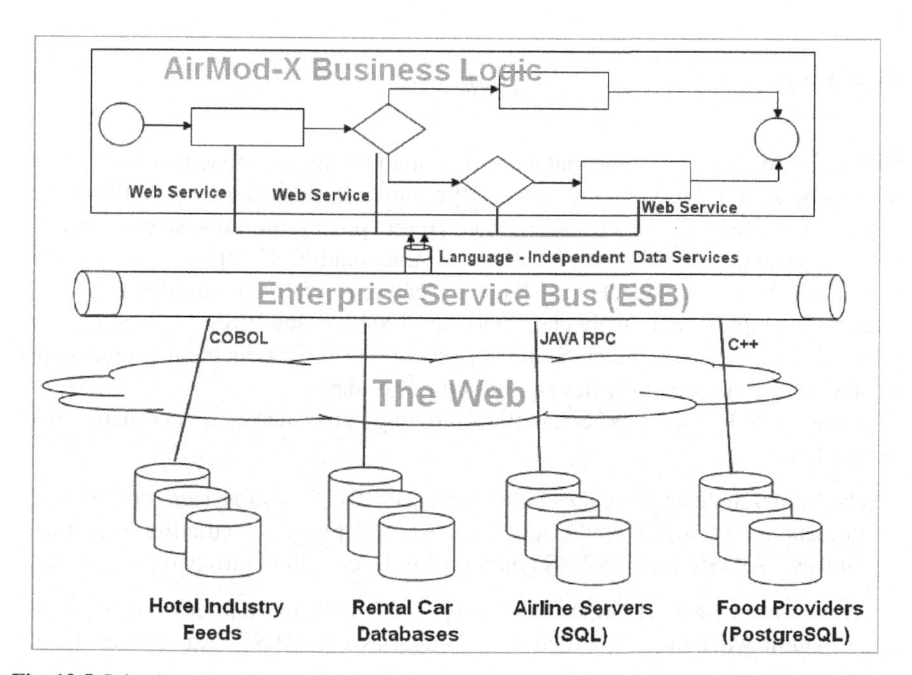

Fig. 19.5 Injection of an Enterprise Service Bus (ESB) to AirMod-X

Let us see how the placement of ESB answers the questions that were encountered in the traditional way of aligning architectures:

- What if a new airline has just begun its operation and it shares data with systems in any new generation language format?
  Answer: With an ESB, the only thing that this new airline needs to do, in order to be compliant to its client base technologies, is that it needs to provide an ESB adapter to its client which is a piece of software that translates one technology to another. This adapter can sit in the space provided on the client's ESB. This way no major shuffle is needed anywhere.
- What if a food provider, previously implementing its data services in C++ now switches over to JAVA without a prior notice?
  Answer: Since JAVA is a fairly common technology, all the airline needs to do is to find out if its ESB already has the JAVA adapter built into it. If it doesn't then it can add it to its ESB. After this step, it is fully ready to tap into the food provider's new JAVA base.
- What if a car rental company shuts down and no longer has data service share?
  Answer: In this case, all the airline needs to do is figure out the extent of space and bulkiness that the adapter for the car rental company technology possesses and accordingly it can either delete this adapter from its ESB or can choose to retain it based on its policies.

This was an example where ESB acted as a uniform gateway to data for a WSS boosting its interoperability and reaction time to change significantly.

19.5.3 SOA and WSS: An Interplay

SOA based WSS are systems that work on collaborating components over the Web whose operation is based on the concept of a unit of business logic capability called as a service. These components represent different processing subsystems inside the support system, or in other words a unit of functionality. Composite components, in which various components are clubbed together to yield a combination of logic modules are highly an unlikely characteristic of SOA-based WSS. Each component is based on a transparent understanding of subsuming and producing message output streams and are highly regardless to the sources of input.

A simple SOA-based WSS has the following subsystems, in accordance with Figure 19.6.

1. The Data Source and Registry Subsystem (DSRS): This subsystem comprises of the data source which could be a database or a registry mediating access to a database. A DSRS for a SWSS could further be classified into two categories:

   (a) Internal DSRS (IDSRS): In this type of DSRS, the data source is located within what may be called the boundaries of a WSS. An internal DSRS

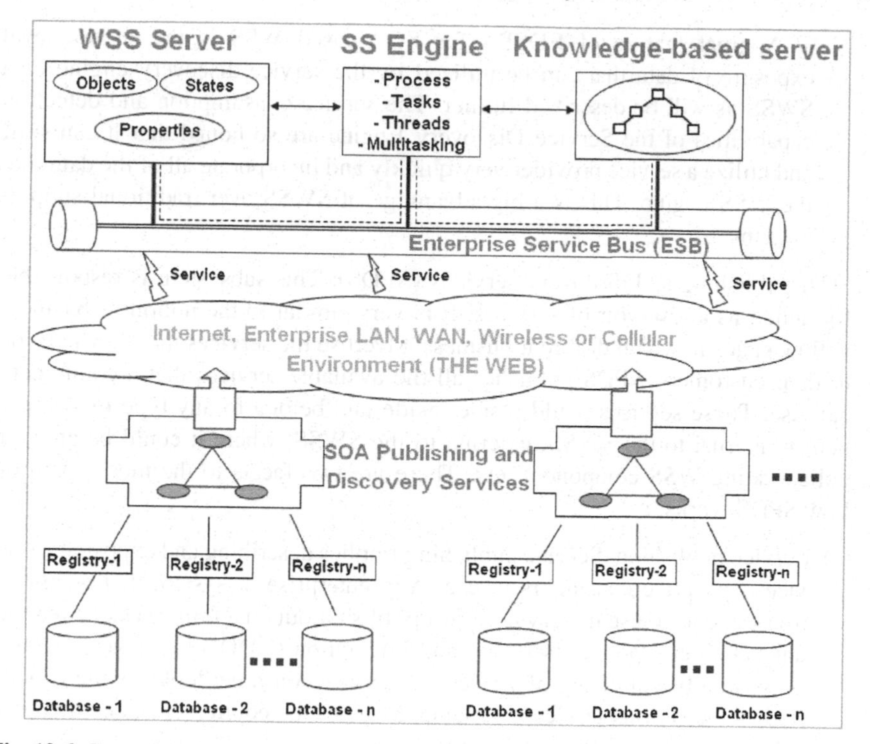

Fig. 19.6 General Architecture of SOA-based WSS (SWSS)

may be a part of a larger composite component of a WSS, coupled together with components like Ontology server and decision support server (DSS) or it could be a stand-alone loosely coupled component that is autonomous and handles channels of data access for all decision-making purposes within a WSS.

One of the salient features of WSS is that due to the prime importance of the Web as an operative medium, an IDSRS could prove less eclectic in terms of the coverage of data as compared to its external counterpart. The Web operative medium really gravitates the need of a rightful balance between the grades of usage depths of IDSRS and External DSRS. A failure to leverage the broad and wide data services of External DSRS could actually lead to a loss of sustainable competitive advantage for an enterprise in a competitive landscape.

(b) External DSRS (EDSRS): This is the most widely used type of data source in SWSS. In this type of data subsystem, efficient and apt use is made of third-party databases and data registries. Potential third-party providers are targeted based on the domain of knowledge base to be exercised in the decision making process of the WSS. Varied classes of third-party knowledge base providers are functional today. From educational institutions to consultancy firms, finding a befitting data source is only a matter of research.

A great feature of EDSRS for SOA-based WSS is the service-level exposure of data that can be utilized by the service discovery engine of a SWSS as will be described further. The service consumption and detection capabilities of the Service Discovery Engine are so honed that it can sniff and utilize a service provider very quickly and incorporate all of the data into the WSS engine. This is a big advantage of SWSS over traditional support systems.

2. SOA publishing and discovery services (SPDS): This subsystem is responsible for acting as a reservoir of services. It is very similar to the notion of business yellow pages in which different business advertise the services that they provide to their customers. SPDS publishes all the available services that a component can use. These services could reside inside and belong to any type of component – internal to the SWSS, external to the SWSS, where it could be another collaborating WSS component, etc. There are two facets to the mechanism of how SPDS works:

   (a) Service publishing: Service publishing implies describing and exposing a service for service consumers to use. Any enterprise or system that wants its usage agents to use its services must publish it out on a framework called the universal description, discovery, and integration (UDDI) framework. UDDI is essentially a directory of services. A service consumer looks up the UDDI, studies the various services available, assesses its requirements, and then decides on employing that service [7].

   In the context of a SWSS, the system could decide that it needs to find a source of data for a medical domain-oriented SWSS. It accesses the UDDI, which has a standard access mechanism and looks up all the medical data services enlisted in the UDDI. For every successful match, the SWSS has the option of invoking that service or not act at all. Two or more collaborating and distributed SWSSs could also decide to utilize each other's business logic for their decision making roles to yield a combined output that is representative of the processing resultant of multiple WSSs. In this case, the collaborating WSSs could publish their services out for each other to use.

   (b) Service discovery: Service discovery is the complement operation to service publishing. Service provider publishes the service and a service consumer discovers a service. The process by which a service consumer discovers a published service and decides to utilize it is called binding between the service consumer and the service provider.

   The above-mentioned two concepts of service discovery and publishing are illustrated in Figure 19.7.

3. The Web subsystem: This is the cardinal subsystem of a SWSS that lends it the paramount potential that it carries. The Web is an ecosystem of knowledge representation over a network called the Internet which effects access to data (text, images, multimedia, other systems, etc.) in a standardized way [19]. The Web is the single most ubiquitous computer network dominating the world today with access to a score of data sources located anywhere across the globe.

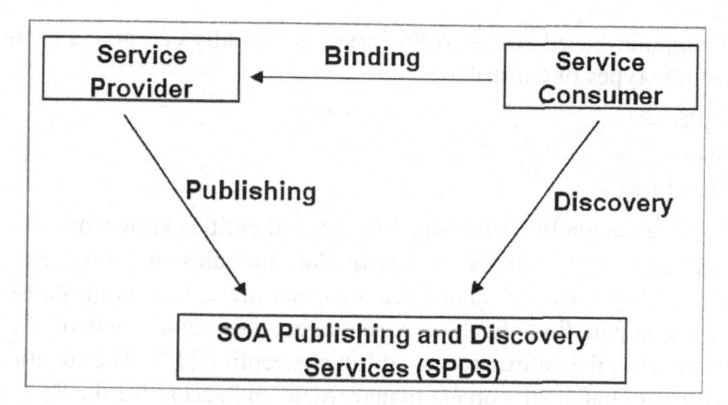

Fig. 19.7 SOA service operations

The Web, in the context of a SWSS, implies a concert of all types of networks that the system can employ in order to service the motion of data across systems between components. Data, for its mobility, needs channels to transport itself. The Web provides that channel. It may comprise of all types of networks available to the SWSS, namely. Internet, Enterprise LAN, WAN, Wireless or Cellular Environment, etc. The entire service implementation of SWSS depends on the Web as a transport layer.

4. Enterprise service bus subsystem (ESBS): ESB, as described earlier in the chapter, constitutes the communication headquarters for a service-oriented system. An ESB has the following functions [14]:

   - Providing connectivity
   - Data transformation
   - (Intelligent) routing
   - Dealing with security
   - Dealing with reliability
   - Service management
   - Monitoring and logging

   At the most basic level, ESB provides logical connectivity for different services to be invoked. An important point to note here is that while the Web provides a physical channel for the functioning of SWSS, an ESB provides a pseudo-physical logical connectivity. Another very important goal of ESB is to ensure interoperability. This was illustrated earlier with the help of the AirMod-X system example.

5. Support system subsystem (SSS): This is the core subsystem representing the engine of the decision-making system that is comprised of models and business logic. The different nodes representing business logic work over the data utilizing the different models to reach a decision and churn out the decision as their output. The SSS is comprised of following three components:

   (a) SS server: A support system server is the host to all the programmatic and logic constructs and agents that actually are symbolic of the execution model of computer-based model. A SS server is typically comprised of the following three types of constructs:

- Objects
- States
- Properties

The server actually handles the life cycle of entities known as objects which are programmatic representation of data modules in a programming language. They have associated features that are called as properties in programming paradigm. Each object changes states representative of the execution stage in the entire process of logic execution [18]. The function of a SS server is to handle the object management and access for the SS Engine.

(b) SS engine: SS engine is the heart of any operational support system. It constitutes the execution sandbox inside which the logic is executed based on the models used. The engine handles all the execution environment management tasks. It comprises of lightweight execution pathways called threads. A single logical unit of execution is called a process. Each process may be comprised of multiple threads. An SS engine may be multithreaded based on the architecture. SS engine heavily utilizes the SS Server to fetch objects and their properties to execute decision model implementations on them using processes.

(c) Knowledge-base server: This server actually resides the SS models that are available for use for the decision-making process. An SS model, at the most generic level, is a what–if analysis rule set that establishes how a support system will model different conditions and produce a decision output based on the data input. Models are the most common instruction assemblies available to a support system that frame a control pattern over the execution of the support system as a whole in producing an optimal output.

19.6 Technologies for SOA Implementation for WSS

Let us try to understand the 'Web' part of (WSS). The Web is a hyperlink-based system, where a hyperlink represents a link to a different location of information (usually a system). This location is named using a special standard called a uniform resource locator (URL), which is further translated into a specific Internet Protocol address (IP Address), a unique identifier for the system or the machine that is being referred to. A request packet is sent to the machine where information to be accessed is located and a return response packet is sent back by the server. In this entire communication scheme, all the data and traffic travel over a communications protocol called the hypertext transfer protocol (HTTP).

Now let us understand the Web service platform. The Web service platform is a collection of interoperable technology standards implemented using a specification called the Extensible Markup Language (XML), which facilitates information interchange across different information systems. XML has gained so much universality that it is the de facto standard for information exchange between diverse technologic platforms. In a SOE, in terms of web services, XML allows the crystallization of specification, discovery, and interchange of service data between the service requester and service provider.

19.6.1 Sample WSS Scenario: AirMod-X

We will consider an example to see how SOA can be implemented in representing a request and response information sets in our previously discussed AirMod-X Web-based airline support system. We consider a used case where a flight for a passenger has been cancelled and a suitable follow-up action is required on the part of airline using AirMod-X. There are two options – book a new flight for the passenger, if available or book a hotel room for the passenger.

This example will be discussed with each technology described below. The following technologies based on XML actually enable the entire implementation of services in a SOA environment:

1. Simple Object Access Protocol (SOAP): SOAP is a standard protocol that lays down a standardized way of exchanging data and messages based on the XML specification. SOAP also utilizes the underlying HTTP protocol to allow the client and server to invoke calls on each other.
   A generic SOAP message, which is just an XML document, contains the following chief parts as shown in Fig. 19.8:

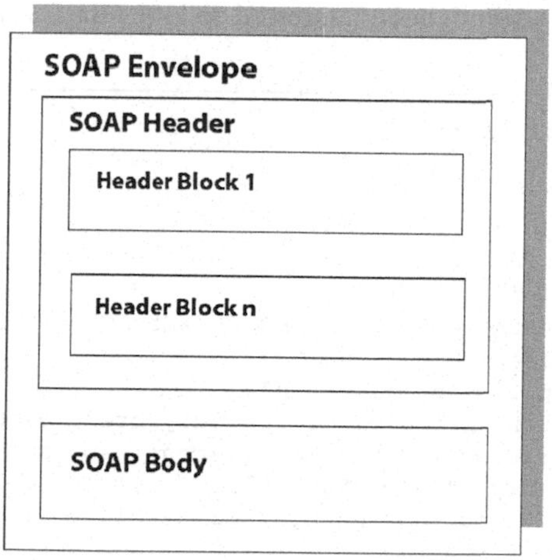

SOAP Envelope
SOAP Header
Header Block 1
Header Block n
SOAP Body

Fig. 19.8 A SOAP message

(a) A mandatory 'Envelope' section that helps recognize an XML document as a SOAP standard message [21]
(b) An optional 'Header' section that contains some lead information
(c) A mandatory 'Body' section that encapsulates the request and response

2. Universal Description, Discovery and Integration Service (UDDI): UDDI is a platform-neutral registry that stores the web services as its directory contents and service henceforth can be published and discovered using this framework. UDDI is the SOA version of the real life yellow pages, white pages and green pages directories.

   UDDI is the nucleus in the process of publishing and discovery of web services in WSS. A support system can, for example, enlist all its processing capabilities as well its data domain descriptions into UDDI for other collaborating support systems to find over the Web and henceforth request access to these services.

   We will now the UDDI aspect to our sample scenario described above. The following query, when placed inside the body of the SOAP envelope, returns details on Hilton Hotel. This search could be used by AirMod-X in the aforementioned scenario when a passenger's flight has been cancelled and hotel room is needed for him:
   Query:

```
<find_business generic="1.0"xmlns="urn:uddi-org:api">
<name>Hilton Hotels</name>
 </find_business>
```

Following is the result. An important point to note is that the following result represents the information that the UDDI entry for Microsoft furnishes out to the service consumer who might be interested to know the services provided by the business.

```
<businessList generic="1.0"
  operator="Hilton Hotels"
  truncated="false"
  xmlns="urn:uddi-org:api">
  <businessInfos>
   <businessInfo
       businessKey="0076B468-EB27-42E5-AC09-9955CFF462A3">
    <name>Hilton Hotels</name>
     <description xml:lang="en">
         Hilton is the proud flagship brand of Hilton
Hotels Corporation and the most recognized name in the global
lodging industry. "Be My Guest" is still the gracious and warm
way we want for our guests to feel at Hilton hotels and resorts
     </description>
      <serviceInfos>
       <serviceInfo
          businessKey="0076B468-EB27-42E5-AC09-9955CFF462A3"
          serviceKey="1FFE1F71-2AF3-45FB-B788-09AF7FF151A4">
```

```
        <name>Web services for smart searching</name>
     </serviceInfo>
    <serviceInfo
        businessKey="0076B468-EB27-42E5-AC09-9955CFF462A3"
        serviceKey="8BF2F51F-8ED4-43FE-B665-38D8205D1333">
      <name>Electronic Business Integration Services</name>
    </serviceInfo>

    <serviceInfo
        businessKey="0076B468-EB27-42E5-AC09-9955CFF462A3"
        serviceKey="611C5867-384E-4FFD-B49C-28F93A7B4F9B">
            <name>Volume Room Booking Program</name>
    </serviceInfo>

    </serviceInfos>
   </businessInfo>
  </businessInfos>
 </businessList>
```

3. Web services definition language (WSDL): WSDL is the default language specification for the service providers. WSDL essentially lists all the necessary parameters that are required by a service consumer to successfully locate and call a Web service [24]. It describes the following vital pieces of information about a service:

   (a) Address information for physically locating a specific service
   (b) Binding information which includes the protocol name and data format type
   (c) Exposed interface information which is basically the collection of signatures of methods that are available to the service consumer and that can be invoked by the consumer on the server

As an example, we will consider the sample scenario that we described previously. Following is the service advertisement for a price and availability quote from Hilton and also a sample request–response snapshot from AirMod-X:

Figure 19.9 describes the entire workflow in a SOA-based WSS with the AirMod-X example. The five steps are as follows:

Step – 1: Upon receiving request form AirMod-X, the SPDS requests for the service business information on Hilton Hotels from the UDDI registry.

Step – 2: The UDDI looks upon its registry and returns the reference information to the SPDS that would allow AirMod-X to discover and find Hilton Web Services.

Step – 3: The SPDS uses the reference information returned by the UDDI to access WSDL documents using the location specified in the reference.

Step – 4: The SPDS reads the WSDL to find out about how and where to access Hilton services. It passes the same to AirMod-X.

1. Overarching paradigm change: For any type of system, SOA represents an immense change in the way systems are organized in information technology architecture. This is primarily because SOA percolates across system boundaries

```
<?xml version="1.0" ?>
<definitions name="Hotel Room Quote"
      targetNamespace="http://hilton.com/roomquote.wsdl"
      xmlns:tns="http://hilton.com/roomquote.wsdl"
      xmlns:xsdl="http://hilton.com/roomquote.xsd"
      xmlns:soap="http://schemas.xmlsoap.org/wsdl/soap/"
      xmlns="http://schemas.xmlsoap.org/wsdl">
<types>
  <schema targetNamespace="http://hilton.com/roomquote.xsd"
      xmlns="http://www.w3.org/1999/XMLSchema">
   <element name="RoomQuoteRequest">
     <complexType>
      <all>
     <element name="city" type="string"/>
     <element name="checkindate" type="date"/>
     <element name="checkoutdate" type="date"/>
      </all>
      </complexType>
   </element>

   <element name="RoomPrice">
      <complexType>
       <all>
     <element name="price" type="float"/>
     </all>
      </complexType>
   </element>
  </schema>
 </types>

 <message name="GetRoomQuoteInput">
     <part name="body" element="xsdl:RoomQuoteRequest"/>
 </message>

 <message name="GetRoomQuoteOutput">
     <part name="body" element="xsdl:RoomPrice"/>
 </message>

 <portType name="RoomQuotePortType">
   <operation name="GetRoomQuoteRequest">
     <input message="tns:GetRoomQuoteInput"/>
     <output message="tns:GetRoomQuoteOutput"/>
   </operation>
 </portType>
```

Fig. 19.9 SOA-based WSS workflow in AirMod-X example

and actually touches the business side of the system. In the WSS context, this represents a change in the way the support systems engines are designed to work and the source orientation that they undergo upon encountering data from diverse sources. Since diverse input sources in SOA are commonplace, a basic interpretation in the engine needs to go into place so as to create synergistic ability of functioning of the support system as a whole.

2. Magnified dependency: SOA also represents a marked augmentation in the dependency of a system on other systems. This is primarily because of the distributed input requirement for a WSS engine. Requiring diverse systems and components' input places a high level of stakes in the input sources particularly the external ones over which the parent system has no control.
3. Complex process lifecycle management: The decision-making process life cycle is heavily based on the nature of the systems that it is deriving inputs and processing power from. A change in the model of access to these auxiliary systems

```
<soap:binding style="document"
transport="http://schemas.xmlsoap.org/soap/http"/>
   <operation name="GetRoomQuoteRequest">
    <soap:operation
soapAction="http://hilton.com/GetRoomQuoteRequest"/>
      <input>
        <soap:body use="literal"
namespace="http://hilton.com/roomquote.xsd"
        encodingStyle="http://schemas.xmlsoap.org/soap/encoding/"/>
      </input>
      <output>
        <soap:body use="literal"
namespace="http://hilton.com/roomquote.xsd"
        encodingStyle="http://schemas.xmlsoap.org/soap/encoding/"/>
      </output>
   </operation>
</binding>

<service name="RoomQuoteService">
   <documentation>My first room booking</documentation>
      <port name="RoomQuotePort" binding="tns:RoomQuoteBinding">
        <soap:address location="http://hilton.com/roomquote"/>
      </port>
</service>

<binding name="RoomQuoteServiceBinding" type="RoomQuoteServiceType">
      <soap:binding style="rpc"
transport="http://schemas.xmlsoap.org/soap/http"/>
        <operation name="getQuote">
          <soap:operation
soapAction="http://www.getquote.com/GetQuote"/>
                                 ..........................................
                                 ..........................................
```

Fig. 19.9 (continued)

requires an accommodation in the life cycle of a decision-making process which is often a complicated task. It entails a thorough sweep of the entire businesses, establishing intertwined dependencies between modules, components, and systems and the working order of the input–output streams. All this requires a very strong and determined effort.

4. Latency for real-time WSS: Real-time WSS are complex and hyperperformance systems that combine several different types of data and processing elements like enterprise data servers, online feeds, legacy systems, and business intelligence technologies. They have very low latency requirement as one of their guiding working principles which may be a little tough to ensure in a SOA-based WSS given that the latency of such a WSS is the aggregate sum of the latencies of the SOA services that the support system consumes for its processing. This can be a big bottleneck given the critical nature of real-time systems.

19.7 Conclusion

The class of computer-based information systems including knowledge-based systems that support decision-making activities has come a long way ever since the advent of the notions of machine 'perception', 'intelligence', and 'mechanical consciousness' [3]. Support systems are 'synthetic' prodigies that when left to operate autonomously in a particular knowledge domain, perceive, and eventually simulate every aspect of learning or any other feature of intelligence.

The doctrines of knowledge representation and knowledge engineering are central to any support system. Most of the problems that machines are expected to address entail extensive, multifarious, and specialized knowledge of the world. For example, objects, states, properties, categories, and relations between objects, events, time, etc. With the birth of the Internet, this requirement of accumulation of extensive and specialized knowledge has been addressed quite successfully. The Internet, by every stretch of imagination, is the most productive invention that complements the field of machine learning and intelligence and ultimately WSS.

The Internet as a knowledge source offers many opportunities as well as presents a lot of challenges. Due to a lack of locus of control governing the content and dissemination of data, some standards, and protocols are necessary that enable support systems to utilize this massive knowledge base to the fullest. The invention of the 'tag' addressed the issue of unstructured data when the association of a tag composed the metadata (data about data) that yielded structure to data. Now remains the idealistic requirement of 'specialized' and 'channelized' data access where a machine could ingest the optimized domain data it needs very rapidly and that utilizes its decision models to exhibit intelligence. This momentous requirement and what was thought to be a pipe dream was satisfied with the rise of SOA. This architectural style of providing data as a 'service' triggered the provisioning of IT infrastructure, whereby each data cluster can be decomposed into small, crisp, digestible, and specialized units that provide live, uninterrupted, and optimized data channels to systems like SS, a facility which is realized by the 'separation of concerns' features of SOA, which essentially derives itself from the decoupling of service representing each knowledge domain. Owing to the Internet, these services constituting SOA operate over the Web and these Web services empower WSS to universally discover these fine-tuned data streams and consume them.

References

1. Bass L, Clements P and Kazman R (2003). Software Architecture in Practice, Addison Wesley. pp. 101–123
2. Brown P C. (2007). Succeeding with SOA: Realizing Business Value Through Total Architecture. Addison Wesley Professional. pp. 45–62
3. Butler S (1872). Erewhon.. Chapter 23: The Book of the Machines http://www.hoboes.com/FireBlade/Fiction/Butler/Erewhon/erewhon23/
4. Champion M, Rerris C, Newcomer E et al (2002). Web Service Architecture. W3C Working Draft, http://www.w3.org/TR/2002/WD-ws-arch-20021114/
5. Chi R T H, Rathnam S, Whinston A B(1992). A framework for distributed decision support systems. Proceedings of the Twenty-Fifth Hawaii International Conference on System Sciences, HI, USA. pp. 345–351
6. Colan M (2004). Service-Oriented Architecture expands the vision of Web services. http://www-128.ibm.com/developerworks/library/ws-soaintro.html, IBM, U.S. pp. 233–250
7. Colgrave J, Januszewski K (2004). Using WSDL in a UDDI Registry, Version 2.0.2. OASIS UDDI Technical Note

8. Erl Thomas (2008). SOA: Principles of Service Design. Prentice Hall, Pearson, Boston. pp. 38
9. Foster H, Uchitel S et al (2003). Model-based Verification of Web Service Compositions. Proceedings of the 18th IEEE International Conference on Automated Software Engineering, pp. 152–161
10. Holsapple C W, Whinston A B (1996). Decision Support System: A Knowledge-Based Approach, West Publishing Company. pp. 161–193
11. Hong T P, Jeng R et al (1997). Internet computing on decision support systems. IEEE International Conference on Systems, Man, and Cybernetics. pp. 106–110
12. Hurwitz J et al. (2006). Service Oriented Architecture For Dummies. John Wiley & Sons. pp. 65–66
13. Jammes F, Mensch A and Smit H (2005). Service-oriented device communications using the devices profile for web services. Proceedings of the 3rd international workshop on Middleware for pervasive and ad-hoc computing, ACM Press. pp. 1–8
14. Josuttis N (2007). SOA in Practice. O'Reilly Publishing. pp. 16–48
15. Krafzig D, Banke K and Slama D (2004). Enterprise SOA: Service-Oriented Architecture Best Practices, Prentice Hall. pp. 189–201
16. Liping S (2005). Decision support systems based on knowledge management. Proceedings of International Conference on Services Systems and Services Management (ICSSSM'05). pp. 1153–1156
17. Louvieris P, Mashanovich N et al (2005). Smart Decision Support System Using Parsimonious Information Fusion. Proceedings of the 7th International Conference on Information Fusion (FUSION). pp. 1009–1016
18. Martin D et al (2004). Bringing Semantics to Web Services: The OWL-S Approach. Proceedings of the First International Workshop on Semantic Web Services and Web Process Composition (SWSWPC 2004) San Diego, California, USA. pp. 26–42
19. Mentzas G et al (2007). Knowledge Services on the Semantic Web. Communications of the ACM Vol 50. pp. 53–58
20. NationMaster: Encyclopedia: Decision Theory. http://www.nationmaster.com/encyclopedia/Decision-theory
21. OASIS Web Service Security, SOAP Message Security 1.0 (WS-Security 2004). http://docs.oasis-open.org/wss/2004/01/oasis-200401-wss-soap-message-security-1.0.pdf
22. Shannon C E (1948). A mathematical theory of communication. Bell System Technical Journal, vol 27:379–423. pp. 623–656
23. The Data Management Association (DAMA International) (http://www.dama.org/i4a/pages/index.cfm?pageid$=$3339)
24. W3C Web Service Definition Language (WSDL) 1.1. http://www.w3.org/TR/wsdl
25. Wikipedia (http://wikipedia.org)
26. Yao J T, Yao Y et al (2004). Web-based Support Systems(WSS): A Report of the WIC Canada Research Centre. Proceedings of the IEEE/WIC/ACM International Conference on Web Intelligence (WI'04). pp. 787–788
27. Yao Y Y (2003). A Framework for Web-based Research Support Systems. Proceedings of the 27th Annual International Computer Software and Applications Conference (COMPSAC'03). pp. 601–606

Appendix A
Contributor's Biography

Ali Babar, Muhammad is an Associate Professor in Software Development Group at IT University of Copenhagen, Denmark. Previously, he was a Senior Researcher with Lero, University of Limerick, Ireland, where he led projects on software architecture and empirical assessment of software development technologies. Prior to that, he was working as a researcher with National ICT Australia (NICTA), where he initiated and led several research and development projects in the areas of software architecture evaluation, architectural knowledge management, and process improvement using empirical methods. During his stay at NICTA, he designed and developed various methods and tools in the area of software architecture knowledge management, which attracted significant industrial interest and funding. He has authored/co-authored more than 100 publications in peer-reviewed journals, conferences, and workshops. He has co-edited a book, software architecture knowledge management: theory and practice. Dr. Ali Babar has also been a co-guest editor of special issues of IEEE Software, JSS, IST, and ESEJ. He has also been involved in academic and professional training by designing and delivering lectures and tutorials since 2000. Recently, he has presented tutorials in the areas of softwarc architecture and empirical methods at various international conferences including ICGSE09, XP08, APSEC07, ICSE07, SATURN07 and WICSA07. Apart from being on the program committees of several international conferences such as WICSA/ECSA, ESEM, SPLC, ICGSE, and ICSP for several years, Dr. Ali Babar is program chair of ECSA2010 and program co-chair of PROFES2010. Prior to joining research and development field, he worked as a software engineer and an IT consultant for several years in Australia. He has obtained an MSc in computing sciences from the University of Technology, Sydney, Australia and a Ph.D. in Computer Science and Engineering from the University of New South Wales, Australia. His current research interests include software product lines, software architectures, global software engineering, and empirical methods of technology evaluation.

Berbers, Yolande is Associate Professor in the Katholieke Universiteit Leuven's Department of Computer Science and a member of the DistriNet research group. Her research interests include software engineering for embedded software, ubiquitous computing, service architectures, middleware, real-time systems, component-oriented software development, distributed systems, environments for distributed and parallel applications, and mobile agents. She received her PhD in computer science from the Katholieke Universiteit Leuven. She is a member of the IEEE.

Carrillo-Ramos, Angela is a Systems and Computing Engineer of the Universidad de los Andes, Bogota, Colombia (1996); Magister in Systems and Computing Engineering of the Universidad de los Andes, Bogota, Colombia (1998); PhD in Informatics of the Universite Joseph Fourier, Grenoble, France (2007); Research Assistant of the Universidad de los Andes (1996–1997)and Assistant professor of the Universidad de los Andes (1998–2003). Currently, she is Associate Professor and Researcher of ISTAR and SIDRe teams of the Pontificia Universidad Javeriana, Bogota, Colombia. Her research work is focused on information systems acceded through mobile devices using software agents. Other research subjects are adaptation (personalization) of information in nomadic environments according to user preferences and context of use, and software engineering.

Curran, Kevin BSc (Hons), PhD, FBCS CITP, SMIEEE, MACM, FHEA Senior Lecturer in Computer Science School of Computing and Intelligent Systems Faculty of Computing and Engineering University of Ulster Magee College, Derry Northern Ireland BT48 7JL.

Domènech, Josep received a BS, MS, and PhD in Computer Science, and an MS in Multimedia Applications from the Polytechnic University of Valencia (Spain), and a BS in Business Administration from the University of Valencia. He obtained the Best University student award given by the Valencian Local Government in 2001. He is currently a lecturer at the Department of Economy and Social Sciences of the Polytechnic University of Valencia. His research interests include Web performance, user and consumer characterization, and Internet architecture.

Fahrenholz, Sally is a publishing and content management professional in Northeast Ohio. She has authored a number of papers and presentations on technology issues specific to supporting both print and online publishing. Sally's publishing career has included management and editorial positions with publishing firms focused on government regulation, legal, scientific, and consumer markets. Past affiliations include Thompson Publishing Group (Washington, DC); West Publishing; and ASM International, where she was actively involved in standards work. While at ASM, Sally chaired the Oasis Open Materials Markup Language Technical Committee and was enthusiastically involved in projects exploring issues involved with materials data exchange using MatML and materials taxonomies. Avidly interested in past and current publishing technologies and practices, Sally has presented papers at a variety of technology conferences, and has most recently focused on best practices and the application of markup languages and taxonomies for online publishing and content discovery.

Fan, Lisa is an assistant professor at Department of Computer Science, Faculty of Science, University of Regina. She received her PhD from the University of London, UK. Dr. Fan's main research areas include Cognitive Informatics, Web Intelligence, and especially web-based learning support systems. She is also interested in intelligent systems applications in engineering (intelligent manufacturing system and intelligent transportation system).

Gensel, Jérôme received his PhD in Computer Science in 1995 and an Accreditation to Supervise Research (Habilitation á Diriger des Recherches) in 2006 at the University Joseph Fourier (Grenoble, France). He is professor in Computer Science at the University Pierre Mendes France (Grenoble, France) since 2007 (he has been Assistant Professor in this University since 1996). JérÔme is member of the STEAMER research team at the Grenoble Computer Science Laboratory (LIG) since its creation in 2007. He has been member of the SIGMA research team at the Software, Systems and Networks Laboratory (LSR-IMAG) from 2001 to 2006, and researcher at INRIA Rhone-Alpes until 2001. His research interests include the representation of Spatio–temporal information, dynamic cartography techniques, geographic information systems, adaptability to users, and ubiquitous Information Systems.

Gil, José A is an associated professor in the field of Computer Architecture and Technology at the Polytechnic University of Valencia, Spain. He teaches at the Computer Engineering School where he is also member of the Staff. Professor Gil obtained his BS, MS, and PhD degrees from the Polytechnic University of Valencia. His current interests include topics related with web systems, proxy cache and distributed shared memory systems. He is joint author of several books on the subject of computer architecture and he has published numerous articles about industrial local area networks, computer evaluation and modeling, proxy cache systems, and web systems. He has participated in numerous investigation projects financed by the Spanish Government and the Local Government of the Cominidad Valenciana and in development projects for different companies and city councils.

Goyal, Madhu is a Lecturer in the Faculty of Engineering and Information Technology, University of Technology, Sydney, Australia. Her research interests lie in Intelligent Agents, Multi-agent Systems and Collaborative Problem Solving.

Hoeber, Orland is an Assistant Professor in the Department of Computer Science at Memorial University of Newfoundland (Canada). He received his Ph.D. from the University of Regina in 2007. His primary research interests include information visualization, Web search interfaces, Web search personalization, human-computer interaction, and the geo-visualization of fisheries and oceanographic data. Dr. Hoeber is one of the principle researchers in the User Experience Lab at Memorial University.

Hsu, D. Frank is the Clavius Professor of Science and Professor of Computer and Information Science at Fordham University. He received a BS from Cheng Kung University (in Taiwan), an MS from the University of Texas at El Paso, and a PhD from the University of Michigan. He has been a visiting professor or visiting scholar at M.I.T., Boston University, Academia Sinica (in Taiwan), Taiwan University, Tsinghau University (in Taiwan), CNRS and the University of Paris-Sud (in France), Keio University (in Yokohama and Tokyo), JAIST (in Kanazawa, Japan), the Beijing International MBA Program (in Beijing University and Fordham University), and DIMACS (at Rutgers University, New Brunswick, NJ). Hsu's research interests are: Combinatorics, Algorithms, and Optimization; Network Interconnections and Communications; Informatics and Intelligent Systems; and Information and Telecommunications Infrastructure. He has authored/coauthored three books, edited/coedited nineteen special issues or volumes, and authored/coauthored over 130 papers in journals and books, and conference proceedings. He has given over 40 keynote, plenary, and special lectures, and over 200 invited lectures at institutions and research organizations in Asia, Europe and the US. Hsu has served as editor for several scientific journals including the *IEEE Transaction on Computers* (IEEE Computer Society), Networks (John Wiley and Sons), the *International Journal of Foundation of Computer Science* (World Scientific Publishing), *Monograph on Combinatorial Optimization* (Kluwer Academic Publishers), the *Journal of Interconnection Networks* (EIC: 2000–06, Editor for Special Issue 2006– *, World Scientific Publishing), and *Pattern Recognition Letter* (2006– *, Elsevier Science). Among the honors and awards Hsu has received are Foundation Fellow (Institute of Combinatorics and it Applications), IBM Chair Professor (Keio University and IBM Japan), Komatsu Chair Professor (Japan Advanced Institute of Science and Technology and Komatsu Corp.), Bene Merenti and Distinguished Teaching Award (Fordham University), best paper award (IEEE AINA05 Conference), Senior Member (IEEE Computer Society), and Fellow (New York Academy of Sciences).

Khan, Sanaullah is a research scholar working in the field of information security and access control. His research interests include distributed information management systems, various technologies in the field of security, especially Trusted Platform Module and mobile platforms. He is also interested in the field of Finance. He holds a Master's degree in Business Administration and has recently joined the Security Engineering Research Group. He has been developing softwares since 2004 for various entrepreneurs mainly in the field of information management.

Khan, Shahbaz is a research scholar working on trusted computing, specifically remote attestation, and secure information flows. His research activities include development of remote attestation techniques based on ontology to attest and control information flows. He is a Master's degree holder in networks and has been working with Security Engineering Research Group since 2008. He has publications in the field of access control and ontological analysis of information. These days he is working on applying his findings on mobile platforms.

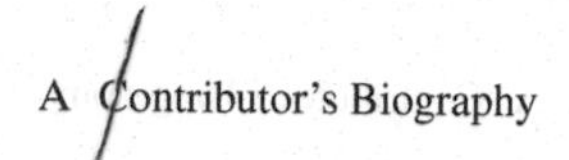

Kim, Dong Won is currently studying for his Master's degree in Computer Science at the University of Regina (Canada). He obtained his Bachelor's degree in Computer Science from Dongguk University (Seoul, Korea). He has over 10-year working experience on IT industry. His research interests include Web-based learning support systems and game-based learning systems. He published a conference paper related to those areas.

Kirchner, Kathrin holds a PostDoc position at the department of Business Information Systems, Friedrich-Schiller-University of Jena, Germany, and works in the fields of (spatial) data-mining, decision support, and knowledge management. She finished her PhD in 2006 with the title "A spatial decision support system for the rehabilitation of gas pipeline networks" in Jena. Since 2001 she has worked as a lecturer in the field of Business Information Systems.

Kirsch-Pinheiro, Manuele Since September 2008, Manuele Kirsch Pinheiro is Associate Professor in the Centre de Recherche en Informatique of the Universite Paris 1 Pantheon-Sorbonne. Previously, she occupied a post-doctoral position on the Katholieke Universiteit Leuven's Department of Computer Science. She received her PhD in computer science form the Université Joseph Fourier– Grenoble I (2006), and her Master's degree from the Universidade Federal do Rio Grande do Sul, Porto Alegre, Brazil. Her research interests include ubiquitous computing, context-aware computing, adaptation (personalization), cooperative work (CSCW), group awareness, and information systems.

Ma, Jun received his Master's and Bachelor's degree in Applied Mathematics from Department of Applied Mathematics, Southwest Jiaotong University, China. He is currently a PhD candidate in the Faculty of Engineering and Information Technology, University of Technology, Sydney, Australia. His research interests lie in automated and approximate reasoning, fuzzy logics, and decision making.

Mustapha, S.M.F.D Syed is currently an Associate Professor in the Faculty of Information Technology, University Tun Abdul Razak. He obtained his PhD and Master in Philosophy from University of Wales, UK, and Bachelor of Science (Computer Science) from University of Texas, USA. He was the Principal Researcher in Asian Research Centre (British Telecom plc). His research interest is mainly on knowledge management, information retrieval, and artificial intelligence. He has published papers related to these areas in international journals and conferences.

Nauman, Mohammad is a research associate working in the field of collaborative systems and usage control. His research interests include remote attestation of distributed systems and intelligent knowledge retrieval systems. He has a Masters in Software Engineering and has been working as a member of Security Engineering Research Group since March 2007. He has several publications in the field of usage control in conferences of international repute. He is also the author of a book on common-sense-based knowledge retrieval systems.

Pont-Sanjuán, Ana received her MS degree in Computer Science in 1987 and a PhD in Computer Engineering in 1995, both from Politechnic University of Valencia. She joint the Computer Engineering Department in the UPV in 1987 where currently she is full professor of Computer Architecture. From 1998 until 2004 she was the head of the Computer Science High School in the UPV. Her research interest include multiprocessor architecture, memory hierarchy design and performance evaluation, Web and Internet architecture, proxy caching techniques, CDNs and communication networks. Professor Ana Pont-Sanjuàn has published a substantial number of papers in international Journals and Conferences on Computer Architecture, Networks, and Performance Evaluation. She has been reviewer for several journals and regularly participates in the technical program committees of international scientific conferences. She also has participated in a high number of research projects financed by the Spanish Government and local Valencian Government. Currently she leads the Web Architecture Research group at the UPV where she has been adviser in several PhD theses and is directly tutoring six more. Since January 2005, she is the Chairperson of the IFIP TC6 Working Group 6.9: Communication Systems for Developing Countries.

Razmerita, Liana is Assistant Professor at Copenhagen Business School. She holds a PhD in computer science from University Paul Sabatier, Toulouse, France. Her PhD thesis is entitled for which User Models and User Modeling in Knowledge Management Systems: an Ontology-based Approach has been awarded in December 2003. Her PhD research work was done at Center of Advanced Learning Technologies, at Insead (European Institute of Business Administration), Fontainebleau, France. Previous publications include more than 40 refereed papers including journal articles, book chapters, and conference papers in domains such as User Modeling, Knowledge Management, e-Learning and e-Government.

Rundensteiner, Elke is Full Professor in the Department of Computer Science at Worcester Polytechnic Institute, and the director of the database systems research laboratory at WPI. Professor Rundensteiner is an internationally recognized expert in databases and information systems, having spent 20 years of her career focusing on the development of scalable data management technology in support of advanced applications including business, engineering, and sciences. Her current research interests include data integration and schema restructuring, data warehousing for distributed systems, analytic stream processing, and large-scale visual information exploration. She has over 300 publications in these and related areas. Her publications on view technology, database integration, and data evolution are widely cited, and her research software prototypes released to public domain have been used by academic and nonprofit groups around the world. Her research has been funded by government agencies including NSF, NIH, DOE, and by industry and government laboratories including IBM, Verizon Labs, GTE, HP, NEC, and Mitre Corporation. She has been the recipient of numerous honors, including NSF Young Investigator, Sigma Xi Outstanding Senior Faculty Researcher, and WPI Trustees' Outstanding Research and Creative Scholarship awards. She is on program committees of prestigious conferences in the database field and has been editor of several journals, including Associate Editor of the *IEEE Transactions on Data and Knowledge Engineering Journal*, and of the *VLDB Journal*.

Sahuquillo, Julio received his BS, MS, and PhD degrees in Computer Science from the Universidad Politénica de València (Spain). Since 2002 he is an associate professor at the Department of Computer Engineering. He has taught several courses on computer organization and architecture. He has published over 70 refereed conference and journal papers. His research topics have included multiprocessor systems, cache design, instruction-level parallelism, and power dissipation. An important part of his research has also concentrated on the Web performance field, including proxy caching, Web prefetching, and Web workload characterization. He is a member of the IEEE Computer Society.

Singh, Vishav Vir is Software Engineer in the Computer Aided Design/Engineering group at Intersil Corporation. He earned his Master's in Software Engineering from San Jose State University, California, USA, in 2008 and his Bachelor's in Technology (BTech) in Computer Science and Engineering from Punjabi University, Patiala, India, in 2005. Prior to joining Intersil, he was with the acclaimed PureXML DB2 Database team at IBM Silicon Valley Laboratory in San Jose, California, USA. He has extensive and substantial experience in the field of service-orientation of technology and service oriented software architectures. At IBM, he was also the chief engineer for the World Wide Web Consortium (W3C) XQuery Test Suite Study and Analysis team that did the analysis of the XML Query Test Suite (XQTS), which assesses the native XML capabilities of IBMs flagship DB2 database. He was awarded the prestigious IBM Thanks award for technical excellence and outstanding commitment for this project. During his Master's degree, he worked in different roles within the N-Tier Client/Server Enterprise Software Technologies Research Lab at San Jose State University. He is a regular speaker at universities and other educational institutions on diverse topics related to service orientation of technology and XML-based support systems. His highlight lectures include a talk at Massachusetts Institute of Technology (MIT) on a service-based XML-based support system and a talk at Stanford University on Service Orientation and Persuasiveness of technology. He is also a regular participant at various IEEE conferences related to the field of technology architectures and support systems. Vishav is a member of Tau Beta Pi Engineering honor society.

Taksa, Isak is an Associate Professor in the Department of Computer Information Systems at Baruch College of the City University of New York (CUNY). His primary research interests include information retrieval, knowledge discovery, and text and data mining. Professor Taksa has published extensively on theoretical and applied aspects of Information Retrieval and Search Engine technology in numerous journal articles, refereed conference papers, book chapters, and a book.

Varde, Aparna is a Tenure Track Assistant Professor in the Department of Computer Science at Montclair State University, New Jersey, USA. She obtained her PhD and MS in Computer Science, both from Worcester Polytechnic Institute in Massachusetts, USA, and her BE in Computer Engineering from the University of Bombay, India. Dr Varde has been a Visiting Senior Researcher at the Max Planck Institute for Informatics Germany; and a Tenure Track Assistant Professor in the Department of Math and Computer Science at Virginia State University, USA. Her research interests span data mining, artificial intelligence and database management with particular emphasis on multidisciplinary work. Dr Varde's research has led to several publications in reputed international journals and conferences as well as trademarked software tools. Her professional activities include serving as a panelist for NSF; a reviewer for journals such as the VLDB journal, IEEE's TKDE, Elsevier's DKE, and IEEE's Intelligent Systems; co-chair of ACM CIKM's PhD workshops; Program Committee Member in conferences such as IEEE's ICDM 2008, Springer's DEXA 2008, EDBT 2009 and ER 2009, and SIAM's SDM 2008. In 2005, she was awarded an Associate Membership of Sigma Xi, the Scientific Research Society, for excellence in multidisciplinary work. Her doctoral work spanned the domain of Materials Science and was funded by the Center for Heat Treating Excellence, an international industry–university consortium. Her current research is supported by collaborative grants through organizations in the USA such as NSF, NIH, and DOE. She has served as an Invited Member of the MatML Committee, an association for the development of the Materials Markup Language, involving organizations such as NIST and ASM International (the Materials Information Society). Dr Varde has worked in the corporate world as a Computer Engineer in multinational companies such as Lucent Technologies and Citicorp. She is currently working on research projects in text mining, scientific data mining, Web databases and machine Learning.

Villanova-Oliver, Marlène received her PhD in Computer Science at the Grenoble Institute of Technology in 2002. She is Assistant Professor in Computer Science at the University Pierre Mendes France (Grenoble, France) since 2003. She is member of the STEAMER research team at the Grenoble Computer Science Laboratory (LIG) since its creation in 2007. She has been member of the SIGMA research team at the Software, Systems, and Networks Laboratory (LSR-IMAG) from 1999 to 2006. Her research interests include adaptability to users and context of use, Web-based information systems, geographic information systems and representation of spatio-temporal information.

Xie, Ermai is a graduate in Computer Science from the University of Ulster, UK. He is presently studying for an MSc in Computing and Intelligent Systems from the University of Ulster. He has worked in the computer industry for a number of years as a network administrator and as a programmer. He has worked for a number of University Departments and is currently engaged on research into novel methods of detecting hand movement in arthritic patients for a local hospital. He has also worked on a diabetes project which utilises mobile phones to accurately gauge the correct amount of insulin to inject. His research interests include ubiquitous computing, neural networks and image processing.

Yang, Xue Dong is currently a Professor and the Department Head of Computer Science at the University of Regina, and an Adjunct Professor with TRLabs, Canada. He received his Ph.D. in Computer Science from Courant Institute of Mathematic Science, New York University in 1989. His research interests include multiresolution methods in computer graphics, visualization, and image processing.

Yao, JingTao joined the University of Regina in January 2002. Before arriving in Canada, he taught in the Department of Information Systems at the Massey University, New Zealand, the Department of Information Systems at the National University of Singapore, Singapore, and the Department of Computer Science and Engineering at Xi'an Jiaotong University, China. He received his PhD degree at the National University of Singapore. He did a BE degree and an MSc degree at Xi'an Jiaotong University. His research interests include softcomputing, data mining, forecasting, neural networks, computational finance, electronic commerce, Web intelligence, and Web-based support systems.

Index